Bioresource and Stress Management

Ratikanta Maiti • Aruna Kumari
Ashok Kumar Thakur
Narayan Chandra Sarkar
Editors

Bioresource and Stress Management

Springer

Editors
Ratikanta Maiti
Facultad de Ciencias Forestales
Universidad Autónoma de Nuevo León
Linares, Mexico

Ashok Kumar Thakur
Department of Seed Science and Technology
Dr. Yashwant Singh Parmar University
of Horticulture and Forestry
Solan, HP, India

Aruna Kumari
Department of Crop Physiology, Agricultural College
Professor Jayashankar Telangana State
Agricultural University
Karimnagar, Telangana, India

Narayan Chandra Sarkar
Department of ASEPAN (Agronomy)
Institute of Agricultural Sciences, Visva-Bharati
Sriniketan, West Bengal, India

ISBN 978-981-10-0994-5 ISBN 978-981-10-0995-2 (eBook)
DOI 10.1007/978-981-10-0995-2

Library of Congress Control Number: 2016945258

Printed on acid-free paper

This Springer imprint is published by Springer Nature
The registered company is Springer Science+Business Media Singapore Pte Ltd.

Preface

In the twenty-first century, new approaches of environmental and social advocacies and bio-resource management have been formulated at the national and international level for preserving the environment and management agendas. It is well understood. One of the most significant needs of the day is to manage the natural resources sustainably vis-a-vis disseminating the knowledge to impart awareness. In this book, we have outlined the consolidated efforts and findings of the researchers, scientists, academicians and scholars engaged in addressing a series of questions and framework on resources management and combating of biotic and abiotic factors for keeping our mother Earth safe for all.

Here, an outline of series of questions, concepts and line-up plans pertaining to the bio-resource and stress management has been drawn. These natural bio-resources have an impact on economic, cultural, aesthetic, scientific and educational segments of the society. Thus, the preservation of this prestigious heritage by conserving or managing in sustainable manner is far more important as these are at danger rather shrinking at a faster rate. The combined and interacting influences of over-exploitation, pollution, modification, destruction or degradation of the native habitats amplify the vulnerability of bio-resource. All these factors are pushing the crease and forcing natural biota to attain a smaller size with every passing day. Immediate attention is next to the door; as the prerequisites are rarely met. The human use of the goods and services provided by nature is necessary; however, continued attempts are to be made to check the genetic erosion. Thus, these alarming notes urge the necessity of adoption of a compromising agenda in management and conservation of bio-resource. These may provide a viable long-term solution, a new paradigm for conservation.

Promoting biodiversity-sensitive management is of equal priority. The land use practices must be compatible with the maintenance of the bio-resource. Nevertheless, the problem of soil fertility degradation may be relieved by the proper use of organic and inorganic fertilizers or by adoption of integrated nutrient management strategies. Some of the chapters enlighten the research contributions encompassing the organic amendments in increasing the nutrient uptakes, microbial activities, etc.

This book not only emphasizes on the general conceptual approaches by different users but also methods on integrated conservation, utility and importance of bio-resource. It has focused with the prime goal of contribution towards the construction of a range of attributes, conservation or management of resources and indicates, at the same time, the areas or topics where further research will be useful under the present scenario of climate change. To elucidate some of the effects and research highlights, an attempt has been made to discuss the wide range of themes for framing out considerable management aspects.

It is also interesting to overview the current perspective to assess the level of depletion or exploitation on bio-resource over the years. The chapters herein encompass the research contributions in the fields of genetics, biotechnology, water conservation, abiotic and biotic stress, seed technology, postharvest physiology, natural resource management, climate change, etc. This has called for the expansion of research over vast corridors looking ahead the necessity of enrichment of inter- and intra-disciplinary research.

Despite the astounding importance of various bio-resources and their abundance on our planet, the degree of their degradation and extinction is elevating owing to the ever-increasing pressure of population growth, urbanization, industrialization, deforestation and other aspects enhancing the level of greenhouse gases. The global impact of these gases is evidenced as the global warming and greenhouse effects. Projections highlight the food crisis to be seen in the forthcoming years with the burgeoning populace, deforestation and land degradation. The large-scale release of these greenhouse gases or burning of fossil fuels, intensive agricultural systems and residual effects of the indiscriminate pesticide use, microbial degradations, industrial effluents and gases into the atmosphere and hydrosphere are exuberating their impact to a much larger extent. Though significant achievement has been obtained in crop improvement, molecular biology and other disciplines, very little progress has been made in increasing crop productivity under sustainable agriculture in the farmers' fields.

This scope of diversity of research activities that prevailed provides an advantage rather than challenges in compiling and editing this book, moreover to indicate a wide range of affordable perspectives. Such typical fields of scientific research endeavour the development of suitable strategies for efficient and reliable management of bio-resource.

The chapters are well organized by broad topics describing the research highlights, with an overall context and concluding summary enlightening the future areas of research. Most of these synthetic chapters are drafted by contributing authors. The editors have stimulated these highly productive research highlights mostly from the 2nd International Conference on Bio-Resource and Stress Management held during 7–10 January 2015 at PJTSAU and ANGRAU, Hyderabad, India. These studies cover a wide scope of disciplinary perspectives, from agronomical to physiological and biochemical right down to the molecular level. Each equipped with their particular set of experimental tools has successfully approached their object of study to provide us with fresh information that will prove useful, in at least two ways. For one thing, those whose interest lies in designing their own research will find in this chapter inspiring ideas on questions of utmost urgency. On the other hand, those who are interested in the general area of stress in plant systems from an ecological point of view – as it may concern global climate change in particular ways in which it threatens ecosystem stability, for example – will find the material most valuable to build on current working models explaining interactions between living organisms and a rapidly changing physical environment and the prospects of evolution of the ecosystems thereupon.

Linares, Mexico	Ratikanta Maiti
Jagtial, Karimnagar, Telangana, India	Aruna Kumari
Solan, Himachal Pradesh, India	Ashok Kumar Thakur
Sriniketan, West Bengal, India	Narayan Chandra Sarkar

Acknowledgements

Many individual researchers and scientists have contributed to the production of this book. We are grateful to all those contributing authors and researchers, especially those who have participated in the 2nd International Conference on Bio-Resource and Stress Management held during 7–10 January 2015 at PJTSAU and ANGRAU, Hyderabad, India, without which this would have been impossible.

We also thank Prof. T.V.K. Singh for sparing his time, in spite of his hectic schedule, to contribute to this symposia and PJTSAU and ANGRAU for offering all facilities to organize this conference to a great success.

We are grateful, in particular, to dozens of scientists, researchers and academicians who kindly acted as referees (anonymously) to several of these chapters. Their critical suggestions, insights and thinking for the improvement of this manuscript are duly acknowledged.

Contents

Contributors

R. V. T. Balazzii Naaiik Electronic Media Wing, ARI, PJTSAU, Rajendranagar, Hyderabad, Telangana, India

Susanta Banik Department of Plant Pathology, School of Agricultural Sciences and Rural Development, Nagaland University, Medziphema, Nagaland, India

Samares Kumar Das Department of Veterinary and Animal Husbandry Extension, Central Agricultural University, Aizawl, Mizoram, India

Aruna Kumari Department of Crop Physiology, Agricultural College, Professor Jayashankar Telangana State Agricultural University, Karimnagar, Telangana, India

Ratikanta Maiti Facultad de Ciencias Forestales, Universidad Autónoma de Nuevo León, Linares, Mexico

A. V. Ramajaneyulu Regional Agricultural Research Station, Professor Jayashankar Telangana State Agricultural University, Palem, Mahabubnagar District, Telangana, India

Jorge Sarquis Ramirez Universidad Veracruzana, Xalapa de Enríquez, Estado de Veracruz-Llave, Xalapa, Mexico

Narayan Chandra Sarkar Department of ASEPAN (Agronomy), Institute of Agricultural Sciences, Visva-Bharati, Sriniketan, West Bengal, India

Somesh Sharma School of Bioengineering and Food Technology, Shoolini University of Biotechnology and Management Sciences, Solan, HP, India

Ashok Kumar Thakur Department of Seed Science and Technology, Dr Y S Parmar University of Horticulture and Forestry, Solan, HP, India

About the Editors

Ratikanta Maiti, Ph.D., D.Sc., is a renowned Botanist of international repute with specialization in crop Physiology and Botany. He worked for 9 years as a senior botanist in Jute Agricultural Research Institute, Barrackpore (I.C.A.R.), Kolkata, West Bengal, and for 10 years as a senior Plant Physiologist in ICRISAT, Patancheru, A.P., India. Then, he worked as Senior Professor and Senior Research Scientist (CONACYT, Mexico, for 18 years) in two Universities in Mexico for 20 years. During his scientific career, he has developed innovative techniques for the screening of salinity and drought of several crops. As author of more than 30 books, he has won several international awards such as international Scientist award offered by IBC, Cambridge and Ehanobotanist Award-2003 sponsored by Friends University, U.S.A. and U.N.D.P. and gold medal for India 2008 offered ABI.

Dr. Ch. Aruna Kumari is working as Assistant Professor, Department of Crop Physiology, at Agricultural College, Jagtial, Professor Jaya Shankar Telangana State Agricultural University. She has 7 years of Research Experience in varied ICAR Institutes and in seed industry (Vibha Seeds). She is the Recipient of CSIR Fellowship, Govt. of India during her doctoral studies. She has been an Author in 2 books viz., Glossary in Plant Physiology, Advances in Cotton, Chilli. Her field of specialization is Seed Dormancy.

Narayan Chandra Sarkar is a faculty member at Visva-Bharati, Department of ASEPAN, Palli Siksha Bhavana (Institute of Agricultural Sciences), Sriniketan, West Bengal, India. He has worked in Vibha Seeds, Syngenta India Ltd and Nagaland University. He has received his PhD in Agronomy from Indian Agricultural Research Institute in 2005. His current research area is soil fertility in rice.

Ashok Kumar Thakur is a faculty member at the Dr. Yashwant Singh Parmar University of Horticulture and Forestry, Department of Seed Science and Technology, in district Solan, Himachal Pradesh, India. He has obtained Ph.D. in 2002 in Horticulture with specialization in Vegetable Breeding from same working University. He has 25 research publications, 04 manuals/books, 06 book chapters and 07 extension publications in the form of popular articles, folders etc. He has experience in working with Vibha Seeds, Bharti Delmonte India Pvt Ltd, ACME Cold Chain Solutions Pvt Ltd. His current area of research is vegetable seed technology.

Social Necessity of an Efficient Management and Conservation of Bio-resource and Stress Management

1

Ratikanta Maiti and Aruna Kumari

Abstract

The growth and economy of a society is highly dependent on profitable utilization of bio-resource, plants and animals for livelihood and survival. This chapter discusses the efficient utilization of native plants, trees and other bio-resources as well as various abiotic stress factors affecting crop productivity under sustainable agriculture. Various innovative techniques, agrotechnology and organic fertilizers need to evolve for increasing productivity under sustainable agriculture. The reserve of organic carbon fixed in the soil by soil microorganisms is the long-term carbon that is released into the environment by increased burning of fossil fuels. It helps to limit the global changes that are continuously increasing due to rapid changes in the amounts of greenhouse gases. Certain tree species have the capacity to capture about 50 % of carbon, viz. *Leucophyllum frutescens*, *Forestiera angustifolia*, *Bumelia celastrina* and *Acacia berlandieri* which may be recommended for planting in highly contaminated areas to reduce pollution and carbon dioxide load in the atmosphere. This chapter urges the necessity for judicious utilization and conservation of native species and also the ethnic knowledge on plant uses inherited from generation to generation.

Nature provides us a variety of goods and services, and for every moment of our life, we use the resources that it has provided. The air encircling the earth enables us to breathe; the plants grown in the soil provide the food to eat, shelter to animals, vegetation and pastures. The energy provided by the sun, coal or fossil fuels are used for several purposes. In a similar manner, the water that falls to the ground in the form of rain is used for irrigation and to meet the various day-to-day needs. It is thus evident that we are dependent on nature to a large extent to meet all of our basic needs – air, food, water, shelter, energy, etc.

The growth and economy of a society is highly dependent on profitable utilization of

R.K. Maiti (✉)
Facultad de Ciencias Forestales, Universidad Autónoma de Nuevo León, Carr. No 85, km 145, NL 67700 Linares, Mexico
e-mail: ratikanta.maiti@gmail.com

A. Kumari
Department of Crop Physiology, Agricultural College, Professor Jayashankar Telangana State Agricultural University, Polasa, Jagtial, Karimnagar 505 529, Telangana, India
e-mail: arunasujanagcjgl@gmail.com

R.K. Maiti et al. (eds.), *Bioresource and Stress Management*, DOI 10.1007/978-981-10-0995-2_1

bio-resource, plants and animals for livelihood and survival. Native plants or trees and crops supply food, medicines, timber for furniture, firewood and various domestic necessities. Besides, we are highly dependent on honeybees for honey, fish and animals for meat, leather, wool, etc. There is a great necessity for efficient utilization of bio-resource and its genetic improvement. Efficient management of bio-resource contributes to improve the economy of mankind worldwide. Therefore, the conservation and efficient utilization of bio-resource is a great necessity for the economy of our society.

With respect to native plant species, there is a great necessity to develop efficient techniques for breaking seed dormancy and propagation of plants and trees. No efficient techniques are available except traditional methods. In this regard, in Northeastern Mexico, few native species such as *Agave lechuguilla*, *Yucca* spp., *Euphorbia antisyphilitica* and wild chilli (*Capsicum annuum*) are overexploited by the semiarid farmers for their economic importance, which may lead to the extinction of these species in the long run. Many cactus species are endangered and becoming extinct. Though simple techniques are available, no attempts are being made by scientists. It urges the necessity for judicious utilization and conservation of native species and also the ethnic knowledge on plant uses inherited from generation to generation.

Since ancient times, many medicinal plants are being used to alleviate various diseases in rural villages as well as in urban areas. This knowledge on the use of medicinal plants is inherited from generation to generation. Many medicinal plants are used to cure a large number of diseases. Systematic studies on the macro- and micronutrient contents of these medicinal plant species are limited. Our research is thus being confined to the selection and identification of the native tree species having a high efficacy of accumulation of macro- and micronutrients into their biomass. These medicinal plants have high nutraceutical values as sources of supplementing the macro- and micronutrients in medicines to the desired in curing a particular disease.

There is a great necessity of interdisciplinary research on medicinal plants as ethnomedicine. In order to have a truly sustainable society, there is a need to understand the impacts of anthropic factors on the nature and the resources that are provided. Decisions are to be made in such a way that the impacts on these bio-resources are minimized, and these which are forming a part of our natural systems are continued to meet the requirements of the future generations too. Scientists, academicians, environmentalists, ecologists and biologists all over the globe are working together in finding better ways of management of the bio-resource as soil, water, vegetation as well as biodiversity.

Soil forms the basis of life and survival of plants. Millions of insects, bacteria, fungi, actinomycetes and algae reside in the soil. Certain fungi as mycorrhiza help plants to retrieve phosphorous from the soil. Certain nitrogen-fixing bacteria such as *Rhizobium*, *Azospirillum* and *Azotobacter* enrich the soil nitrogen content. Certain phosphorous-solubilizing bacteria such as *Pseudomonas* and *Bacillus* and fungi *Aspergillus* and *Penicillium* enrich the soil phosphorus contents. Certain bacterial species such as *Pseudomonas* are useful in phytoremediation in removing the oil spills and other metal contaminants from the soil and water surfaces. Likewise, some bacterial and fungal species like *Bacillus* and *Trichoderma* are effective biocontrol agents against a number of harmful pests. Certain nematodes and mites also reduce the pest species by feeding on them.

Developed and developing worlds are facing ever-increasing problems. Overcropping or abuse of mineral fertilization and pesticides is leading to the destruction of our soils, a valuable resource which is scarce. Nevertheless, this problem of soil fertility degradation may be relieved by the proper use of organic and inorganic fertilizers or by adoption of integrated nutrient management strategies. The organic residues not only enrich the soil microbial population but also influence the soil structure and the nutrient turnover, increase biodiversity and influence many physical, chemical and biological parameters. Nowadays, a very interesting subject of investigation is the use of organic amendments so that the biological properties of the soil are properly maintained in enabling in the maintenance of soil functioning and structure. Indiscriminate use of fertilizers and pesticides are degrading the soil environment leading to its pollution. The soil biological activity is being lost due to significant amounts of contaminants that are accumulated in the soils. Organic manures or organic amendments can enrich the soil microflora and lead to a build-up of microbial biomass and increased microbial activity. The soil enzymes are very important and several researches have highlighted the increase in the soil enzymatic activities with the addition of organic amendments. In dry areas, the presence of soil fauna, litter decomposition and soil chemical and biochemical properties are very important in the organic resource management.

Under the present conditions, there is a need for a transition towards durable and sustainable relations between humans and nature. An advance in scientific knowledge has given a platform for the efficient management of our natural bio-resource such as soil, water and biodiversity. An important agricultural and environmental component of soil is soil organic matter. It helps in the maintenance of plant nutrition, protects the soil quality and also controls the fate of contaminants in the soil. It is the only reservoir of organic carbon. The largest amounts of organic carbon on the earth's surface are reserved in the soil. Thus, this reserve of organic carbon in the soil is potentially capable of fixing the long-term carbon that is released into the environment by increased burning of fossil fuels. It helps to limit the global changes that are continuously increasing due to rapid changes in the amounts of greenhouse gases. Thus, the science of sustainable management of bio-resource offers enormous opportunities for scientific development for the base interventions or formulation of policies for the conservation of these valuable resources or the adoption of a green technology, and sustainability of agro-environment is the need of the present day apart from meeting the demands of the increased population.

Climate change is occurring due to increase of CO_2 and other greenhouse gases, which have a direct impact on the growth, development and survival of tree species of a forest. The climatic factors prevailing in different semiarid, tropical and temperate regions affect the growth of trees. The change in climatic factors related to atmospheric condition are the solar radiation, light, wind, temperature, precipitation, relative humidity and intensity of light. Climate determines the distribution of vegetation in a forest ecosystem. There exists a good relation of the climate with the conservation and development of forest. It is essential for the foresters to have good understanding of the climate changes and its impact on forest productivity and to take necessary measures to protect it. It is difficult to determine the changes of climate on plant growth and human activity and to adopt effective measures to mitigate climate changes. Global warming leads to melting of glaciers and swelling of sea levels thereby causing climatic disasters.

The earth receives radiant energy from solar radiation for its utilization by the plants for photosynthesis and other human activities. Short wave solar energy (visible) received from the sun passes through the atmosphere, thereby warming the earth's surface. Long wave thermal radiation is absorbed by a number of greenhouse gases (GHGs). These greenhouse gases accumulate in small amounts in the atmosphere and reflect long wave thermal radiation in all directions. The greenhouse gases cause the retention of heat to the lower atmosphere due to absorption and reradiation by clouds and other gases. Some of the radiation is directed towards the earth's surface. The amount of GHG in the atmosphere influence global temperature. The greenhouse gases affecting climate change in the earth surface are water vapour (H_2O), carbon dioxide (CO_2), nitrous oxide (NO_2), ozone (O_3), carbon monoxide (CO) and chlorofluorocarbons (CFCs). With an increase of agriculture, animal husbandry, overgrazing and an increase in human population indirectly the levels of these greenhouse gases are increasing. Increased global warming is endangering the security of persistence of life of humans, plants, animals and so on. Over and above, an incessant logging of trees for timber has direct impact in the increase of greenhouse gases. An increase in the accumulation of GHG has direct effect on forest growth and productivity.

In this context, increasing global warming owing to an increase of greenhouse gases associated with drought and other abiotic stresses are causing great concern in endangering plants, animals and other beneficial native microorganisms. This in turn is affecting crop productivity under sustaining agriculture affected with drought, salinity and other abiotic stresses. This is associated with tremendous increase of human population leading to increased hunger and poverty. Under this situation, there is a great necessity of research on the conservation of native animal and plant species or trees with special reference to trees and economic plants, on the one hand, and adopting suitable technologies to increase crop productivity in stress-prone areas under sustainable agriculture, on the other hand. Increasing global warming has direct impact on wild animals endangering their survival. High temperature has detrimental effects on the growth of plants, animals and their survival. Efficient utilization of animal or plant resources is essential for strengthening rural economy. This urges the necessity of protection, conservation and judicious utilization of plant and animal resources.

Trees in the forests are saviours of our lives. These give us food, shelter and daily necessities and supply oxygen for our respiration. Protecting forests is a great necessity for our existence. We are very much concerned about how the human activities such as the burning of fossil fuels, conversion of forests to agricultural lands and other illegal activities cause significant increase of carbon dioxide and other greenhouse gases in the atmosphere. On the other hand, deforestation and the increased use of forest products and burning of fossil fuels are contributing to the gradual increase of greenhouse gases in the atmosphere. Fortunately, the trees and forest with their ability to fix CO_2 and carbon into their biomass provide an opportunity to mitigate climate change through carbon sequestration.

Forest tree species have the capacity to fix CO_2 load from the atmosphere, thereby reducing the exposure to noxious CO_2 gas by humans protecting our health and saving our lives. This in turn leads to the accumulation of carbon, the source of energy for fuel after combustion. This is important for wood industry of high commercial importance. The tree species having high capacity of carbon sequestration (carbon fixation) could have high potential for reducing the carbon dioxide load through the process of photosynthesis.

A large variability in macro- and micronutrient contents in leaves of several native woody tree species of semiarid Mexico has been observed. These species had the ability to capture high carbon of about 50 % and were found to be efficient carbon sequestrants (carbon fixers). The species capturing high carbon about 50 % were *Leucophyllum frutescens*, *Forestiera angustifolia*, *Bumelia celastrina* and *Acacia berlandieri*. These species may be recommended for planting in highly contaminated areas to reduce pollution. Plant species with high carbon sequestration could be planted during town planning to reduce carbon dioxide load in the atmosphere. In developed countries, this excessive carbon dioxide load build-up in the atmosphere, owing to continuous emission from burning of fossil fuels, is captured and then injected in the much deeper soil impervious layer, thereby reducing the carbon dioxide load from the atmosphere.

We have hypothesized that trees with open canopy are expected to be more efficient in photosynthesis for their capacity in the capture of solar radiation and greater carbon fixation. We have selected few tree species with open canopy with about 50 % carbon sequestration. Thus, these tree species with high carbon fixation capacity could be effectively used in planting in highly contaminated areas, factory sites and cities with high carbon load in the atmosphere.

In view of the above discussion, the following options may be effective for reducing the carbon load from the atmosphere: (1) the selection of native tree species with high capacity of carbon sequestration and the promotion of their plantation in polluted areas, in development planning of townships; (2) selection of legumes, *Acacia*, and C_4 plants; and (3) preferably the selection of fast-growing species in particular for heartwood species. We feel that these potential lines of research to reduce carbon dioxide load from the atmosphere and concerted research inputs need to be directed in this direction. At the same time, incessant logging and other human activities and expansion of agriculture in logged areas need to be rigorously controlled by forest authorities. Rigorous training needs to be given to the forest rangers to control this menace. Conservation of forest health and its growth can save our lives from the menace of contamination in the atmosphere. There is necessity to develop efficient techniques to propagate native plant species and trees through seed germination.

With respect to crop productivity, thanks to the efforts of plant breeders, plant protection scientists and biotechnologists for developing high-yielding crop cultivars with increased substantial yield under high-input situations, but little progress has been achieved with respect to increased yield under sustainable agriculture where several abiotic stresses affect the crop productivity drastically. High-input cultivars cannot adapt under this condition. It may be mentioned here that biotechnologists contribute a lot in the development of crop cultivars resistant to insects such as Bt cotton and other crop species, but they failed to increase productivity under sustainable agriculture especially in poor farmers' field in rural areas where high-input crops fail to adapt. There is a great necessity in the transfer of technology from the lab to the farmers' field and to convince farmers with the benefit of the technology better than their traditional methods. We cannot impose our technology, rather convince them with the results and try to modify their traditional procedures.

In this venture, there is a necessity of concerted research activities in multidisciplinary team including breeders, physiologists, plant protection specialists, soil scientists, agronomists and biotechnologists to address this problem in an attempt to increase crop productivity under these stress-prone areas. Crop cultivars tolerant to salinity, drought and heat stress have great potential to increase productivity under these stress-prone areas. In this aspect, we adopted a novel strategy to screen high-yielding crop cultivars which were tested over multilocation trials for their adaptation and good yield by the seed companies and research institutes and then select crop cultivars for tolerance to salinity and drought with good success. Various resources management as land water, plants are particularly focused on strategies of management in such a way that the quality of life of the present and the future generations is not affected. The long-term implications of actions should take into account the future prospects while making any decisions about the land or agriculture or bio-resource that are being managed. The main ultimate goal of bio-resource management is to have sustainability – balancing social, economic and environmental factors in making sure that our future generations can equally be benefited from the goods and services of the various bio-resources. Our economical, social and environmental well-being is dependent on the sustainable management of these resources.

In this respect, crop physiologists may contribute to develop low-cost technology to screen high-yielding pipeline crop cultivars and select crop cultivars tolerant to these stresses. Simple low-cost technologies have been developed by us to screen and select crop cultivars tolerant to drought, salinity and other abiotic stresses. We developed these technologies and were successful in several field crops such as cotton, sunflower, maize, pearl millet, rice, castor and few vegetable crops. We feel that using these low-cost technologies is possible to increase crop productivity under sustainable agriculture in collaboration with agronomists, soil scientists, breeders, etc. who can contribute their innovative technologies in this venture.

We also developed simple priming technique in India for enhancing flowering and increasing the productivity of several vegetable crops such as tomato, chilli, bottle gourd, watermelon, cucumber, bitter gourd and other cucurbits. This needs to be confirmed in other countries.

In this juncture, we are aware that strenuous efforts of the breeders, molecular biologists and other scientists have contributed to the significant achievements in attaining increased crop productivity under high-input situations, but a little progress has been achieved in increasing productivity under sustainable agriculture in the farmer's fields. High-input, high-yielding crop cultivars cannot thrive well under low-input situations. In this respect, we adopt a strategy to develop simple low-cost technology to screen a large number of pipeline varieties or hybrids of few field and vegetable crops and finally select cultivars for tolerance to salinity and drought. Later, these cultivars could be handed over to molecular biologists for its validity. There is a great necessity of an interdisciplinary research team consisting of breeders, physiologists and soil scientists to address the issue and develop a strategy to increase crop production under sustainable agriculture in the farmers' fields. There is a great necessity of breeding and selection of crop cultivars tolerant to abiotic stresses such as salinity, drought and heat stress, etc. which have high potential in increasing crop productivity under sustainable agriculture in the farmer's field.

In the Second International Conference on Bio-resource and Stress Management, ANGRAU & PJTSAU held in Hyderabad, India, 7–10 January, 2015, a large number of papers (980) are presented dealing with management of bio-resource and stress management. Brief account of the research highlights in the bio-resource and stress management are presented herewith: Bio-resource and Stress Management: A Perspective.

Socioeconomy of Bio-resource and Stress Management

2

Samares Kumar Das and Ratikanta Maiti

Abstract

The chapter discusses the role of plants and animals in the socioeconomic condition of human beings. The economy of the society depends on proper utilisation of bio-resources, such as food crops, medicinal plants, fibres for clothes, firewood, etc. which are affected by abiotic and biotic stresses. Emphasis has been made on the role of women in society. Technology transfer for upliftment of socioeconomy and utilisation of various natural resources by human beings and also suggests specific strategy for rural development.

2.1 Role of Bio-resource in Socioeconomy of Rural People

Plants and animals play an important role in the socioeconomic condition of human beings. The well-being of the society depends on proper utilisation of bio-resources, such as food crops, medicinal plants, fibres for clothes, firewood, etc. In the Second International Conference on Bio-resource and Stress Management, several papers were presented. Here, we give a brief summary of these papers.

Manipur harbours a rich diversity of different plant species which also includes several dye-yielding species. The study was undertaken during 2010–2012 in different parts of Manipur, and 30 species were recorded for dye-yielding plants. Kikim et al. (2015) documented 30 species used for dye-yielding plants in Kangchup hills of Senapati district in Manipur state (India). Meiteis and Meitei Pangal (Muslim) communities have been using species like *Strobilanthes flaccidifolius*, *Lithocarpus dealbata*, *Bixa orellana*, *Tectona grandis*, *Parkia javanica*, *Osbeckia chinensis*, etc. traditionally in combination with other plants for extraction and preparation of dye-utilising indigenous process.

We depend on various plant resources, food crops and vegetables for our livelihood and economy which are affected by abiotic and biotic stresses. We are mentioning here a few of biotic

S.K. Das (✉)
Department of Veterinary and Animal Husbandry Extension, Central Agricultural University, Aizawl, Mizoram, India
e-mail: samcau.d1@gmail.com

R.K. Maiti
Facultad de Ciencias Forestales, Universidad Autónoma de Nuevo León, Carr. No 85, km 145, NL 67700 Linares, Mexico
e-mail: ratikanta.maiti@gmail.com

R.K. Maiti et al. (eds.), *Bioresource and Stress Management*, DOI 10.1007/978-981-10-0995-2_2

and abiotic stresses affecting crop productivity and the economy of rural people.

Nandeshwar et al. (2015) studied the economics of vegetable production and marketing in Akola district of Maharashtra (India). Channels of vegetable marketing were (i) producer-wholesaler-retailer-consumer, (ii) producer-retailer-consumer and (iii) producer-consumer. Producer's share in consumer's rupees was highest in channel (iii). However, channel (i) was found to be the most important channel of distribution.

Chakma and Sharma (2015) conducted a study for the assessment of production and marketing of cabbage in the district of West Tripura (India). Sixty farmers were sampled following multistage-stratified random sampling. In the case of the selected farmers of Teliamura Block, farmers were found well familiar with the cabbage production technology knowledge; however, they lacked the knowledge regarding seed treatment methods, seed rate in nursery technology, biofertilisers in fertiliser technology, etc.

Diseases reduce crop productivity owing to climatic effects. Tomato is an important vegetable whose productivity is affected by disease infection. Bhat et al. (2015) observed prevalence of tomato diseases across semiarid regions of Karnataka. High rainfall and humidity coupled with moderate temperature triggered late blight. The amount of rainfall and number of rainy days also affected the appearance and severity of bacterial spot, buckeye rot, powdery mildew and *Septoria* leaf spot.

Abiotic stress such as salinity reduces crop productivity in rural villages, thereby affecting rural economy. Bhagyashree et al. (2015) did a comparative study of sugar beet and fodder beet in salinity stress tolerance. Salinity stresses were imposed in these two varieties after the attainment of two-leaf stage. They inferred that fodder beet variety CALIXTA is much more tolerant to salinity stress than sugar beet variety SZ-35; hence, fodder beet can be grown much more efficiently in saline track.

Insects attack food products, thereby reducing their quality; the study by Deepthi et al. (2015) aimed to carry out the feeding aversion behaviour of rose-ringed parakeet (*Psittacula krameri*) 'by choice' and 'no-choice' method to different concentrations of various plant products. Crude extract of *Annona squamosa* showed promising results. Therefore, it may be used to reduce crop damage by rose-ringed parakeet in crops like sunflower, sorghum and other minor millets.

A study was undertaken by Meena (2015) in *Acacia* for understanding germination, establishment and juvenile growth during the natural regeneration of vegetation. It is concluded that seedling establishment of *Acacia* is technically feasible, and it provides a viable methodology useful for restoration of barren and degraded lands.

Chauhan et al. (2015) successfully reared bumble bees (*Bombus haemorrhoidalis* Smith) without undergoing hibernation for more than 15 months in Himachal Pradesh (India).

2.2 Role of Women in Society

In rural society, women play an important role both in the domestic and outside field and construction works. Women work hard in carrying heavy load leading to skeletal disorder and other problems. The most common of these were a lack of sleep, back problems, worrying about work, irritability and feeling down. This may be either due to incorrect design of the equipments used or due to improper design of the workstations. Agriculture is the highest employer of women's labour to the extent of 76 % in India. Work-related musculoskeletal disorders are the most prevalent illnesses among informal sector workers in India. Based on the feedback received from the farm women through participatory rural appraisal, Tiwari et al. (2015) carried out research to reduce the drudgery of farm women by introducing women-friendly farm equipment.

Kale et al. (2015) studied constraints faced by farm women in dairy farming and was undertaken in Kalmeshwar Tahsil of Nagpur district in Vidarbha region of Maharashtra State with sample size of 100 dairy farm women from

15 villages. In India, women's involvement in livestock management is a long-standing tradition, and dairy farming has been an integral part of homestead farming system. Constraints are the circumstances or the causes which prohibit the dairy farmers from adoption of the improved management practices. The constraints faced by farm women while dairy farming were high cost of quality concentrate feeds (98.00 %) and high cost needed for purchasing crossbred animals and veterinary medicine (95.00 %), followed by decline in performance of exotic animals due to the high temperature (94.00 %), low water profile, lack of irrigation facilities (85.00 %), inadequacy of green fodder round the year (82.00 %) and nonremunerative price for milk (80.00 %).

Thakur (2015) emphasised mainstreaming of gender concerns in agriculture for inclusive growth of the society at large. Women are the backbone of agricultural-based economy owing to the fact that men mostly migrate to the urban areas for search of better jobs. Worldwide, women play a vital role in food production in the developing world. They contribute as much as 60 % of labour on family farms, but they often have no control over farm income or agricultural resources such as seeds, fertiliser and land. Evidence shows that if women farmers across the developing world had the same access as men do to resources such as land, improved seed varieties, new technologies and better farming practices, yields could increase by as much as 30 % per household, and countries could see an increase of 2.5–4 % in agricultural output. In order to ensure inclusive growth of the society at large, mainstreaming of gender concerns in agriculture should form the core area of the developmental process so that the world could be freed from the scourge of global hunger.

Chavan et al. (2015) assessed training needs of farm women in storage of food grains and their constraints selecting randomly 120 respondents from Raigad district of Konkan region. More than half of the respondents (55 %) expressed training need in the 'use of fumigants at stored place' followed by the 'use of preventive measures at stored place' (46.67 %), followed by 'identification of stored grain pest and its nature of damage and control measures' (40 %). Major constraints in food grain storage practices were high cost of improved storage structure (85 %), lack of knowledge in identification of stored grain pest (69.16 %) and unavailability of different storage structures (56.66 %).

Kumari et al. (2015) interviewed randomly selected 23 women representatives from two blocks of Samastipur district in Bihar (India) to assess the extent of empowerment and associated problem of elected women members in Panchayati Raj institutions. The findings revealed that 21.73 % of respondents were with high empowerment in respect of health welfare and development of public property. The association between empowerment and age was negative but highly significant at 1 % level of probability.

2.3 Technology Transfer for Upliftment of Socioeconomy

Kadam et al. (2015) suggested strategies for documentation and validation of indigenous technological knowledge (ITK) in bio-resource management. Four steps were identified by them for inclusion of ITK in technology generation, assessment and adaptation process. They are documentation, validation, refinement and integration. Story, survey, participatory rural appraisal (PRA) or rapid rural appraisal (RRA), observation, documentary evidence, etc. were considered by them for documentation and validation of ITK.

Lotha and Sharma (2015) studied 60 households engaged in tea cultivation in six villages of Dimapur district in Nagaland (India). It was found that tea growers were well aware of the importance of planting methods, improved variety, land preparation, manures and fertilisers, plant protection measures, plucking and processing. They adopted the recommended practices for tea plantation. Formation of cooperative societies may help tea growers obtain higher price of produce and accelerate the adoption of recommended technology as well. Emphasis may be given to reduce the gap between the farming system and extension system in order to enhance the dissemination of information on recommended technology and better farming practices. Department of Horticulture and Agriculture may give high priority to provide the farmers of the study area more technical guidance through conducting demonstrations, training, field days, seminars, workshops, etc. on plant protection measures and processing. The most prominent constraints perceived by the respondents were lack of proper credit facilities, transportation and proper marketing facilities. In order to overcome such problems, government and concerned sectors may give emphasis on establishment of cooperative societies and extension of credit. Formation of cooperative societies in the study area may support the tea growers for obtaining higher price of the produce; it may also accelerate the adoption of recommended technology as well.

Reddy and Reddy (2015) studied the economics of direct-seeded rice (DSR) in Andhra Pradesh (India). About 40 % labour saving was observed at field in DSR compared to traditional transplantation method. Additional revenue due to adoption of DSR was much higher in unfavourable season compared to normal season.

Ajrawat et al. (2015) identified potential of information technology (IT) in agriculture and rural sector. Most applications of ITs in agriculture and rural sector were related to trade of inputs and outputs through *e-Choupal*, extension and training activities for rural people, advantages of agritourism, knowledge transfer from cities to villages through e-kiosks and geographical information system for management of natural resources.

2.4 Natural Resource as Source of Livelihood

Loktak Lake, the largest freshwater wetland and a Ramsar Site located in Manipur, is the source of livelihood and the lifeline for more than 50 villages located in and around the lake. The lake is the habitat of the endemic brow-antlered deer or Sangai deer (*Rucervus eldii eldii*) protected under Schedule I of Indian Wildlife Protection Act, 1972, and listed in endangered category in IUCN Red List of Threatened Species. The lake is rich in bio-resource which provide livelihood to the local population. Dey and Laishram (2015) conducted a questionnaire survey to understand the participation of women living in and around the lake in sustainable livelihood and also the level of conservation awareness among them.

In Cameroon, day by day the natural resources which are highly essential in agriculture like water, land, animal and vegetation are being degraded, eroded and dwindling. Therefore, the farmers should be trained to judicially utilise these resources. Management of natural resources beyond watershed areas, particularly in irrigated areas, is very important due to the nonavailability of a proper scale to measure watershed farmer's attitude towards natural resource management; it was thought necessary to construct a scale for the purpose. Paul and Marie (2015) described entrepreneurship development programme followed in rural Cameroon. The programme has two components: 12-module classroom course and follow-up work.

2.5 Strategy for Rural Development

Well-defined strategy needs to be formulated for the development of agriculture and economy of rural people. Sharma (2015) mentioned policy for development of agriculture in Northeast India. An analysis has been taken into consideration of the regulation of agricultural markets, introduction of legislative measures, maintenance of demand – supply balance, price support, and regulation of external trade. However, owing to the skewed nature of agricultural development underdeveloped regions such as the northeastern region remained neglected and became reliant upon the developed regions for requisite supplies of the basic commodities. To provide a big push to agricultural development and hence upliftment of the region, the role of the government and the corporate sector has great significance, while the nongovernment organisations (NGOs) voluntarily can come forward to fabricate its social network owing to cultural diversity and inherited backwardness that pose big barriers on the individuals' mindsets.

Shuya et al. (2015) conducted a survey on 50 piggery farmers in Nagaland who availed loan from the commercial banks. Major problems faced by the bankers were nonrepayment, supervision, uneven distribution of the borrowers and misutilisation of funds. While major problems faced by the borrowers were cost and availability of piglet, feed, interest rate, lack of scientific knowledge, disbursement of loan, supervision, certificates and guarantor, lengthy and slow bank procedures, marketing, disease and loan amount.

Sangeetha et al. (2015) studied productivity and economics of rice (*Oryza sativa* L.) and black gram (*Vigna mungo* L.) in rice-black gram cropping sequence as influenced by organic manures conducted at Tamil Nadu Agricultural University, Coimbatore, during *rabi* and summer seasons of 2007 and 2008. They studied the effect of organic sources of nutrients (enriched FYM compost, vermicompost, FYM + neem cake, enriched FYM compost + vermicompost + FYM, composted poultry manure and enriched poultry manure compost) and recommended NPK fertilisers on productivity and economics of rice and black gram in rice-black gram cropping sequence. The results revealed that the application of enriched poultry manure compost on equal N basis (2.3 t ha^{-1}) recorded higher yield attributes and grain yield of rice (4,675 kg ha^{-1} in 2007 and 4,953 kg ha^{-1} in 2008), which was however comparable with composted poultry manure.

References

Ajrawat B, Singh AP, Vaid A Kumar A et al (2015) Application of information communication technologies (ICTs) in agriculture development in India. In: Compendium of abstracts of the 2nd international conference on bio-resource and stress management, ANGRAU & PJTSAU, Hyderabad, India, 7–10 January 2015

Bhagyashree A, Nilima K, Rameshchand AA et al (2015) Comparative study of sugar beet and fodder beet in salinity stress with special reference to antioxidant system. In: Compendium of abstracts of the 2nd international conference on bio-resource and stress management, ANGRAU & PJTSAU, Hyderabad, India, 7–10 January 2015

Bhat NM, Vennila S, Sardana HR et al (2015) Scenario of tomato diseases in the semi-arid region of Karnataka as influenced by rainfall. In: Compendium of abstracts of the 2nd international conference on bio-resource and stress management, ANGRAU & PJTSAU, Hyderabad, India, 7–10 January 2015

Chakma A, Sharma A (2015) Production and marketing of cabbage cultivation in West district of Tripura state. In: Compendium of abstracts of the 2nd international conference on bio-resource and stress management, ANGRAU & PJTSAU, Hyderabad, India, 7–10 January 2015

Chauhan A, Rana BS, Katna S (2015) Successful rearing of bumblebee, *Bombus haemorrhoidalis* Smith, year round in Himachal Pradesh, India. In: Compendium of abstracts of the 2nd international conference on bio-resource and stress management, ANGRAU & PJTSAU, Hyderabad, India, 7–10 January 2015

Chavan KM, Mehta PG, Kale RA (2015) Training needs of the farm women in storage of food grains and their constraints. In: Compendium of abstracts of the 2nd international conference on bio-resource and stress management, ANGRAU & PJTSAU, Hyderabad, India, 7–10 January 2015

Deepthi KC, Rao VV, Reddy VR (2015) Effectiveness of plant products in feeding aversion of birds under captive studies. In: Compendium of abstracts of the 2nd international conference on bio-resource and stress management, ANGRAU & PJTSAU, Hyderabad, India, 7–10 January 2015

Dey M, Laishram J (2015) Women's participation in sustainable livelihood: a study on selected villages in and around Loktak Lake of Manipur, Northeast India. In: Compendium of abstracts of the 2nd international conference on bio-resource and stress management, ANGRAU & PJTSAU, Hyderabad, India, 7–10 January 2015

Kadam RP, Ahire RD, Pawar GS (2015) Strategy for documentation and validation of indigenous technological knowledge in bio-resource. In: Compendium of abstracts of the 2nd international conference on bio-resource and stress management, ANGRAU & PJTSAU, Hyderabad, India, 7–10 January 2015

Kale RA, Tikale VS, Tambe DR et al (2015) Constraints faced by farm women in dairy farming. In: Compendium of abstracts of the 2nd international conference on bio-resource and stress management. ANGRAU & PJTSAU, Hyderabad, India, 7–10 January 2015

Kikim A, Khan MR, Yadava PS (2015) Traditional dye yielding plants used by different communities of Manipur, Northeastern India. In: Compendium of abstracts of the 2nd international conference on bio-resource and stress management, ANGRAU & PJTSAU, Hyderabad, India, 7–10 January 2015

Kumari AR, Singh M, Suhane RK (2015) Extent of empowerment of elected women Panchayat Samiti members in Panchayati Raj system. In: Compendium of abstracts of the 2nd international conference on bio-resource and stress management, ANGRAU & PJTSAU, Hyderabad, India, 7–10 January 2015

Lotha N, Sharma A (2015) Production and marketing of tea cultivation in Dimapur district of Nagaland state. In: Compendium of abstracts of the 2nd international conference on bio-resource and stress management, ANGRAU & PJTSAU, Hyderabad, India, 7–10 January 2015

Meena V (2015) Tool for restoration of forest and degraded land in Rajasthan – taxonomy of *Acacia* sp. through seedling morphology. In: Compendium of abstracts of the 2nd international conference on bio-resource and stress management, ANGRAU & PJTSAU, Hyderabad, India, 7–10 January 2015

Nandeshwar NS, Ganvir BN, Tayade PM (2015) Economical analysis of vegetables. In: Compendium of abstracts of the 2nd international conference on bio-resource and stress management, ANGRAU & PJTSAU, Hyderabad, India, 7–10 January 2015

Paul FDZ, Marie TJ (2015) Rural entrepreneurship for sustainable development. In: Compendium of abstracts of the 2nd international conference on bio-resource and stress management, ANGRAU & PJTSAU, Hyderabad, India, 7–10 January 2015

Reddy KG, Reddy KK (2015) Economics of climate resilient practices in a case of direct seeded rice in Andhra Pradesh, India. In: Compendium of abstracts of the 2nd international conference on bio-resource and stress management, ANGRAU & PJTSAU, Hyderabad, India, 7–10 January 2015

Sangeetha SP, Balakrishnan A, Devasenapathy P (2015) Productivity and economics of rice (*Oryza sativa* L.) and black gram (*Vigna mungo* L.) in rice-black gram cropping sequence as influenced by organic manures. In: Compendium of abstracts of the 2nd international conference on bio-resource and stress management, ANGRAU & PJTSAU, Hyderabad, India, 7–10 January 2015

Sharma A (2015) Policy for agricultural development in northeast India. In: Compendium of abstracts of the 2nd international conference on bio-resource and stress management, ANGRAU & PJTSAU, Hyderabad, India, 7–10 January 2015

Shuya K, Nakhro R, Sharma A (2015) Micro-financing for socio-economic development of piggery farmers in Nagaland. In: Compendium of abstracts of the 2nd international conference on bio-resource and stress management, ANGRAU & PJTSAU, Hyderabad, India, 7–10 January 2015

Thakur N (2015) Mainstreaming of gender concerns in agriculture. In: Compendium of abstracts of the 2nd international conference on bio-resource and stress management, ANGRAU & PJTSAU, Hyderabad, India, 7–10 January 2015

Tiwari R, Tomar DS, Dixit AK et al (2015) Impact of advanced transport machinery for reducing drudgery and work-related stress of farm women. In: Compendium of abstracts of the 2nd international conference on bio-resource and stress management, ANGRAU & PJTSAU, Hyderabad, India, 7–10 January 2015

Climate Change: Its Impact on Bio-resource and Sustainable Agriculture

3

Aruna Kumari and Ratikanta Maiti

Abstract

Climate change is a complex alteration of climate, which is subtle and continuous, yet extremely important through its consequences for vegetation of various types that thrived under constant or relatively unchanged climates. Potential adaptation strategies for management of the impact of climate change—viz developing cultivars tolerant to heat and salinity stress and resistant to flood and drought, modifying crop management practices, improving water management, adopting new farming techniques such as resource-conserving technologies, crop diversification, improving pest management, better weather forecasting and crop insurance, and harnessing the indigenous technical knowledge of farmers—are briefly discussed. The chapter makes a brief assessment of research undertaken on the effects of global warming and climate change on various aspects—(1) impact of climate on agricultural production and forestry; (2) crop production; (3) impact of an increasing level of carbon dioxide on security of life; (4) impact of climate change on food inflation; (5) suggestion of various mitigation strategies for climate change; (6) carbon sequestration technology to reduce carbon pollution; (7) climate-smart agriculture; (8) conservation practices under rain-fed agriculture; (9) intercropping; (10) genotype × environment; and (11) impact of climate on livestock production—and discusses technologies that need to be adopted to combat climate change.

3.1 Introduction

Global warming, or climate change, has been strongly attributed to greenhouse gases (GHGs) in the Earth's atmosphere, like carbon dioxide (CO_2), methane (CH_4), nitrous oxide (NO_2) and chlorofluorocarbons (CFCs). These GHGs absorb the thermal radiation emitted by the Earth's surface. Thus, the rising concentrations of GHGs in the atmosphere could lead to a change

A. Kumari (✉)
Department of Crop Physiology, Agricultural College, Professor Jayashankar Telangana State Agricultural University, Polasa, Jagtial, Karimnagar 505 529, Telangana, India
e-mail: arunasujanagcjgl@gmail.com

R.K. Maiti
Facultad de Ciencias Forestales, Universidad Autónoma de Nuevo León, Carr. No 85, km 145, NL 67700 Linares, Mexico
e-mail: ratikanta.maiti@gmail.com

R.K. Maiti et al. (eds.), *Bioresource and Stress Management*, DOI 10.1007/978-981-10-0995-2_3

in energy balance and eventually the world's climate. CO_2 is by far the largest contributor to the man-made enhanced greenhouse effect (IPCC 2007) cited in Crutzen and Wacławek (2015).

Increasing global warming is associated with an increase in abiotic and biotic stresses, which have direct negative impacts on plants, animals, soils and crop productivity.

According to Crutzen and Wacławek (2015), anthropogenic actions have increasing effects on the environment on all scales, in a lot of ways overcoming natural processes. During the last 100 years the human population went up from little more than one billion to six billion and economic activity increased nearly ten times between 1950 and the present time. In the last few decades of the twentieth century, anthropogenic chlorofluorocarbon release caused a dramatic decrease in levels of stratospheric ozone, creating an ozone hole over Antarctica. Thus as a result UV-B radiation from the sun increased, causing an enhanced risk of skin cancer. Release of more greenhouse gases by mankind (such as CO_2, CH_4 and NO_2) to the atmosphere increases the greenhouse effect. Atmospheric greenhouse gas concentrations would continue to rise and remain high for hundreds of years, thus warming Earth's climate. Warming temperatures raise the sea level by melting mountain glaciers and ice caps, due to which portions of the Greenland and Antarctic ice sheets melt or flow into the ocean. Ice loss from the Greenland and Antarctic ice sheets could contribute an additional 19–58 cm of sea level rise, hinging on how the ice sheets react.

3.2 Impact of Climate Change on Agricultural Production and Agroforestry

Global climate change leads to changes in the long-term weather patterns that characterize the regions of the world. Modern agriculture, food production and distribution are the major contributors of greenhouse gases. Agriculture is directly responsible for 14 % of total greenhouse gas emissions. Broader rural land use decisions have an even larger impact. Deforestation currently accounts for an additional 18 % of emissions. In the long run, climatic change could affect agriculture in several ways such as the quantity and quality of crops in terms of productivity, growth rates, photosynthesis and transpiration rates, and moisture availability. Climate change is likely to have a direct impact on food production across the globe. Drivers of climate change through alterations in atmospheric composition can also influence food production directly by their impacts on plant physiology.

Coping with the impact of climate change on agriculture will require careful management of resources like soil, water and biodiversity, while the extension strategies to combat the impact of climate change on agricultural production and agroforestry (Srilathavani and Preethi 2015) have emphasized that to cope with the impact of climate change on agriculture and food production, India will need to act at the global, regional, national and local levels. Extension can help farmers prepare for greater climate variability and uncertainty, create contingency measures to deal with exponentially increasing risks and alleviate the consequences of climate change by providing advice on how to deal with droughts, floods and so forth. Further, extension can also help with mitigation of climate change. This assistance may include providing links to new markets (especially carbon), information about new regulatory structures and new government priorities and policies. Innovating and reinstating effective extension services will become more important than ever in a changing climate.

Rani et al. (2015) discussed the impact of changing climate on agriculture. According to them, climatic change and increasing climatic variability are likely to aggravate the problems of future food security by exerting pressure on agriculture. However, there are a lot of uncertainties about the assessment of the impact of climate change on agriculture. Global climatic changes affect agriculture through their direct and indirect effects on crops, soils, livestock and pests. The increase in temperature, depending upon the current ambient temperature, can reduce crop duration, increase crop respiration rates, change photosynthate partitioning to economic products, affect the survival and distribution of pest populations and hasten nutrient mineralization in soils, decrease fertilizer-use efficiencies and increase evapo-transpiration rates. Indirectly, this may lead to considerable effects on land use due to snow melt, availability of irrigation water, frequency and intensity of droughts and floods, soil organic matter transformations, soil erosion and changes in pest profiles, thereby reducing arable areas due to submergence of coastal lands and availability of energy.

In order to manage the impact of climate change, the potential adaptation strategies are developing cultivars tolerant to heat and salinity stress and resistant to flood and drought, modifying crop management practices, improving water management, adopting new farming techniques such as resource-conserving technologies, crop diversification, improving pest management, better weather forecasting and crop insurance, and harnessing the indigenous technical knowledge of farmers.

Climate change (and agricultural climate change) is a complex alteration of climate, which is subtle and continuous, yet extremely important through its consequences for vegetation of various types that thrived under constant or relatively unchanged climates. Some of the main effects of climate change with specific reference to agriculture and food production especially during the last decade are attributed to increased occurrence of storms and floods, increased incidence and severity of droughts and forest fires, steady spreading out of frost-free intervals and potential growing seasons, increased frequency of diseases and insect pest attacks, and vanishing habitats of plants.

In the face of growing climatic variability impacting the existing ecosystem, it is important for the international scientific community (Kumawat et al. 2015) to use all accessible knowledge to stop or reverse this trend to the maximum extent possible. There have been various attempts and viable measures in the past to bring down atmospheric GHGs to slow down climate change. The early impacts of climate change already are being felt worldwide. Future impacts will affect a broad array of human and natural systems, with consequences for human health, food and fibre production, water supplies, and many other areas vital to economic and social wellbeing.

Moderation of greenhouse gas emission is required to combat the effect of climate change. Rice–wheat cropping system agricultural interventions contribute towards the greenhouse gas (GHG) emissions responsible for global warming that have put forward changes in climatic variables. Efforts should be made to reduce GHG emissions for restraining global warming. The feasibility of different management practices for mitigating the greenhouse gas emissions from rice–wheat cropping sequences evaluated by Choudhury et al. (2015) revealed that out of the three different crop establishment methods as the main plot treatments, M_1 (rice: SRI, wheat: conventional tillage), M_2 (rice: puddled transplanted rice, wheat: conventional tillage + 30 % residue incorporation), M_3 (rice: DSR, wheat: zero tillage + 30 % residue retention) and four nutrient management sub-plot treatments—viz S_1 (100 % of recommended dose of fertilizer (RDF) through inorganic sources), S_2 (75 % of RDF through inorganic sources + 25 % N of RDF through organic sources), S_3 (50 % of RDF through inorganic sources + 50 % N of RDF through organic sources) and S_4 (S1 + Moong bean as green-manure) DSR—emitted lower CH_4 (1.39 mg m^{-2} h^{-1}), CO_2 (0.57 mg m^{-2} h^{-1}) and N_2O (0.36 mg m^{-2} h^{-1}) at the maximum tillering stage of rice. The same trend was followed under zero tillage in wheat with lower emission ranges of all three gases, i.e. 0.95, 1.29 and 0.58 mg m^{-2} h^{-1}, respectively. The lowest emissions of CH_4 and CO_2, with values of 1.87 and 1.24 mg m^{-2} h^{-1}, respectively, from rice and 1.57 and 3.23 mg m^{-2} h^{-1}, respectively, from wheat, were observed under 100 % RDF through inorganic fertilization, whereas N_2O emissions were in reverse to the emissions of CH_4 and CO_2. Crop establishment is good through minimum soil disturbance along with 50 % supplementation of inorganic fertilizers; organics could be an option to lower GHG emissions.

3.3 Impact of Climate Change on Productivity of Crops

Climate change is a burning issue and one of the major problems at the present time in the agriculture field. Climate change is both harmful and beneficial for some plants. An increased level of CO_2 is beneficial for CAM and C_4 plants where the higher CO_2 levels are used for increased photosynthesis and growth rates but in cereals and other crops this climate directly affects growth, reducing growth and production. Increased CO_2 in soil affects the growth of soil micro-organisms and root growth. Increases in temperature result in increasing evapo-transpiration rates, less water availability and high demands for water for growth. Flowering and fruiting rates are adversely affected and plants are unable to survive in high temperatures. If the temperature rises 10 °C, the yield of cereal crops decreases by 4–5 million tonnes per year. Increasing temperatures cause reductions in the abundance of soil micro-organisms, which are beneficial for nutrient mobilization and fixation. They also adversely affect nitrogen availability. Climate change also accelerates the emergence of insect pest populations, raising the chance of crop infestation and resulting in immortality. There is a need for some effective policy and application of better scientific management for regulation of climate in a sustainable way with betterment of biodiversity and crop production which helps in reduction of poverty and maintenance of the national economy (Panigrahi and Das 2015).

Climate change will be an additional challenge to produce enough food grain for the ever-growing population. In this direction, annual, monthly and season-wise weather data for 25 years (1986–2010) have been analysed using the Weather Cock Analysis tool developed by CRIDA, Hyderabad, for establishing the effect of climate change trends and their effect on crop growth and development in the North Eastern dry zone of Karnataka. The North Eastern dry zone of Karnataka belongs to the semi-arid tropic zone and in a normal year receives around 650–750 mm of rainfall. The 25-year rainfall data showed that for 10 years there were no meteorological droughts, for 12 years mild droughts occurred and in the remaining 3 years moderate droughts occurred, whereas with respect to agriculture drought, this occurred for a period of 2 weeks during the rainy seasons in 1986, 1993, 1994, 2006 and 2008 and for a period of 1 week in 1999, 2001–2005 and 2009 (Shanwad et al. 2015).

The 25-year data indicated that the highest rainfall of 1054.2 mm was recorded during 1998 and the lowest rainfall of 435 mm was observed during 2003. The highest rainfall (249.6 mm) on a single day, which was recorded on 2 October 2009, caused heavy damage to human life and agricultural crops in the North Eastern parts of Karnataka and the Telangana regions of Andhra Pradesh. An annual maximum temperature of 35 °C was recorded during 2002 and 2007 and an annual minimum of 18.9 °C was recorded during 2010 (Shanwad et al. 2015) in studies of climate change trends and their effects on agriculture crops in the North Eastern dry zone of Karnataka, which also indicated that in the past 75 years the lowest temperature (7.3 °C) was recorded on 19 December 2010 and on 22 December 2010. These variations in weather parameters will definitely affect crop growth and development. The crop losses due to climate variability will vary from region to region depending on regional climate, crops and cropping systems, soils and management practices. Further, studies using advanced 'temperature gradient tunnels' (Shanwad et al. 2015) with different dates of sowing revealed that a temperature increase of 1–2 °C reduced the yield of cotton from 48 to 14 q ha^{-1}, pigeon pea from 15 to 7 q ha^{-1} and quality protein maize from 18.5 to 8.5 q ha^{-1}. However, increases in yields of all crops were noticed with a CO_2 increase from 390 to 550 ppm (Shanwad et al. 2015). But the interaction of temperature and CO_2 showed mixed results.

Climate change is associated with slight shifts in seasons and the occurrence of erratic and untimely rains in most parts of the country. Berseem (*Trifolium alexandrium* L.) is an important winter forage crop in Northern India, occupying the second largest area under fodder crop cultivation. The merit of this annual legume crop lies in its multi-cut nature (five to seven cuts), a long period of fodder availability (from November to April) and high green fodder yield (80–120 t ha^{-1}). The influence of weather on its seed yield (Vijay et al. 2015) indicated that there was induction in the number of flowers (21–25 %) followed by reduced seed set (20–50 %) during 2013–2014 compared to 2012–2013. This reduction transformed into a severe yield loss (47–78 %) during 2013–2014. The analysis of weather parameters pinpointed the occurrence of winter rains and increased RH during the pre-flowering and flowering periods during 2013–2014. Thus, these findings showed that weather conditions, particularly rainfall during the flowering season, play a pivotal role in seed yield in insect-pollinated crops like Berseem and necessitates more study on it in the era of climate change.

Post-rainy season *rabi* sorghum genotypes in response to changing climate (Alse et al. 2015) indicated that sowing of *rabi* sorghum in the first week of October produced higher grain yields (3552 kg ha^{-1} in 2012–13, 2473 kg ha^{-1} in 2013–14, pooled mean 3013 kg ha^{-1}) and Stover yields (10,408 kg ha^{-1}, 6718 kg ha^{-1}, 8563 kg ha^{-1}, respectively) than sowing in the first or third week of September, the third week of October or the first week of November. The grain and Stover yields were higher due to ideal atmospheric conditions for crop growth in the first week of October. Early sowing dates were found to invite pests (shoot fly and stem borer), disease (rust) and birds. Late sowing suffered from moisture stress during the flowering and grain-filling stages and led to lower grain and Stover yields.

During the first year of experimentation, the sorghum genotype Phule Revati produced significantly higher grain (3131 kg ha^{-1}) and Stover yields (9091 kg ha^{-1}) than CSH 15R, CSV 22R and Phule Anuradha; however, contrasting results were observed during the second year of investigation and in the pooled mean, where CSH 15 R was found to be significantly superior, and it was on a par with Phule Revati and CSV 22R over the rest of the genotypes in respect of grain yield. The Stover yield was significantly higher under CSV 22R, which was comparable with Phule Revati (Alse et al. 2015).

Heavy rain impact on vertisols during the maturity period of certain spices, viz coriander, fenugreek and ajwain grown in Andhra Pradesh, were assessed by Kalidasu et al. (2015). Coriander, fenugreek and ajwain are the important seed spices grown in Andhra Pradesh. The crop growth period of these spices mostly spans the entire *rabi* period starting from October. The harvesting window of short-duration seed spices (less than 100–120 days) falls between January and March in South India and between February and March in North India. The long-duration seed spices (more than 150 days) come to harvesting during February to March, if they are *rabi* sown. The harvesting window occasionally meets with unforeseen rain in many places and, depending on the intensity of rainfall, these crops get affected. The effect of such undesirable rainfall is less documented and information is scanty. Heavy rainfall during vegetative stages led to temporary wilting to premature death of coriander and ajwain plants. Incessant rains during the maturity period not only affect grain quality through discolouration but also lead to grain shattering and vivipary in coriander and fenugreek. Varied responses were also observed in moisture retention and vivipary in coriander genotypes. As climate change is inevitable, data collection regarding damage due to unforeseen climatic conditions is vital for better management of seed spice crops in future.

In coriander (*Coriandrum sativum* L.), several issues like lack of a unified protocol, delineation of suitable test environments and difficulties in selection of a genotype over a wide range of environments are becoming difficult in selection of cultivars. A study undertaken in 13 coriander genotypes to assess their performance in multi-environments (Giridhari et al. 2015) by GGE biplot analysis revealed that in the mega-environment analysis (Genotype by Environment Data analysis) the genotype DH-220 was the highest yielder in the test locations of Jabalpur, Jobner, Hisar, Raigarh and Pantnagar, and RKD-18 in the test environments of Kota, Jagudan, Ajmer, Navsari and Lam. The AEC view (Average Environment View) of the individual GGE biplot of the two mega-environments helped to rank genotypes within those environments, respectively. LCC-236 and DH-220 were found to be promising in the respective mega-environments. The evaluation of test environments revealed mega-environments as ideal test environments among the environments that were evaluated.

Micro-analysis of all the environments revealed two diverse mega-environments, i.e. niche 1: Lam, Ajmer, Navsari, Jagudan and Kota; niche 2: Pantnagar, Raigarh, Dholi, Hisar, Jobner and Jabalpur. Among the 11 test environments, Dholi, Pantnagar, Kota and Jagudan were the most discriminative of the tested genotypes. A simplistic environment model from a study by Giridhari et al. (2015) indicated that for reasonable discrimination of a large number of genotypes, only three ideal test environments (Pantnagar, Dholi and Jagudan) can be deployed, thus saving resources and energy. For further discrimination among elite ones, full set environments may be utilized. Similarly, robust and effortless screening of a few elite genotypes can be made in the aforementioned three environments. However, these propositions need further validation with existing and new multi-environment data (MET) to draw more accurate and robust conclusions.

The Karl Pearson's correlation and multiple linear regression correlation analysis carried out in a study on the impact of change in weather on productivity of paddy (Rao et al. 2015) revealed that productivity of paddy is positively correlated with rainfall (0.44), followed by maximum temperature (0.07), but negatively correlated with minimum temperature (−0.42), followed by evaporation (−0.36), etc. Among the weather parameters, the significantly highest positive correlation was found between minimum temperature and evaporation (0.75) while the highest significant negative correlation was between maximum temperature and afternoon relative humidity (−0.55). Regression estimates indicate that a unit increase in rainfall results in a 0.01-unit increase in yield, whereas a unit increase in evaporation decreases the yield by 0.41 units.

In sweet corn hybrid Sugar 75, weather parameters during *kharif*, *rabi* and summer had a significant influence on its growth (Lakshmi et al. 2015a) and yield. Results in individual seasons and pooled over the seasons, i.e. *kharif*, *rabi* and summer, showed positive and significant phenotypic and genotypic correlations among all the traits during the *kharif* season of 2009, except for plant height. Plant height showed a significant correlation with dry matter content but showed a non-significant correlation with the remaining traits in 2009. Environmental correlations were non-significant for the majority of the traits in all the seasons and pooled over the seasons. Pooled data on correlation of weather parameters with yield and yield-contributing characters showed that in *kharif* (the rainy season) maximum temperatures had positive significant correlations with all of the traits except plant height, 100 seed weight and seed weight per cob. Rainfall, rainy days, RH 1 and RH 11 had negative significant correlations with all the traits. In *rabi*, relative humidity had positive and significant correlations with dry matter, cob length, cob girth and kernels per row, cob weight with husk, green cob and fodder yields. Further, in the summer season, minimum temperature had positive and significant correlations with plant height, dry matter, cob length, cob girth, kernel rows, kernels per row, fodder yield and LAI. Weather parameters like maximum temperature in *kharif*, relative humidity in *rabi* and minimum temperature in the summer seasons had profound effects on yields of sweet corn.

Cotton production in Andhra Pradesh is mostly rain fed and predictions show that the precipitation in the region due to climate change may benefit cotton farmers if the monsoon comes at a proper time. Six cotton entries (DHH 543, G Cot H 8 BGII, G Cot H12 BGII, Ankur 3028 BGII and LH 144/12) in black cotton soils under rain-fed conditions were studied for the impact of weather on plant growth, disease incidence and seed cotton yield (Kumari et al. 2015). Of the entries tested, the entries NCS 145 BGII, Ankur 3028, BGII and LHH 144/12 recorded significantly superior seed cotton yield under both normal and delayed sowing conditions, whereas under delayed sowing conditions, the entries NCS 145 BGII and Ankur BGII gave significantly higher seed cotton yields. A significant variation was also noticed in the boll number due to dates of sowing. Under normal sowing, LH 144/12 produced higher numbers of bolls per plants, whereas under delayed sowing conditions, the highest number of bolls was recorded by NCS 145 BGII. The highest seed cotton yield under late-sown conditions with NCS 145 BGII and Ankur 3028 BGII was associated with higher flower numbers, % fruit set and boll numbers of plants. Regarding disease incidence, the entries Ankur 3028 BGII, NCS 145 BGII and LH 144/12 recorded lower incidence rates of leaf spot diseases, especially NCS 145 BGII with regard to *Alternaria* leaf spot and the other two entries with regard to *Cercospora* leaf spot, *Myrothecium* leaf spot and grey mildew.

The sub-mountainous Himalayan region of Uttarakhand is characterized by events like prolonged droughts and flash floods in the hills. Most of the farms in this region are small and marginal, therefore they suffer the most. Finger millet, along with other millets, has been termed as climate resilient and one of the hardy crops that can withstand changing climatic conditions more efficiently. There is a necessity to understand the relationship of climatic factors to different crop stages over the years. Sood et al. (2015) studied the effect of changing climatic conditions on grain yield of finger millet. The years were grouped into two categories, namely group A and group B, for each variety. Group A had all the years where grain yield was above the average and group B had all the years where grain yield was below the average of 11 years. The different meteorological parameters of group A and group B were averaged separately for valid interpretations. There was a decreasing trend in minimum and maximum temperatures and rainfall in the *kharif* seasons. For those years when the average yield was higher (Group A), the mean maximum, minimum and mean temperatures and sunshine hours for August to September were higher, and rainfall, as well as the numbers of rainy days, for August and September were lower in comparison to the years where the yield was lower than the mean yield (Group B).

It was observed that rainfall during June–July enhanced and high rainfall during August–September reduced the yield level of finger millet crop in the hills. This was found to be true for all the three varieties but the long-duration variety PR 202 failed to produce a higher yield due to low temperatures in October, affecting grain filling. It indicated clearly that early and medium-duration varieties of finger millet are suitable in hill ecosystems and are resilient to changing environmental conditions in the hills.

3.4 The Impact of Increasing Carbon Dioxide and Its Future

An increase in atmospheric carbon dioxide levels will improve the fertilization effect on crops with the C_3 photosynthetic pathway and thus will promote their growth and productivity. Atmospheric concentrations of carbon dioxide have been steadily rising, from approximately 315 ppm in 1959 to a current atmospheric average of approximately 385 ppm. The knowledge of plant responses to future CO_2 concentrations rests on the results of experiments on free-air carbon dioxide enrichment (FACE). FACE experiments by Aatla (2015) have been developed to understand the effects of elevated levels of CO_2. Natural or agricultural ecosystems were allowed to be fumigated with elevated concentrations of CO_2 in the field without use of chambers.

Across a range of FACE experiments, with a variety of plant species, growth of plants at elevated CO_2 concentrations of 475–600 ppm increased the leaf photosynthetic rates. Carbon dioxide concentrations were also found to be important in regulating the openness of stomata, allowing CO_2 to diffuse into leaves for photosynthesis, but also providing a pathway for water to diffuse out of leaves. As CO_2 concentrations increase, plants can maintain high photosynthetic rates with relatively low stomatal conductance. In FACE experiments, growth under elevated CO_2 decreased stomatal conductance of water, resulting in a decrease of overall plant water use.

A secondary effect of elevated CO_2 is increased dry matter production. Elevated CO_2 also leads to changes in the chemical composition of plant tissues. Leaf nitrogen concentrations in plant tissues typically decreased in FACE under elevated CO_2, with decreased nitrogen per unit of leaf mass. Further, there was a decrease in protein concentrations in grains under elevated CO_2. Crop concentrations of nutritionally important minerals including calcium, magnesium and phosphorus also decreased under elevated CO_2. These studies indicated that single-cell photosynthesis with a substantial increase in the concentration of CO_2 around the carboxylating system(s), by engineering C_4 genes into C_3 plants, could lead to producing a C_4-like environment in the plants.

3.5 Intricacies of Climate Change

Climate change threatens to increase soil erosion, reduce soil quality, lower agricultural productivity, and negatively impact food security and global sustainability, making it one of the most severe challenges that are to be faced in the twenty-first century.

Mondal and Thokchom (2015) conducted a study to assess the perception of climate change and its effect on the environment as perceived by the scientists of the Central Rice Research Institute, Cuttack. Data were collected from 40 scientists by a questionnaire method in the months of October–December 2011. For this purpose, five important sectors (viz the environment, agriculture and allied, coastal zone, forest and wild life, hydrology and water resources) were identified and presented to the scientists in ten pairs. It was found that due to climate change the environment will be severely damaged as perceived by the scientists of CRRI, followed by agriculture and the allied sector, coastal zone, hydrology, water resources, forest and wild life.

3.6 Impact of Climate Change on Food Inflation in India

Food inflation affects everyone, as food is a basic necessity for health. The poor people in a society spend a large part of their income on food, thereby exhibiting the greater impact of food inflation on them. Food production in India depends mainly on the weather and the amount and pattern of rainfall received. With the changing climate, temperate and rainfall are changing. This will impact agriculture, leading to uncertainty in production; any changes in production will translate into price changes affecting every consumer.

Bhat and Swamy (2015) discussed the impact of climate change on food inflation in India and possible measures for adaptation. It is important to analyse food inflation in India in the context of climate change. An attempt has been made to understand the issues related to food inflation in India and the methodologies followed in India to measure inflation and differences between wholesale and consumer price indexes. The share of food was higher in rural areas and also with the rise in income, the share of food expenditure was reducing in both urban and rural areas. This exhibits the significance of food for lower-income groups and also for the people living in rural areas. Inflation in different food commodities and components has been compared with non-food inflation.

Trends in food inflation in India have been compared with other developing nations and with developed nations. To control food inflation, several initiatives have been undertaken in India. The Central Government has advised State Governments to delist fruits and vegetables from the Agriculture Produce Marketing Committee Act. This can help to reduce the confusion among traders, creating competition among them. This will also lead to a unified national market allowing free movement of fruits and vegetables across the country.

The Essential Commodities Act 1955 and the Prevention of Black-Marketing and Maintenance of Supplies of Essential Commodities Act 1980 help the Government to take action against unlawful hoarding and black marketing. In India, the period from September to January coincides with the festival season and the lean period in terms of arrivals of vegetables. Traders and middlemen take advantage of this situation by hoarding produce. Under the circumstances of changing climate, suitable measures for adaptation have been suggested for controlling food inflation in India.

3.7 Mitigation Strategies for Climate Change

Management of climate change mitigation and adaptation is key for environmental conservation, sustainability of cropping systems, soil and water quality, and food security. Certain management decisions regarding conservation practices, such as no-till, conservation of agriculture and returning crop residue to the field to increase nutrient cycling, can contribute to carbon sequestration and help us mitigate and adapt to climate change.

Additionally, management of grasslands, restoration of degraded or desertified lands, nitrogen management to reduce greenhouse gas emissions, precision conservation management at a field and/or watershed level, and other management alternatives can also help us mitigate and/or adapt to climate change.

Different aspects of carbon management, nitrogen management, manure management, management in low-input systems (sustainable agriculture) and grazing land management were discussed in Climate Change Mitigation, Adaptation And Sustainability In Agriculture: Issues And Intricacies (Verma et al. 2015), and indicated that certain management decisions that reduce soil erosion, increase carbon sequestration to improve soil function, soil quality and soil health, and contribute to the resilience of soils and cropping systems will be needed to respond to climate change and related challenges such as food security. They suggest that without management decisions that increase soil and water conservation, food security for the world's growing population will be harder to achieve.

3.8 Carbon Sequestration

Various mitigation practices are being adopted to reduce the negative effect of climate change on plants and crops. Carbon dioxide sequestration is an effective strategy to reduce carbon load from the atmosphere. Carbon sequestration means removal of CO_2 from the atmosphere by plants, to be converted into organic matter by microbes and stored in the soil. Coconut plantations or farm ecosystems could be used (Swetha and Manthri 2015) in many ways to reduce CO_2 emissions via carbon capture or sequestration in the crop-soil system: (1) substitution of fossil fuel with biodiesel or biomass from coconut oil; (2) sequestration of C in coconut plantations, as a mono-crop or with intercrops; (3) enhancing C sequestration through coconut plantation management; and (4) conserving C sinks in coconut farms. The coconut tree, a woody perennial with a single main stem, meets the FAO criteria for a "forest". The Philippine Department of Environment and Natural Resources has included coconut as a reforestation crop. Coconut and other plantation crops contribute significantly to the ecosystem by sequestering CO_2.

Carbon sequestration in rice soils under long-term integrated nutrients on inceptisols has been studied to assess the long-term effects of chemical fertilizers alone and in combination with organic sources on sustainability and soil fertility under a rice–rice cropping system (Sridevi et al. 2015) with 12 treatments comprising chemical fertilizers alone and in conjunction with FYM, paddy straw and *Gliricidia* at 50 % and 25 % substitution levels of nitrogen during *kharif* and with 75 % and 100 % RDF in *rabi*. Carbon sequestration was studied as changes in fractions of oxidizable soil organic carbon (SOC) and profile carbon stocks at the end of the 24th crop cycle (2011–2012), by collecting depth-wise soil samples at 15 cm intervals from 0 to 60 cm. There was significant variation in the labile and non-labile fractions of SOC in the top layer (0–15 cm) while at lower depths (15–60 cm) such a variation due to the nutrient supply system was not significant at the end of 24 years. The amount of total SOC sequestered over 24 years at the 0–60 cm depth was lowest with control and was highest with 50 % RDF + 50 % N substitution through FYM. The total amount of carbon sequestered at the 0–60 cm depth was increased by 18–26 % with nitrogen substitution through glyricidia, 31 % with recommended doses of fertilizers, 35–38 % with nitrogen substitution through paddy straw and 44–48 % with nitrogen substitution through FYM over control. But when the treatments were assessed with reference to 100 % NPK for their carbon sequestration potential, paddy straw substitution contributed to an extent of 24–31 % additional total SOC and FYM substitution resulted in 56–58 % higher total SOC sequestration over chemical fertilization, while *Gliricidia* green leaf manuring did not contribute to carbon sequestration. Passive SOC, a stable form, was highest with 50 % RDF + 50 % N substitution through FYM, followed by 75 % RDF + 25 % N substitution through FYM, contributing to 41.4 and 35.8 % increases over control and 15.7 and 11.0 % increases over 100 % NPK at the 0–60 cm depth.

An assessment of soil C was made by assessing the soil carbon pools by various crop genotypes adopted to mitigate climate change (Sarangi et al. 2015) in two drought-prone districts, Sonepur (Badma) and Jharsuguda (Bhoimunda). Green gram variety Durga, black gram variety Prasad, cowpea variety Utkal Manic, groundnut variety Smruti and maize hybrid MH-9468 at Sonepur, and green gram variety TARM-1, black gram variety PU-31, cowpea variety YB-7, groundnut variety Devi and maize hybrid P-30R77 at Jharsuguda were taken up as interventions. Technological interventions through respective Krishi Vigyan Kendra's were practised in farmers' fields. The soil organic carbon (SOC) and its pools easily available SOC fraction (EAC), slowly available SOC fraction (SAC) and durable SOC fraction (DC) were determined after the second year from 420 post harvest surface soil samples from a depth of (0–15 cm) were analysed. The easily available SOC fraction was about 10 % of the SOC, the slowly available SOC fraction was about 50 % of the SOC and the durable SOC fraction was the rest. Among all the interventions, the improved genotypes produced higher SOC than their local genotypes with hybrid maize P-30R77 being the highest (1.5925 %) followed by groundnut variety Debi (1.12 %) recorded at Jharsuguda. A maize crop had the highest durable SOC fraction. Among legumes, ground nut followed by cowpea produced higher DC, and green gram was higher than black gram in terms of the durable SOC fraction.

Carbon sequestration studied through coconut-based cropping systems (Swetha and Manthri 2015) revealed that coconut-based cropping systems can be a mitigation strategy for reducing carbon dioxide load in the atmosphere, which is claimed to result in global warming, unpredictable rainfall patterns, floods and droughts. It has a serious impact on food security and agricultural productivity as climate affects the capacity of a country to feed its people and generate an adequate quantity and acceptable quality of crops in a sustained way. In response to the global warming crisis, the Kyoto Protocol has introduced ground-breaking concepts on carbon credits, carbon footprint and emissions trading. One carbon credit permits 1 tonne of carbon dioxide or a corresponding amount of other greenhouse gases to be discharged in the air. Thus, a new commodity was created in the form of emission reductions or removals. Since carbon dioxide is the principal greenhouse gas, people speak simply of trading in carbon. Carbon is now tracked and traded like any other commodity. This is known as the "carbon market". India and China are likely to emerge as the biggest sellers of carbon credits through carbon sequestration.

Though preliminary studies have been done on carbon sequestration of coconut in some countries, more detailed studies on C sequestration with reference to different agro-climatic regions of coconut are yet to be carried out. Efforts are to be made to include coconut and other plantation crops in carbon trading or clean development mechanisms (CDMs) at the highest policy-making level. Carbon sequestration below ground in the form of organic carbon and microbial dynamics in coconut and coconut-based cropping or farming systems should be studied in detail.

3.9 Climate-Resilient Technologies

The impact of climatic factors is more pronounced in the case of rain-fed ecosystems. Rain-fed ecologies are characterized by their proneness to various production constraints provoked by the prevailing environmental conditions. In rain-fed areas, mainly the problems of land degradation, soil erosion and moisture stress lead to low-input responses in agriculture and thus low productivity. Singh et al. (2015) discussed the potentials of climate-resilient technologies to revamp agriculture in fragile rain-fed ecosystems in a study conducted, under the National Initiative on Climate Resilient Agriculture (NICRA), in Said-Sohal village in the Kathua district of Jammu and Kashmir state. The potentials of various climate-resilient technologies to insulate agriculture against weather abrasions were assessed by performing activities towards natural resource management, crop production, livestock, institutional arrangements, capacity building and other extension activities. All the activities in integration had a positive effect in providing climate proofing for agriculture and indicated that these activities could successfully address the problem of land degradation and soil erosion, enhancing the access to life-saving irrigation and green fodder availability to a certain extent. Further, these integrated activities led to enhanced productivity of crops and livestock, and made the village self-reliant in oilseed and pulses. Besides this, the institutional arrangements, capacity-building programmes and extension activities had positive commutative effects in ushering the capacity, as well as the access, of the farmers to technological provisions leading to better agriculture under environmentally fragile conditions.

3.10 Sustainable Agriculture

With the development of human civilization on the planet Earth, agriculture has been an integral part of human needs and has subsequently developed as a basic human right. This led to the enactment of a Food Security Bill in India in the year 2013. Nowadays, boosted by new studies indicating that agriculture can be a cost-effective emissions reduction strategy and identifying many synergies between adaptation and mitigation, climate-smart agriculture is becoming the new paradigm. FAO (2010 cited in Meena et al. 2015b) has defined agriculture as "Agriculture that sustainably increases productivity, resilience (adaptation): reduces or removes GHG (mitigation) and enhances achievement of national food security and development goals". It is composed of three main pillars: sustainably increasing agricultural productivity and incomes; adapting and building resilience to climate change; and reducing and/or removing greenhouse gas emissions, wherever possible (Meena et al. 2015b). These are the needs of the day and future climate-smart agriculture.

Organic farming helps to improve soil productivity and attain sustainable yield in various crops. Chandrappa et al. (2015) discussed the development of organic farming practices for system-based high-value crops by inclusion of eight treatments: combination of recommended NPK + 50 % N as FYM, 10 kg $ZnSO4$ ha^{-1} based on soil tests of recommended NPK + 50 % N as FYM, 10 kg $ZnSO4$ ha^{-1} based on soil tests, organic sources of $N^{1}/_{3}$ on an equivalent basis either through FYM + vermicompost + non-edible oil cake intercropping or addition of trap crops (location-specific in each season); weed and pest control as per recommendations (no inclusion of chemicals, pesticides or herbicides); rock phosphate to substitute the P requirement of crops + phosphate-solubilizing bacterial culture (PSB); biofertilizer containing N and P carriers; and 100 % NPK with secondary and micronutrients + 10 kg $ZnSO4$ ha^{-1}. The maximum grain yield of maize (6022.6 kg ha^{-1})

was obtained in kharif through application of 50 % recommended NPK + 50 % N as FYM + inorganic sources of micronutrients + 10 kg $ZnSO_4$ ha−1. Similarly, the highest pod yield (2096.10 kg ha−1) was registered in a groundnut crop with application of 100 % NPK + secondary and micronutrients based on soil tests. These studies concluded that both organic and inorganic sources of nutrients provide better nutrient supply for obtaining sustainable yields.

3.11 Intercropping

Investigations carried out in two agroforestry models, viz Aswagandha and *Andrographis* intercropped agri-horticultural systems in 4-year-old Amla and *Terminalia* plantations with nutrient management practices, indicated that among the different cropping situations studied (Loria et al. 2015), growth parameters like plant height, dry matter production and leaf area per plant were markedly higher under a sole cropping situation when compared to an intercropping situation both in anola and *Terminalia*. The profit per rupee investment also showed the same trend under the three cropping situations with the adoption of integrated nutrient management practices in both the agroforestry models. Land equivalent ratio (LER) or income equivalent ratio (IER) values were also maximum under the intercropping situation in *Terminalia* as compared to anola intercropping.

3.12 Conservation Practices Under Rain-Fed Rice Fallows

Chickpea productivity was enhanced through the adoption of a strategy of conservation practices under rain-fed rice-fallow regions (Kumar et al. 2015). A considerable area of rice (11.7 m ha out of 42.56 m ha) remains fallow after a rice harvest. The major parts lie in Eastern and Central India. Soil- and water-related factors like low moisture content in the soil after a rice harvest followed by a fast decline in the water table with advancement of crop seasons and mid- and terminal droughts at the flowering and pod-filling stages are the reasons for fallowing after rice.

During the post-rainy season (winter season), due to lack of irrigation facilities, second winter crops depend on the effective utilization of carry-over residual soil moisture. The inherited soil physical and biological constraints of rice fallow affect seed germination, seedling emergence and crop establishment due to disruption of soil structure, poor aeration and mechanical impedance of the seed zone. Short-duration low-water-requiring pulses combined with suitable crop management techniques for utilization of residual soil moisture offer an excellent opportunity for utilization of carry-over residual soil moisture in rice fallows (Sireesha et al. 2015).

Rice crop residue mulch can be an effective measure for reducing soil evaporation, increasing water storage and minimizing weed infestation. Thus, it facilitates more retention of soil moisture, helps in control of temperature fluctuations and improves physical, chemical and biological properties of soil, as it adds nutrients to the soil and ultimately enhances the growth and yield of crops. Further, it boosts the yield by 50–60 % over no-mulching under rain-fed situations.

3.13 Bio-resource Conservation

Bio-resource conservation is important under climate change conditions. Several beverage crops such as coffee, cocoa and tea form important bio-resources for conservation under climate change (Vinutha and Swamy 2015). Coffee, cocoa and tea are major agricultural commodities in the world. The waste produced from these industries disposed to the environment and to near water bodies causes severe health hazards such as giddiness, skin irritation, stomach pain, nausea and breathing problems.

Coffee pulp has caffeine and polyphenols, which are anti-physiological factors for animals. Coffee pulp disposed to nature without any treatment causes putrefaction of organic matter in soil. Hence, it is essential to manage this preferably by organic means. Coffee pulp is rich in organic matter. It is an ideal substrate for microbial processes. On the other hand, coffee, cocoa and tea waste has many health benefits, such as soluble dietary fibre, and coffee pectin boosts the levels of high-density lipids and helps in locking bile lipids. Mucilage of coffee and cocoa contain anthocyanin, paranthocyanin, bioflavonoids and antioxidants. The by-products produced will provide additional income for the farmers and improve the economic status of the farming community. Coffee pulp is a waste material from the coffee industry. It can replace up to 20 % of commercial concentrates in dairy cattle feeding, with no adverse effects and a 30 % cost saving. The water drained from coffee cherry extract is another potential source of biogas production. Coffee pulp is turned over every few days in a heap preserved for a few years as in conventional compost making. However, it is a good source of humus and organic soil carbon. One kilogram of coffee bean product produces 1 kg of coffee husk in the wet process, while the dry process produces 0.18 kg of coffee husks (Vinutha and Swamy 2015).

Pelletized dried cocoa pod husks can be used as animal feed. Cocoa pod husk ash is used mainly for soft soap manufacture. Pelletized dried cocoa pod husks can be used as animal feed. Cocoa pod gum is extracted from cocoa pod husks by alcohol precipitation. Cocoa pod gum can be used as a binder in the food and pharmaceutical industries for binding pet foods, emulsifiers, pharmaceutical pills, etc. Cocoa pod husks could also be used as fertilizer for food crop production. Cocoa bean shells can be used as organic mulch and soil conditioner. Products from cocoa pulp are soft drinks; fresh cocoa pulp juice is collected, sterilized and bottled. For the production of alcohol or alcoholic drinks, the fresh juice is boiled, cooled and fermented with yeast. Pectin for jam, jelly and marmalade is extracted from the sweatings by precipitation with alcohol (Vinutha and Swamy 2015).

The waste from tea factories amounts to about 2 % of total production. Tea factory waste (TFW) has been successfully used for the removal of various toxic metals (Zn, Cr, Ni, Cu, Cd and Pb). TFW characterization shows a high surface area for adsorption. Currently it is either returned to the field as fertilizer or sold at a very low price and is used to produce caffeine, polyphenol pigments (edible colours), polymers, animal feed, foaming agents, vinegar, and tea seed oil as an alternative to groundnut and olive oils for cooking, cosmetics and pharmaceuticals. Refuse tea is a by-product of the tea industry removed during tea processing. It comprises mainly fibre and stalks removed from winnowers, dryers and colour separators in the factory (Vinutha and Swamy 2015).

3.14 Precision Sprayers for Sustainable Agriculture

Precision canopy sprayers can effectively be used to protect the environment for sustainable agriculture (Chowda and Swamy 2015). It has been emphasized that pesticide spraying generally provides effective crop protection, but this is very much dependent on the correct type of equipment and its usage, though some wastage of pesticide to the surroundings either by missing or overdosing is inevitable. The use of chemicals has contributed to weed control and prevention of biotic stresses such as diseases and insect infestation. An estimate of crop and crop produce loss shows that annually about one third of total food production is being destroyed by pests. Spraying machines applying pesticides have also given rise to concern as they frequently create drift, which reduces efficiency, affects off-target crops and creates environmental pollution and operator contamination. A sprayer is a machine to apply the fluids in the form of droplets. Precision spraying of the canopy allows applying pesticides only to the target areas, in correct quantities according to the canopy size, growth stage and season. It helps to apply products in an economic and environmentally sound manner. Weed control, disease infection, environmental pollution and insect damage are significant issues in agricultural crop production. Weeds compete with crop plants for moisture, nutrients and sunlight. Some insects that attack crops, including aphids, spider mites and whiteflies, feed on the underside of leaves or beneath the plant canopy. Disease infection and insect damage have detrimental effects on crop yield and quality. Nearly 60 % of the total losses in grape production worldwide are due to virus diseases. The use of canopy sprayers and chemicals has contributed to weed control and prevention of biotic stresses, such as diseases and insect infestation, and to protecting the environment.

3.15 Climate-Smart Agriculture

Climate change has direct impacts on agriculture. Under these circumstances it is essential to adopt climate-smart agriculture. In the face of the depletion of fossil fuel resources on a worldwide basis there is an urgent search for alternative energy sources to meet the present day demands. Solar energy is a clean, inexhaustible and environmentally friendly potential resource among the renewable energy options. Sinha et al. (2015) undertook a study of solar energy for parameterization of a 100 kWp solar power plant for electricity generation. Energy generation by a proposed grid-connected SPV power plant was calculated. They estimated the potential of grid-quality solar photovoltaic power on a college campus. They also developed a system based on the potential estimations made for a chosen area of $29 \times 65\ m^2$.

3.16 Effects of Soil Moisture Regimes and Organic Manures on Dehydrogenase Activity

In inceptisols soils, the soil moisture regimes and organic manures had a substantial influence on soil enzymatic activities. The soil dehydrogenase enzyme activity was found to increase with increasing moisture levels up to field capacity (Hagavane et al. 2015). Moisture levels above or below the field capacity led to a reduction in the activity of soil dehydrogenase. Among the sources of organic manures, vermicompost increased the activity of dehydrogenase enzymes in soil. The highest dehydrogenase activity (0.134 mg NH_4–N g soil 24 h) was recorded in moisture at field capacity with vermicompost on the 30th day of incubation.

3.17 Ion Balance with Reference to Organic Acids

Salinity stress has negative impacts on agricultural yield throughout the world, affecting production whether it is for subsistence or economic gain. Soil salinization and alkalization is rapidly increasing on a global scale. Beetroot variety PAC60008 (Nilima et al. 2015) subjected to salinity stress of 0.2, 4, 6, 8 EC after the growth of the two-leaf stage to the first physiological growth stage indicated that under salinity conditions, oxalic acid and succinic acid were dominant compounds. Further, as the salt stress increased, the concentration of oxalic acid also increased in the shoots. Though succinic acid was initially higher, it showed a gradual decrease in its concentration. At higher salinity of 8 ds per m, succinic acid levels were not detected, but higher levels of increased malionaldehyde contents were detected. Lipid peroxidation at high salinity levels led to the accumulation of malionaldehyde. Increased highest total antioxidant activity was seen at salinity levels of 2 EC only. There was also synthesis of proteins of small molecular weight with the increase in the intensity of salinity stress. This indicates that in higher salinity stress where there is more accumulation of positive ions in the plant body, the positive ions are balanced by the negative ions of organic acids. These results showed that there are different adaptive strategies for *Beta vulgaris* seedlings in organic acid metabolism under salt and alkali stress.

3.18 Climate Change Affects Livestock Production

Climate change affects the productivity of animals and their behaviour. Methane has a negative impact on shrimp culture. Methane, a greenhouse gas emitted from aquatic sediments, is mediated by methanogens, a diverse group of micro-organisms dominantly present in anaerobic sediments. High numbers of methanogens are detrimental as methane gas is a potential contributor to global warming. Shrimp cultured pond sediments are also anaerobic and the presence of methanogens result in the production of methane. The present study has given us a clear insight into the diversity of methanogens in shrimp pond sediment. Further, studying the activity of individual bacteria will provide valuable information for taking mitigation measures to decrease the emission of greenhouse gases (Thulasi et al. 2015).

Global warming is increasingly associated with an increase in sea levels. Sea surface across the Indian Ocean may have a significant influence on the summer monsoon over India. For example, the onset of the monsoon was on 6 June in 2014 and the monsoon rainfall was unusually in deficit (more than 30–50 % in some districts across the Peninsular India) till the first week of July and picked up later, finally ending with a deficit of 12 % across the country while several regions in the south experienced unusual deficits in monsoon rainfall. In contrast, heavy floods were noticed in Orissa, Jammu and Kashmir, and several other meteorological subdivisions across the country in 2014. The unprecedented cyclone 'Hudhud' in the second week of October 2014 devastated the three northern districts of Andhra Pradesh. Such monsoon aberrations and intensity of cyclones are expected to frequently be due to global warming and climate change. Another important aspect of the monsoon in India is that whether the monsoon is normal, excessive or in deficit, as was noticed in 2012, 2013 and 2014, one region or another experiences floods or droughts and the agriculture sector, including animal agriculture, is adversely affected (Rao and Sejian 2015).

Climate change and global warming are a burning global issue affecting various sectors of livelihood, mainly agriculture including livestock production and reproduction. The livestock sector is increasingly organized in long market chains, which employ at least 1.3 billion people globally and directly support the livelihoods of 600 million poor small-holder farmers in the developing world. Livestock provides a more reliable source of income to the farmers throughout the year. In the changing climate scenario, livestock are exposed to different types of stresses such as heat, nutritional and walking stresses. The cumulative effect of these multiple environmental stresses impacting livestock production and reproduction may be more detrimental than the stresses that would occur in isolation. Studies should focus on the impact on growth, production, reproduction and immunity against diseases (Nair et al. 2015).

3.19 Climate Change Impacts the Livestock Sector

Considerable mortality is likely to occur during the occurrence of cyclones, floods and droughts. Lack of fodder and drinking water are the constraints in feeding animals during prolonged drought periods and thereby mortality is noticed. Cold and heat waves are not uncommon in India and they adversely affect poultry and dairy production to a considerable extent. The classic examples were the cold waves in 2002–2003 and 2005–2006 and the heat wave in 2003–2004. Poultry farming was adversely affected due to the heat wave in May 2003 across the State of Andhra Pradesh (Rao and Sejian 2015). Therefore, suitable housing and other management techniques are essential to minimize the adverse impact of heat and cold waves in poultry and dairy farming. The year 2010 was the warmest year in India, followed by 2009, and the very first decade of this century was recorded as the warmest, since 8 out of the 10 years were identified as warm years. The increase in the mean annual surface air temperature across India was 0.54 °C during the last 110 years, while the rate of increase was high in post-monsoon and winter seasons.

The rate of increase in temperature was high across the West Coast, North East and the Western Himalayas of India when compared to that of other zones within the country. Reports on global warming projections indicate that increases in temperature will be restricted to around 2–3 °C by the year 2080 (Rao and Sejian 2015) through concerted efforts globally. A decline in rainfall was also noticed during the monsoon season in many parts of the country over the last 50–60 years. Availability of water will be a constraint in the coming decades under the projected climate change scenario. To meet the requirement of egg, milk, meat, fish and food grain production by 2020 in tune with the growing population in warming India, there is an urgent need for climate change adaptation and mitigation strategies in animal agriculture for sustenance of rural livelihoods.

3.20 Impact of Climate Change on Pests and Diseases

Chilli is an important vegetable, as well as a spice crop, in India and occupies an area of 7.92 lakh ha with a production of 13.76 lakh tones (Indian Horticulture Database 2013 cited in Lakshmi et al. 2015b). Though chilli crops are attacked by a large number of pathogens, heavy crop losses are attributed to viruses. Among these viruses, chilli leaf curl virus, causing leaf curl disease, is a prime limiting factor in chilli production in Andhra Pradesh, causing crop losses of up to 100 %. Chilli leaf curl virus belongs to a single-stranded DNA group in the family Geminiviridae and genus *Begamovirus*, and is transmitted by whitefly (*Bemisia tabaci*).

Lakshmi et al. (2015b) investigated the impact of climate change on the incidence of chilli leaf curl virus. According to them, in chilli variety CA960 (Sindhur) the correlation and regression of 3-year data results revealed that there was a significant negative correlation with minimum temperature (2011: $r-0.32$; 2012: $r-0.57$; 2013: $r-0.26$) and afternoon relative humidity (2011: $r-0.51$; 2012: $r-0.55$; 2013: $r-0.50$) in all the years. The maximum temperature, morning relative humidity and whitefly population showed a positive correlation with the disease incidence, and rainfall was negatively correlated.

In 2011–2012, during the month of January, the whitefly population washed out because of rainfall (141.9 mm) and there was no disease spread for 1 month. The coefficient of multiple determination (R^2) values were also significant (2011: 0.782; 2012: 0.862; 2013: 0.635) and implied that 0.7597 (the mean R^2 value for the 3 years) of the variation in the development of the chilli leaf curl virus was explained by five independent variables (minimum and maximum temperatures, morning and afternoon relative humidity levels, and rainfall) and one dependent variable (whitefly population) (Lakshmi et al. 2015b). Therefore, it indicated that the incidence of chilli leaf curl virus and the whitefly population were highly influenced by climate change.

Variability in climate influences the incidence of insects affecting crop production. The mango defoliator *Cricula trifenestrata* Helfer completes four generations in a year. It is therefore crucial to identify the critical stage of development for management intervention. The bio-ecology of mango defoliator in the Terai region of West Bengal (Bera et al. 2015) indicated that all the developmental stages varied over seasons. The gradient of temperature and RH also played a significant role in the different life stages. The larval duration was found to be longest during January–April (45.30 days) and shortest in March–May (36.40 days). A wide variation in the pupal period was also recorded in different seasons (24.40–89.40 days for males and 25.74–93.20 days for females). The life cycle depended mainly on the pupal stage and was completed in 77.00–146.00 days. Therefore, management of this pest should be directed towards killing of summer-diapausing and early winter-hibernating pupae resting on damaged twigs.

The variation in the seasonal incidence of fruit flies (*Bactrocera cucurbitae* Coq.) on cucumber was studied from 2008 to 2010 (Kukanur et al. 2015). Peak fruit fly populations were noticed during the 48th standard week in 2008 and the 49th week in 2009. However, a gradual increase was observed from September to December and thereafter decreased during both the years. Though the fly population was positively correlated with the maximum temperature, minimum temperature, rainfall and evening relative humidity, it had a negative relationship with morning relative humidity.

Global climate change is the current burning issue around the world. A study of the impact of climate change on the biology of two BPH (*Nilaparvata lugens*) populations from West Godavari, Andhra Pradesh and Punjab, studied under three different graded levels or treatments of CO_2 (elevated CO_2 at 550 ppm with temperature control, elevated CO_2 at 550 ppm without temperature control and ambient CO_2 and temperature) in the susceptible variety TN1 (Sunil et al. 2015) revealed that there was a significant difference between the three treatments and populations of first-, second- and third-instar nymphs. The nymphal duration was prolonged in the Punjab population (15.9 ± 0.4*) at elevated CO_2 at 550 ppm without temperature control, while it was shortest in the West Godavari population at elevated CO_2 at 550 ppm with temperature control (12.5 ± 0.2*). A higher percentage of females emerged at elevated CO_2 at 550 ppm without temperature control (76.2 ± 0.1) in the West Godavari population than in the Punjab population (38.1 ± 0.1). Adult longevity was greater in the West Godavari population (12.2 ± 0.3) than in Punjab (10.5 ± 0.5). The total life span ranged from 21 to 25 days. A shorter life span was observed in the West Godavari population (22.8) than in the Punjab population (24.3). Higher fecundity was observed in elevated CO_2 at 550 ppm with temperature control (277 eggs per female). Honeydew excretion did not differ significantly in adults (range 184–264 mm^2) but was significant in third-instar nymphs with higher excretion observed in the Punjab population (39.5 mm^2). A higher survival rate was observed in elevated CO_2 at 550 ppm without temperature control on the sixth and ninth days (69.0 and 47.0). Emergence of macropterous adult males was higher in the West Godavari population (36.4*) than in Punjab (27.5).

3.21 Genotype × Environment Interaction

Genotype × environment interaction is one of the most important steps to encourage utilization of the most stable genotypes by growers. Numerical classificatory or pattern analysis methods have been applied more widely in comparing the responses of cultivars and/or breeding lines across environments. Eswarareddy and Nandan (2015) made an analysis of the genotype × environment interaction of winged bean yield (*Psophocarpus tetragonolobus* L.) DC in the North East Plain zone. The present research provided additional information concerning the behaviour of a set of 15 selected winged bean lines over different years grown in the North East Plain zone. These winged bean lines were evaluated for the stability of seed yield, 100-seed weight, flowering time, plant height and biomass. Significant differences existed between years, lines and lines × year interactions for all traits except for 100-seed weight.

Two methods of multivariate analysis cluster and principal components were utilized firstly to determine whether a pattern existed among lines in their response across years, and secondly to examine the relationships among them. In both analyses, each line was presented as a vector whose elements were given by the performance of lines in each year. The analyses that were used arranged the lines into groups that were differentiable in terms of performance and stability. These results (Eswarareddy and Nandan 2015) provided useful information to aid the choice of winged bean lines in the North East Plain zone. In fact, the effect of the year was much more important than the other effects at least for the flowering time, seed yield and biomass. Clear differences for these traits between the winged bean lines across growing seasons were evident. Conversely, the effect of the line on the 100-seed weight was clearly most important, while the year and year × line interaction effects were of little importance. This suggested that seed size is a stable trait in the tested grass pea lines, which have shown good potential to respond better in most favourable growing conditions. Utilizing principle component and cluster analyses to examine genotype performance, the advanced winged bean lines that had a significant stability over different years were identified (Mysore Local, TMV Local and IC95227). These lines were relatively indifferent to environmental variation and always had good performance.

3.22 Improvement in Fruit Quality by Bagging

In central Uttar Pradesh, India, various polyethylene covers of different colours (silver, white, yellow, green, black and pink) were used to study the bagging effect on fruit quality of guava cv. Lalit (Meena et al. 2015a). The fruit bagging improved the growth and quality of guava fruits. The fruit size, weight and pulp content increased due to fruit covering, and bagging with yellow polythene gave the best result. Fruit bagging also improved the fruit quality in terms of TSS, total sugars and the TSS-to-acid ratio and maximum values (14.25 °B, 11.14 % and 30.07, respectively) were found under yellow polythene bagging. Vitamin C content (171.14 mg 100 g^{-1}) was found to be maximum in white polythene bagging. Thus, among the various fruit-covering materials, bagging with yellow coloured polythene was found to be the best for overall improvement of the physicochemical quality of winter season guava cv. Lalit under central Uttar Pradesh (i.e. Lucknow) conditions.

References

Aatla HB (2015) Rising atmospheric carbon dioxide: plants face: the future. In: Compendium of abstracts of the 2nd international conference on bio-resource and stress management. ANGRAU & PJTSAU, Hyderabad, 7–10 Jan 2015

Alse UN, Mishra JS, Waskar DP et al (2015) Response of *rabi* sorghum genotypes to changing climate. In: Compendium of abstracts of the 2nd international conference on bio-resource and stress management. ANGRAU & PJTSAU, Hyderabad, 7–10 Jan 2015

Bera C, Chaudhuri N, Ghosh J et al (2015) Bio-ecology of mango defoliator *Criculatrifenestrata* Helfer (Lepidoptera: Saturnidae) in terai region of West Bengal, India. In: Compendium of abstracts of the 2nd international conference on bio-resource and stress management. ANGRAU & PJTSAU, Hyderabad, 7–10 Jan 2015

Bhat S, Swamy BKN (2015) Impact of climate change on food inflation in India: issues and measures for adaptation. In: Compendium of abstracts of the 2nd international conference on bio-resource and stress management. ANGRAU & PJTSAU, Hyderabad, 7–10 Jan 2015

Chandrappa H, Chandravamsi P, Kumara O et al (2015) Development of organic farming practices for system based high value crops. In: Compendium of abstracts of the 2nd international conference on bio-resource and stress management. ANGRAU & PJTSAU, Hyderabad, 7–10 Jan 2015

Choudhury SR, Das A, Sharma RP et al (2015) Moderation of greenhouse gas emission to combat the effect of climate change in rice-wheat cropping system. In: Compendium of abstracts of the 2nd international conference on bio-resource and stress management. ANGRAU & PJTSAU, Hyderabad, 7–10 Jan 2015

Chowda RM, Swamy BKN (2015) Precision canopy sprayer to protect environment for sustainable agriculture. In: Compendium of abstracts of the 2nd international conference on bio-resource and stress management. ANGRAU & PJTSAU, Hyderabad, 7–10 Jan 2015

Crutzen PJ, Wacławek S (2015) Atmospheric chemistry and climate in the anthropocene/chemia atmosferyczna i klimat w antropocenie. Chem Didactics Ecol Metrol. doi:10.1515/cdem-2014-0001

Eswarareddy G, Nandan R (2015) Analysis of the genotype × environment interaction of winged bean yield (*Psophocarpus tetragonolobus* (L.) DC) in North East Plain Zone. In: Compendium of abstracts of the 2nd international conference on bio-resource and stress management. ANGRAU & PJTSAU, Hyderabad, 7–10 Jan 2015

Giridhari K, Surya Kumari S, Nirmal Babu K et al (2015) Biplotevaluation of test environments and cultivar selection of coriander (*Coriandrum sativum* L.) in India. In: Compendium of abstracts of the 2nd international conference on bio-resource and stress management. ANGRAU & PJTSAU, Hyderabad, 7–10 Jan 2015

Hagavane SR, Chaudhari RD, Shinde RB (2015) Influence of soil moisture regimes and sources of organic manures on dehydrogenase activities in inceptisols College of Agriculture, Dhule-424004, Maharashtra, India. In: Compendium of abstracts of the 2nd international conference on bio-resource and stress management. ANGRAU & PJTSAU, Hyderabad, 7–10 Jan 2015

Kalidasu G, Suryakumari S, Naramnaidu L (2015) Impact of heavy rains during growth and maturity period in certain seed spices in vertisols. In: Compendium of abstracts of the 2nd international conference on bio-resource and stress management. ANGRAU & PJTSAU, Hyderabad, 7–10 Jan 2015

Kukanur VS, Naganagoud A, Waseem MA (2015) Seasonal fluctuations of melon fruit flies in cucumber. In: Compendium of abstracts of the 2nd international conference on bio-resource and stress management. ANGRAU & PJTSAU, Hyderabad, 7–10 Jan 2015

Kumar N, Singh SS, Ghosh PK et al (2015) Enhancing chickpea productivity through conservation practices under rainfed rice-fallow regions of India. In: Compendium of abstracts of the 2nd international conference on bio-resource and stress management. ANGRAU & PJTSAU, Hyderabad, 7–10 Jan 2015

Kumari SR, Bharathi S, Sreelakshmi B et al (2015) Influence of weather on growth, disease incidence and seed cotton yield under rainfed conditions in black cotton soils. In: Compendium of abstracts of the 2nd international conference on bio-resource and stress management. ANGRAU & PJTSAU, Hyderabad, 7–10 Jan 2015

Kumawat A, Kumar R, Yadav RS (2015) Climate change and agriculture. In: Compendium of abstracts of the 2nd international conference on bio-resource and stress management. ANGRAU & PJTSAU, Hyderabad, 7–10 Jan 2015

Lakshmi YS, Sreelatha D, Pradeep T et al (2015a) Impact of weather parameters on sweet corn yield. In: Compendium of abstracts of the 2nd international conference on bio-resource and stress management. ANGRAU & PJTSAU, Hyderabad, 7–10 Jan 2015

Lakshmi TV, Lakshmi PV, Rajani A et al (2015b) Impact of climate change on the incidence of chilli leaf curl virus. In: Compendium of abstracts of the 2nd international conference on bio-resource and stress management. ANGRAU & PJTSAU, Hyderabad, 7–10 Jan 2015

Loria N, Verma KS, Bhardwaj SK (2015) An assessment of the environmental impacts of catchment management on the status of pong wetland in District Kangra of Himachal Pradesh, India. In: Compendium of abstracts of the 2nd international conference on bio-resource and stress management. ANGRAU & PJTSAU, Hyderabad, 7–10 Jan 2015

Meena KR, Maji Sutanu, Kumar S et al (2015a) Effect of bagging on fruit quality of guava under central Uttar Pradesh. In: Compendium of abstracts of the 2nd international conference on bio-resource and stress management. ANGRAU & PJTSAU, Hyderabad, 7–10 Jan 2015

Meena RS, Kote P, Hashim M (2015b) Need and future of climate smart agriculture. In: Compendium of abstracts of the 2nd international conference on bio-resource and stress management. ANGRAU & PJTSAU, Hyderabad, 7–10 Jan 2015

Mondal S, Thokchom R (2015) Impact of climate change as perceived by scientists of Central Rice Research Institute (CRRI): Cuttack, Odisha. In: Compendium of abstracts of the 2nd international conference on bio-resource and stress management. ANGRAU & PJTSAU, Hyderabad, 7–10 Jan 2015

Nair AB, Sejian V, Prasada Rao GSLHV et al (2015) Impact of climate change on livestock production and reproduction. In: Compendium of abstracts of the 2nd international conference on bio-resource and stress management. ANGRAU & PJTSAU, Hyderabad, 7–10 Jan 2015

Nilima K, Patel S, Adbhai A et al (2015) Maintenance of ion balance in beta vulgaris: special reference to organic acids. In: Compendium of abstracts of the 2nd international conference on bio-resource and stress management. ANGRAU & PJTSAU, Hyderabad, 7–10 Jan 2015

Panigrahi TK, Das SP (2015) Climate change: impact on crop treasure. In: Compendium of abstracts of the 2nd international conference on bio-resource and stress management. ANGRAU & PJTSAU, Hyderabad, 7–10 Jan 2015

Rani KS, Shankar R, Padmaja G (2015) Impact of agriculture in a changing climate. In: Compendium of abstracts of the 2nd international conference on bio-resource and stress management. ANGRAU & PJTSAU, Hyderabad, 7–10 Jan 2015

Rao GSLHVP, Sejian V (2015) Climate change impacts in livestock sector. In: Compendium of abstracts of the 2nd international conference on bio-resource and stress management. ANGRAU & PJTSAU, Hyderabad, 7–10 Jan 2015

Rao SG, Rao IVYR, Rao HS (2015) Impact of change in weather on paddy productivity in Visakhapatnam district of Andhra Pradesh, India. In: Compendium of abstracts of the 2nd international conference on bio-resource and stress management. ANGRAU & PJTSAU, Hyderabad, 7–10 Jan 2015

Sarangi C, Mohapatra SC, Mishra SN et al (2015) Assessment of soil carbon pools by various crop genotypes adopted to mitigate climate change. In: Compendium of abstracts of the 2nd international conference on bio-resource and stress management. ANGRAU & PJTSAU, Hyderabad, 7–10 Jan 2015

Shanwad UK, Srinivas AG, Shankergoud I et al (2015) Climate change trends and its effect on agriculture crops in North Eastern Dry Zone of Karnataka. In: Compendium of abstracts of the 2nd international conference on bio-resource and stress management. ANGRAU & PJTSAU, Hyderabad, 7–10 Jan 2015

Singh AP, Vaid A, Kumar A et al (2015) Potentials of climate resilient technologies to revamp agriculture in fragile ecosystems. In: Compendium of abstracts of the 2nd international conference on bio-resource and stress management. ANGRAU & PJTSAU, Hyderabad, 7–10 Jan 2015

Sinha P, Pandey VK, Lakra A et al (2015) Study of solar energy for parameterization of 100 kWp solar power plant for electricity generation. In: Compendium of abstracts of the 2nd international conference on bio-resource and stress management. ANGRAU & PJTSAU, Hyderabad, 7–10 Jan 2015

Sireesha A, Rao PC, Ramalakshmi CHS (2015) Mobility of oxyfluorfen in vertisol and alfisol. In: Compendium of abstracts of the 2nd international conference on bio-resource and stress management. ANGRAU & PJTSAU, Hyderabad, 7–10 Jan 2015

Sood S, Panday SC, Gupta A et al (2015) Effect of changing climatic conditions on grain yield of finger millet in sub-mountainous Himalayan region of Uttarakhand. In: Compendium of abstracts of the 2nd international conference on bio-resource and stress management. ANGRAU & PJTSAU, Hyderabad, 7–10 Jan 2015

Sridevi S, Venkata Ramana M, Madhavi M et al (2015) Carbon sequestration in rice soils under long-term integrated nutrient management. In: Compendium of abstracts of the 2nd international conference on bio-resource and stress management. ANGRAU & PJTSAU, Hyderabad, 7–10 Jan 2015

Srilathavani C, Preethi M (2015) Extension strategies to combat impact of climate change on agricultural production and agroforestry. In: Compendium of abstracts of the 2nd international conference on bio-resource and stress management. ANGRAU & PJTSAU, Hyderabad, 7–10 Jan 2015

Sunil V, Sampath Kumar M, Chiranjeevi K et al (2015) Effects of elevated CO_2 on biology of brown planthopper, nilaparvatalugens (Stal). In: Compendium of abstracts of the 2nd international conference on bio-resource and stress management. ANGRAU & PJTSAU, Hyderabad, 7–10 Jan 2015

Swetha S, Manthri K (2015) Carbon sequestration through coconut bases cropping systems: a mitigation strategy for reducing carbon dioxide. In: Compendium of abstracts of the 2nd international conference on bio-resource and stress management. ANGRAU & PJTSAU, Hyderabad, 7–10 Jan 2015

Thulasi D, Muralidhar M, Vinayakumar K et al (2015) Abundance and diversity of methanogenic archaea in shrimp. In: Compendium of abstracts of the 2nd international conference on bio-resource and stress management. ANGRAU & PJTSAU, Hyderabad, 7–10 Jan 2015

Verma AVN, Alluri NR, Rao JSP (2015) Climate change mitigation, adaptation and sustainability in agriculture: issues and intricacies. In: Compendium of abstracts of the 2nd international conference on bio-resource and stress management. ANGRAU & PJTSAU, Hyderabad, 7–10 Jan 2015

Vijay D, Maity A, Malaviya DR (2015) Influence of weather onberseem (*Trifolium alexandrinum* L.) seed production. In: Compendium of abstracts of the 2nd international conference on bio-resource and stress management. ANGRAU & PJTSAU, Hyderabad, 7–10 Jan 2015

Vinutha SN, Swamy BKN (2015) Beverage crops generated waste for bio-resource conservation under present climate change. In: Compendium of abstracts of the 2nd international conference on bio-resource and stress management. ANGRAU & PJTSAU, Hyderabad, 7–10 Jan 2015

4 Natural Resource Management

Ratikanta Maiti and Aruna Kumari

Abstract

Natural resources play an important role in the livelihood of human beings and animal kingdom. The chapter makes a brief review of researches underway on various aspects of natural resource management. This includes heritage of Deccan region, use of medicinal and aromatic plants, geranium as source of oil, contract farming for sustainable agriculture, water resource management and water quality, and research measures for maintaining water quality, conservation of fossil fuels, soil resource management, integrated nutrient management, and organic agriculture and soil amendments. Besides various other technologies in the management of natural resources such as bioresource management in a society, resource use efficiency in fish production is discussed. Research on the use of sorghum and other crops as biofuels for sustainable agriculture and conservation of natural resources is mentioned.

4.1 Biodiversity

Natural resources are the gifts of God. Nature nourishes us with sources of food, shelter, fiber, energy, coal, timber, and domestic necessities of our daily needs. We need to conserve and exploit them judiciously in order to conserve these treasures for the future generations. Natural resources or their products are managed in different manners. A few aspects pertaining to the management and utilization of the existing natural resources are mentioned herewith.

Nature provides us a repository of resources for various activities. These are in the form of crop plants, medicinal plants, trees, forests, water, air, minerals, etc. There is an urgent need to utilize each of these resources sustainably and preserve them for the benefit of humankind.

R.K. Maiti (✉)
Facultad de Ciencias Forestales, Universidad Autónoma de Nuevo León, Carr. No 85, km 145, NL 67700 Linares, Mexico
e-mail: ratikanta.maiti@gmail.com

A. Kumari
Department of Crop Physiology, Agricultural College, Professor Jayashankar Telangana State Agricultural University, Polasa, Jagtial, Karimnagar 505 529, Telangana, India
e-mail: arunasujanagcjgl@gmail.com

R.K. Maiti et al. (eds.), *Bioresource and Stress Management*, DOI 10.1007/978-981-10-0995-2_4

4.2 Repository and Heritage of Deccan Region

The University campus of Professor Jayashankar Telangana State Agricultural University is spread over 7890 ha. It consisted of varied habitat types of open scrub, rocky out crops, mixed forests, wetlands, and agricultural landscapes. This area has rich diversity of native flora and fauna. The floral diversity included 424 species belonging to 252 genera and 64 families. The faunal diversity recorded in this region includes a total of 412 species belonging to 43 orders and 151 families. Out of 412 species, 19 species of mammals, 172 species of birds, 57 species of herpetic fauna, and 164 species of invertebrates inhabit in different habitats of the area. Among the total birds, three species are near threatened category namely: Black headed ibis (*Threskiornis melanocephalus*), Painted stork (*Mycteria leucocephala*), and Oriental darter (*Anhinga melanogaster*) are found to be habituating in this geographical area. Out of the 57 species, two species were considered as near threatened category namely: Red sand boa (*Erix johinii*) and Indian black turtle (*Melanoche lystrijuga*); and three RARE species which includes Indian chameleon, Green keel back (*Macropisthodonplum bicolour*), and Indian flap shell turtle (*Lissemys punctata*) are also finding their habitat in this location. Several conservation measures were undertaken to maintain this unique habitat features for sustainability of the biodiversity. Swamy et al. (2015) discussed about the highlights of success in preserving the repository and heritage of Deccan region.

4.3 Prevalence of Fruit Bats

The status and distribution studies (Venkateswarulu et al. 2015) of five species of pteropodid bats, namely Fulvous fruit bat (*Rousettus leschenaultii*), Indian flying fox (*Pteropus giganteus*), Greater short-nosed fruit bat (*Cynopterus sphinx*), Lesser short-nosed fruit bat (*Cynopterus brachyotis*), and the Dawn bat (*Eonycteris spelaea*) in Ranga Reddy district of Telangana State, India, has shown that a total of 18 species of plants were identified as potential roosting sites for all the five species of bats. Most of the roosts were predominantly located within the towns in the vicinity of old buildings near water areas and other plantations. Among the species, Indian flying fox highly preferred large canopy trees and the other species showed variations in selection of trees. Different sites were marked and developed spatial distribution maps in relation to habitat features and the feeding patterns. The preference for fruits by pteropodids varied according to the development stages of fruits namely, immature, unripe and ripe. There was a relationship between the foraging activities of bats and the moon phase. Bats exhibited a varied foraging pattern and flight height.

Lake Kolleru is the largest fresh water wetland ecosystem in South India. This lake is the home for 189 species of birds, including the rare and endangered Grey Pelican. Lake Kolleru has been designated as a RAMSAR site.

4.4 Medicinal and Aromatic Plants

Nature provides us different groups of plants. Medicinal plants are valuable resources of a wide array of chemicals. These chemicals are used in the preparation of medicines to cure several diseases. The importance of medicinal plants dates back to our civilization. Many of the tropical and temperate forests have a repository of these medicinal plants where in some of these are still unexploited. Deforestation being caused by anthropic factors is leading to a loss of these valuable repositories. There is an urgent need to conserve medicinal plants that are already on the verge of extinction.

The nutrient management influence growth and development of these plants. The nutrient management studies on macro and micronutrients availability of inceptisols under Java citronella (*Cymbopogon winterianus*) at Nagarjun Medicinal Plants Garden, Dr. P.D.K.V, Akola during *kharif* 2009–2010 and 2010–2011 (Nandapure et al. 2015) revealed that in *Cymbopogon winterianus* the maximum availability of macronutrients (N, P, K, S) was obtained through the application of 10 t FYM + 140:40:80 kg NPK per ha and the highest micronutrients availability of Fe, Mn, Zn, and Cu with the application of 100 kg N through FYM. Further it could be concluded that the application of FYM at 10 t per ha in combination with chemical fertilizer at 140:40:80 kg NPK per ha was found to be more beneficial way of nutrient management to improve the availability of macro and micro nutrients status of soil under Java citronella.

4.5 Geranium as Important Source of Oils

Rose-scented Geranium (*Pelargonium graveolens*) is source of one of the most important essential oils. Geranium oil is widely used in soaps, perfumery, and cosmetics industries. It is also used as therapeutic (aromatherapy) to combat menopausal problems, skin, stress, and anxiety. In India, it is grown in Deccan plateau located in Kerala, Andhra Pradesh, Maharashtra, and Tamilnadu. In recent years, the oil production has gone down due to significant reduction in cultivated area in the high altitude areas of Tamilnadu. Large quantities of geranium oil are being imported to India to meet the requirements. The composition of essential oils is highly dependent on the genetic structure of the plant but may be strongly influenced by environmental factors, such as water availability, soil structure and composition, temperature and ontogenetic factors as harvest period, plant age, flowering, and others. Geranium responds well to input of organic matter. Studies have shown the influence of fertilization on yield and content in the composition of essential oils in medicinal and aromatic species, and this effect was directly related to the influence of mineral nutrition on the metabolism of the plant. Reddy and Susila (2015) studied the effect of spacing and integrated nutrient management on herbage yield of Geranium (*Pelargonium graveolens*). The results revealed that planting of Geranium at 45 × 45 cm spacing produced highest herbage yield (37.66 t ha^{-1}). Among the different manures studied, the application of 50 % RDF and Neemcake 1.0 t ha^{-1} produced highest herbage yield (40.615 t ha^{-1}).

4.6 Viable Options for Conservation of Biodiversity

Organic manurial addition to the soil has several advantages apart from maintaining the soil fertility levels. One such organic manure is donkey manure. Proper spreading of this manure and its incorporation in the soil resulted in only 5 % loss. In an animal operated spreader (Reddy et al. 2015a) that consisted of three volumes of manure box and three openings of sliding plate for spreading the donkey manure, about 21 % of its nitrogen was lost to the atmosphere in stack piled manure, whereas animal spreader accounted for only 5 % loss of nitrogen to the atmosphere.

The maximum field capacity of donkey manure spreader was 0.31 ha hour^{-1} and lowest 0.21 ha h^{-1}. Further, the donkey manure application to field was observed as a best viable option to conserve biodiversity.

4.7 Contract Farming for Sustainable Horticulture

Under the present climate change, contract farming of medicinal and aromatic plants for sustainable horticulture (Vinutha and Narayana Swamy 2015) are very important. The World Health Organization (WHO) has emphasized the need for utilization of the indigenous system of medicine based on the locally available medicinal plants in the developing countries. Medicinal and aromatic plants (MAPs) and their derivatives would play a major role in the medical therapy in spite of advances in chemical technology and appearance of cheap synthetic substitutes. The MAPs are emerging as the industrial crops that require comparatively less inputs but gives better returns to the growers.

Contract farming varies depending on the nature and type of contracting agency, technology, nature of crop or produce, and the local and national context (Vinutha and Narayana Swamy 2015). The MAPs have gained importance in horti-business and development of entrepreneurship. The experience of horticulture development in India has demonstrated that the existing system of delivery of horticulture inputs and purchase and use of horticulture output have not been efficient in reaching the benefit of better linkage between horticulture and horti-processing industry to the farmers or the horti-industry. Given the nature of modern farming involving tremendous amount of technological input and market orientation which require capital resources, it is inevitable to involve private corporate business interests in horticulture development through contract farming system. In fact, all over the world contracting of some kind is a necessity for many or most forms of modern commercial horticulture.

4.8 Water Resource and Its Quality

Annual rainfall is the main source of water that is stored in lakes, ponds, and rivers for human needs as well as for the growth of the plants and crops. The abundance of the plants is highly influenced by the amount of rainfall. The quality of water is affected owing to the pollutants derived from industrial and human activities.

4.9 Water Management

Water management studies on finger millet (*Eleusine coracona*) in Alfisol of mid-land situation in Chhattisgarh plains (Bisen et al. 2015) has shown that soil mulch with 50 % soil moisture depletion not only gives higher yield (16.71 q ha^{-1}) but also leads to higher water use efficiency (WUE) (3.81 kg ha^{-1} mm) followed by soil mulch with 60 % soil moisture depletion (15.97 q ha^{-1}). The cumulative yield and growth parameters data reviewed indicated that there was significant reduction in the yield of finger millet after 70 % depletion of soil moisture. The depth of water was estimated to bring the field from its deficit moisture status to its field capacity. On the basis of this, a simple guideline based on the moisture depletion condition was developed for the farmers so that they can analyze their field situation and irrigate ragi for better production.

4.10 Water Quality Assessments and Restoration Measures

Lakes are unique gift of god, diverse, productive, and interactive ecosystems in the world. One of the National Lake of India is the Rewalsar Lake in the historic holy town of Rewalsar, district Mandi (H.P), India. This lake comes under the category of sacred lakes and is unique in terms of religious and ecological significance. Kashyap et al. (2015) studied water quality assessment and suggested restoration measures for it. The lake water quality when analyzed for physicochemical characteristics viz., pH, electrical conductivity, transparency, and total dissolved solids (TDS): alkalinity, acidity, calcium, magnesium, sodium chloride, potassium, phosphate, sulphate, biological oxygen demand (BOD), and chemical oxygen demand (COD) on a monthly basis over a period of 6 months in year 2011 revealed that it is alkaline in nature. The TDS and transparency values were high. The sites that showed greater pollution had higher levels of BOD and COD. Every parameter showed a significant increase with increased lake water pollution during the period of the annual Baisakhi fair. This lake is touching the critical stage of water pollution and is dying slowly due to massive anthropogenic pressure.

There is an urgent need of restoration of its ecosystems and suitable measures should be taken at the earliest. Some of the remedial measures for its restoration suggested include.

- Canalization of rills-gullies
- Construction of check dams
- De-weeding and dredging of slit
- Establishment of water treatment plant
- Proper outlet for water from lake
- Creation of public awareness

Lake Kolleru is the largest fresh water wetland ecosystem in South India. This lake is the home for 189 species of birds, including the rare and endangered Grey Pelican. Lake Kolleru has been designated as a RAMSAR site. The spatial and temporal evaluation of water quality of Keller Lake and Upputeru Wet Land Ecosystem, Andhra Pradesh, India, in monthly collected water samples from selected 30 benchmark sites (Sreenivas et al. 2015) has shown that the charge balance error %age indicated by the balance between the total cations (TZ^+) and total anions (TZ^-) is between ± 1 and $\pm 10\,\%$. The major cation that was detected was sodium ion (61.5 %) followed by magnesium (19.9 %), calcium (16.7 %), and potassium (1.9 %). Among the anions, chloride was the major anion (47.4 %) slightly followed by bicarbonate (40.4 %), sulphates (11.8 %), and carbonates (0.5 %). The spatial variability was known through the preparation of monthly water quality maps by the use of Surfer software. Further, the temporal variability was assessed by cations and their ratios. Salinity in the system was found to increase gradually from January to May and reached its highest salinity level. The salinity level gradually decreased in the month of December. Apart from salinity, Mg:Ca ratios were also taken into consideration for the evaluation of the system. The fresh environment in the lake and river is evidenced during August–November. September month recorded the lowest salinity in the system. Increased salinity during summer months was attributed to reduced inflows and more evaporation and more backflow of salts from the sea.

Land uses and season have an impact on the physicochemical characteristics of surface water. This was evidenced in the water quality studies conducted in water samples collected from 39 sampling sites under different land uses (agriculture, forest, and urban). The physicochemical characteristics viz., pH, electrical conductivity (EC), temperature, total dissolved solids (TDS), calcium (Ca), magnesium (Mg), nitrate (NO_3^-), chloride (Cl^-), chemical oxygen demand (COD), and biological oxygen demand (BOD) assessed (Chauhan et al. 2015) revealed the maximum pH (7.69), EC (266.39 μS cm^{-1}), TDS (297.70 mg l^{-1}), Ca (66.08 mg l^{-1}), temperature (19.25 °C), BOD (1.45 mg l^{-1}), and COD (18.33 mg l^{-1}) in the surface water under urban land use, whereas Mg (12.90 mg l^{-1}), Cl^- (3.05 mg l^{-1}), and NO_3^-(4.19 mg l^{-1}) were maximum under agriculture land use. Maximum pH (7.79), temperature (22.58 °C), BOD (1.71 mg l^{-1}), COD (17.41 mg l^{-1}), Ca (64.61 mg l^{-1}), Mg (13.87 mg l^{-1}), and Cl^- (3.39 mg l^{-1}) were recorded during summer season, whereas EC (264.75 μS cm^{-1}), TDS (297.58 mg l^{-1}), and NO_3^- (3.91 mg l^{-1}) were maximum during rainy season. The water quality index (WQI) for selected land uses ranged from 19.89 to 25.12 whereas seasonwise water quality index (WQI) ranged from 18.77 to 27.62, which showed that the water quality falls between good and excellent and small variations in water quality indices were seasonal. These results clearly indicated that the surface water quality of Kandaghat block of Solan district of Himachal Pradesh was affected by changing land use patterns during different seasons.

4.11 Natural Resource

4.11.1 Conservation of Fossil Fuels

The world is confronted with the twin crises of fossil fuel depletion and environmental degradation. The indiscriminate extraction and consumption of fossil fuels have caused a reduction in petroleum reserves. These finite reserves are highly concentrated in certain regions of the world. It causes a severe foreign exchange problem for the countries importing petrol from the other countries. Hence, it is necessary to look for alternative fuels (Jadhav et al. 2015), which can be produced from materials available within the country.

4.11.2 Castor Bean as a Source of Biodiesel

Castor bean (*Ricinus cummunis*) is a potential plant species for biodiesel production in fragile and marginal environments. Although castor is probably indigenous to Eastern Africa but today castor is distributed worldwide. Castor establishes itself easily as a "native" plant and can often be found on wasteland and near rail roads and has recently also been used extensively as decorative plant in parks, etc. Castor seed contains 40–60 % oil, which is rich in triglycerides, mainly ricinolein, and provides a great natural resource for biodiesel production. Global castor seed production is around one million t $year^{-1}$. Leading producing areas are India, China, Brazil, and the former USSR. There are several active breeding programmes for castor.

Castor bean in arid or semi arid regions using local varieties can provide a valuable resource for biodiesel production. This type of holistic agricultural approach provides the regional farmers and local biodiesel industry with two strategic benefits: (1) For the regional farmers, the planting and harvesting of local varieties of castor bean on marginal lands provides the farmers with not only additional income but also provides stabilization of marginal lands that are susceptible to erosion through either water or wind. (2) The local biodiesel industry benefits in that they are guaranteed in continuous supply of raw material for the production of biodiesel which in turn gives viable economics to the overall local biodiesel industries.

Castor bean does not compete with food crops, as it can be grown on marginal lands. Castor bean is not competing with food grade oils. Its toxicity is sensed by animals and therefore not foraged on. It is high yielding, yielding as much as 350–650 kg of oil per hectare when no maintenance is applied to the crop i.e fertilizers, to a high-end yield where annual rainfalls happen at the right times, (1000 l ha^{-1}). It has a very high oil content of approximately 50 %. The oil seed is collected by hand, by picking the ripe pods from the plant. It requires only moderate rainfall (approx. 600 mm) and can withstand long periods of drought, but will thrive under higher rainfall. It is an uncomplicated crop that requires little attention during its growing periods. Due to its low demand on soil fertility, it is ideal to replant marginal lands to prevent desertification and erosion.

An unintended but important advantage to a castor bean project is that the plants absorb carbon dioxide, thereby reducing greenhouse gas accumulations in the atmosphere. The estimated carbon dioxide absorption level of castor bean plants is 34.6 tonnes per hectare, with two growing cycles per year. Production of castor oil worldwide is 0.5 million tonnes per annum. Consumption of petro diesel per day is approximately 10 million tonnes. If the entire petro diesel is to be replaced by castor biodiesel it needs to produce 7000 times the castor oil that is being produced today. It has a lot of industrial usages and therefore market is already in existence. The biodiesel prepared from castor oil has certain properties that are attractive particularly for cold climate. It may be mentioned that it has flash point of 190.7 °C which is much higher than petro diesel and other vegetable oil biodiesel. The oil is stable at low temperature and makes it an ideal combustible oil for region of extreme seasonal weather.

India has enjoyed a period of intense economic growth in last several years. Securing future energy supplies in a shrinking energy market is vital if this is to continue. After a decade in which oil reserves have shown a pronounced drop, it has become necessary to find new sources. However, the chances of finding oil wells of considerable size are becoming more and more limited. The benefits of growth however have failed to filter down to the many millions of its population still living in rural poverty. Despite the reforms which went some way to open up the economy, India still considers it a welfare state with development planning taking a key role. In its approach to biodiesel production, we can see the state's attempt to address both the issue of energy security in relation to liquid fuels and its obligations to provide for the rural poor (Jadhav et al. 2015).

4.11.3 Sorghum a Potential Source of Energy

Sorghum is an annual C_4 crop in the grass family, having high photosynthetic efficiency with considerable variability in growth characteristics. It contains both cultivated and wild races and possesses a significant amount of genetic diversity for traits of agronomic importance and has the potential to be an excellent diversified biofuel crop which is able to fill the needs of multiple bioenergy conversion process. The production of biofuel from plant-based biomass is becoming an important alternative to non-renewable energy sources.

India has witnessed substantial gains in the renewable energy domain, the country still needs to make sustained and focused efforts to develop its renewable sector and realize its vast potential. Northern India, with only 25 % of realized renewable energy potential, offers substantial opportunities for the country to meet its energy needs through renewable energy. To achieve this, all the stakeholders, including the Central and State Governments, the public and private sectors, and research and scientific institutions, need to make a concerted effort to help India secure its energy needs without endangering the country's environment and thereby actively contribute to balanced regional development (Mundiyara et al. 2015). He reports that sorghum is a potential source of renewable energy in India. The fuel ethanol from sorghum is the best choice to be implemented in India regarding both economic and environmental considerations. Because sweet sorghum has higher tolerance to drought, water logging, salt, alkali, and aluminium soils, it may be harvested 3–4 months after planting and planted 1–2 times a year (in tropical areas). Its energy output or fossil energy input is higher than sugarcane, sugar beet, corn, wheat, etc., especially in tropical areas like India. Sorghum juice is assumed to be converted to ethanol at 85 % theoretical, or 54.4 l ethanol per 100 kg fresh stalk yield. Ethanol production directly from juice is about 3000 l ha^{-1}. Sorghum has high amount of sucrose and invert sugar which are easily converted to ethanol; therefore, it seems that sorghum is the most suitable crop for biofuel production in arid regions of the world.

4.11.4 Soil Resource Management

Soil fertility levels and nutrient management influence the growth and yield of crops. In cluster bean *Cyamopsis tetragonoloba* L., studies on the fertility levels and agrochemical on its growth and yield (Yogi et al. 2015) have shown that the variety RGC-1066 was superior to RGC-1038 in respect of plant height, dry matter accumulation, pods $plant^{-1}$, seeds pod^{-1}, and test weight. Similarly, the seed yield (1300 kg ha^{-1}), stover yield (3297 kg ha^{-1}), and biological yield (4597 kg ha^{-1}) were also higher with RGC-1066. The maximum values of nitrogen and phosphorus content in seed and stover and protein content in seed, total uptake of nitrogen, phosphorus and potassium, net returns (INR 34,395 ha^{-1}), and B:C ratio (1.41) were also recorded with variety RGC-1066. Results further revealed that the application of 100 % RDF significantly increased the plant height at 60 DAS and at harvest, dry matter accumulation per meter row length at all the growth stages, chlorophyll content, pods $plant^{-1}$, seeds pod^{-1} and seed, stover and biological yield, protein content in seed and total uptake of nitrogen, phosphorus and potassium, net returns (INR 39,937 ha^{-1}) and B:C ratio (1.58) over preceding levels. Results have also shown that among the agro-chemicals, application of thiourea at 500 ppm as foliar spray being at par with TGA at 100 ppm, increased the plant height and dry matter accumulation at 60 DAS and at harvest, number of pods per plant, number of seeds per pod, test weight, seed, stover and biological yield, N and P content in seed and stover and protein content in seed and total potassium uptake, net returns (INR 35,617 ha^{-1}), and B:C ratio (1.45) over control. However, in terms of total uptake of nitrogen and phosphorus and dry matter accumulation at 30 DAS, chlorophyll content, the treatment 500 ppm thiourea proved significantly superior.

4.11.5 Baby Corn

Maize is a crop with increasing production every year, especially in the state of Andhra Pradesh. During rainy season about 15 varieties of maize viz., Vivek-9, Vivek-11, Vivek-15, Vivek-17, Baby corn-1, Ashwini, Bioseed-1, Maduri, Harsha, Him-129, BH-2187, Varun, HQPM-1, African Tall, and J-1006 developed for different use patterns, i.e., grain varieties, green cob cultivars, baby corn as well as fodder varieties were tested for baby corn potential and their subsequent utilization as fodder (Shanti et al. 2015). Highest yield of baby corn was observed in Harsha i.e. 68.96 q ha^{-1}. Most of the varieties exclusively released for baby corn gave highest baby corn yields ranging from 66.04 q ha^{-1} (Vivek-11) to 68.96 q ha^{-1} (Harsha). Traditional maize varieties like Ashwini (50.43 q ha^{-1}), Maduri (50.75 q ha^{-1}), Varun (49.18), and BH-2187 (49.04) recorded commendable yields next to baby corn varieties.

The study showed that there was a gradual decrease in crude protein content of husk from first pick to third pick i.e., from 14.88 % (first pick), 10.06 % (second pick), and 8.75 % (third pick) as in case of Vivek-11 and so in other varieties Vivek-17 and Baby corn-1. On the contrary, the crude protein % of cob (dehusked) increased gradually from first pick to third pick in the three baby corn varieties tested with mean values of 8.97 % (first pick) <11.38 % (second pick) <12.40 % (third pick). This indicates that the protein content of husk is being channelized to cob with advancing age of cob. This also infers that with advancing age of crop (with each pick) the channelization of protein is increasing to the reproductive parts of plant (dehusked cob). This also highlights that the husk of cobs can also be efficiently used as fodder owing to its high crude protein content.

4.11.6 Integrated Nutrient Management

Integrated use of inorganic fertilizer coupled with organic manure, phospho compost, and crop residue i.e. cotton stalk on yield, nutrient uptake, and their residual effect on soil fertility in inceptisol of Akola, Maharashtra, (Shinde et al. 2015) revealed that the application of 100 % Recommended dose fertilizer (RDF) + 5 t of FYM ha^{-1} resulted in highest NPK uptake as well as fertility status by use of 100 % organics in the form of FYM and resulted in higher yield. Integrated use of organic along with optimum doses of chemical fertilizers not only produced highest crop yield but also increased the uptake and fertility status of experimental soil.

Different sources of nitrogen, organic, and inorganic sources their combinations were evaluated in chilli (*Capsicum annuum* L.) for better growth, flowering, and fruit yield of chilli (Pariari et al. 2015). The growth parameters like plant height and number of branches per plant were recorded maximum with the combination of vermicompost and urea at 50 %. Flowering was delayed with increase in inorganic nitrogen level (25, 50 and 75 %) in the growing media. The yield attributes including fruit yield was found highest also with nitrogen received from vermicompost and urea at 50 % level. The qualitative aspects were increased with the application of neem cake compared to other inorganic sources.

4.11.7 Organic Fruit Production

Modern agriculture as well as horticulture has been heavily depending on the fossil fuel based inputs such as inorganic fertilizers, pesticides, herbicides, and energy intensive farm machinery. The indiscriminate use of pesticides and chemicals has caused serious damage to the ecosystem. Accumulation of nitrates has caused water pollution leading to carcinogenic effect on human body. It is therefore necessary to use ecofriendly methods for sustainable fruit production of guava. Dutta et al. (2015) studied the effect of biofertilizer on organic fruit production of guava grown in new alluvial zones of West Bengal. Results revealed that different treatment of biofertilizer increased the fruit weight and other biochemical composition of fruit. Among different treatments, *Azospirillum brasilense* + VAM gave the best result. Biofertilizer namely *Azospirillum brasilense* + VAM can be applied for sustainable fruit production of guava which are possibly safe and residues free.

4.11.8 Organic Agriculture

On sandy loam soils in cowpea *Vigna unguiculata* (L.) Walp, the application of vermicompost increased the plant height, number of branches per plant, number of pods per plant, number of seeds per pod, seed, straw and biological yield, N, P, and K content in seed and straw, chlorophyll content in leaves, number of total and effective root nodules, leg haemoglobin content in root nodules and up to 4 t ha^{-1} (Khan et al. 2015). The organic carbon, available nitrogen, phosphorus, and potassium in soil increased significantly with increasing levels of vermicompost at harvest of the crop. Seed inoculation with *Rhizobium* + PSB significantly increased the plant height; number of branches per plant; number of pods per plant; number of seeds per pod; seed, straw, and biological yield; N, P, and K content in seed and straw; chlorophyll content in leaves; and number of total and effective root nodules. Leg haemoglobin content in root nodules and the test weight and harvest index were found non-significant. However, in respect of phosphorus content uptake in seed and straw with PSB alone proved superior to control and *Rhizobium*. The organic carbon, available nitrogen, phosphorus, and potassium in soil significantly increased with inoculation of *Rhizobium* + PSB at harvest stage of the crop (Khan et al. 2015).

The effect of different forms of organic manures and systems of planting on growth of scented rice (*Oryza sativa* L.) cv. Pusa Basmati studied (Dipali et al. 2015) studied has shown that amongst the forms of organic manures, FYM (12 t ha^{-1}) + Panchgavya (3 %) 4 foliar spray in combination with system of rice intensification (SRI) was found to be the best for obtaining higher growth and growth attributing parameters viz., plant height (cm), number of tillers $hill^{-1}$, plant dry weight (g), CGR (g m^{-2} day^{-1}), and RGR (g g^{-1} day^{-1}) in rice.

Maize (*Zea mays* L.) response studies to organic manures and *Azospirillum* in foothills of north-eastern India (Kharutso et al. 2015) has shown that the plant height, stem thickness (cm) and leaf area index were highest with application of FYM at 10 t ha^{-1} followed by FYM at 10 t ha^{-1} + *Azospirillum* at 20 g kg^{-1} seed. FYM at 10 t ha^{-1} has also produced highest yield attributing characters like cobs weight (120.51 g), number of seed rows (34.84), length of cob (15.29 cm), grain weight (81.66 g), grain yield (18.16 q ha^{-1}), and B: C ratio 2.5.

4.11.9 Organic Farming

Intensive cultivation with excess and imbalance use of chemical fertilizers cause reduction in soil fertility status may also reduce the availability of micronutrients and yield, also the concerns about the increased pest and disease infestation and increased cost; organic farming seems to be an alternative, which may not be acceptable on short-term basis, but may be viable on long-term basis. Some organic manures, viz. farmyard manure, vermicompost, and green manure, are important constituents of the organic farming system which may not only improve the soil health but also enhance the quality of the produce.

The performance of rice to different organic manures (Vineethaa et al. 2015) studied revealed that highest grain yield of 6307 kg ha^{-1} was obtained when karanj cake was applied at 3 t ha^{-1} which was on par with RDF (160: 60:40 kg N: P_2O_5:K_2O ha^{-1}) followed by poultry manure, FYM, vermicompost, press mud, and neem cake which were in turn on par with each other. This might be due to number of panicles hill^{-1} and number of filled grains panicle^{-1} formed due to the application of karanj cake.

Praveen and Swamy (2015) studied retrospect and prospects of organic fertilizer for enhancing the quality of life. Organic farming is basically a holistic management system, which promotes and improves the health of the agro-ecosystem related to biodiversity, nutrient biocycles, soil microbial, and biochemical activities. Organic farming emphasizes on management practices involving sustainable use of organic manures, green manuring, organic pest management practices, and so on. It has also come to mean that it is a system of farming that prohibits the use of artificial fertilizers and synthetic pesticides.

The adverse effects of modern agricultural practices are not only on the farm but also on the health of all living things and thus on the environment. These have been well documented all over the world. Application technology, particularly in terms of the use of chemical fertilizers and pesticides all around us, has persuaded people to think aloud. Their negative effects are manifested through soil erosion, water shortages, salinization, soil contamination, genetic erosion, etc. Further, in organic farming all practices are interrelated and the end product will be decomposable one. So there is no cause for environmental pollution in organic farming, ideal combination of agronomical, physical, and biological measures bring down the population harmful microbes and do not release any residues in soil and crop environment.

The conventional farming had helped India not only to indiscriminate use of chemical fertilizers, pesticides, weedicides, etc. over a period of time, resulted in environmental pollution and hazards in animals and human health. The biological science embraces the entire significant outcome from the various disciplines to enhance the quality of life. However, among the available alternatives, organic agriculture is a visible option to enhance the quality of life of human beings in the coming decades for better health, wealth, and prosperity.

The plant parts from organically managed soils are intuitively rich and safe crops which are grown under balanced nutritive approach. Crop uptake nutrients as and when required in entire growing period convert the absorbed nutrients properly into sink. When the food materials from organically managed soils are consumed, natural immunity of the human beings is developed. Organic input materials are less costly source, readily available at the door step and very easy to apply. The nutrients are available in entire crop period in a balanced way. It provides optimal environment to the crop growth and taps the full genetic potential of crops to provide satisfactory level yield and quality. The end product is superior in nutritive quality than chemical farming and gets higher offer in the competitive market offered premium price. In organic farming products, presence of potassium improves the keeping quality and also nutritive value in balanced way. It offers much premium price in market and it enhances the quality of life of human beings both in producing as well as in consuming sector.

Hence, relatively high success of organic farming in some countries are due to the high awareness of the health problems caused by the consumption of contaminated food products, the ill effects of environment degradation, and appropriate supports by the government and organizations which promotes the organic farming in the country as well in the global levels.

4.11.10 Organic Amendments

Soils fix huge amounts of phosphorus significantly and limit its availability to plants. Only a part of fertilizer phosphorus is readily soluble in water; thus, the water solubility of phosphorus is not always the best estimate of phosphorus availability to the plants. The major problem is phosphorus fixation in the soil due to inorganic elements like Al, Fe, and Ca.

Organic amendments applied through FYM, vermicompost, poultry manure, and paddy straw (Kumar et al. 2015) affected the availability of different chemical fractions of phosphorous. All organic amendments application increased the

Olsen's P. Availability of phosphorus increased in the order of paddy straws ‹ FYM ‹ vermicompost ‹ poultry manure, respectively. FYM at 20 t ha^{-1} increased the saloid-P, Al-P, and Fe-P. Poultry manure at 20 t ha^{-1} also increased all fractions of phosphorus. During the incubation period increase in P availability up to initial 28 days was observed but at later stage, P availability decreased. Available P is significantly correlated with saloid-P (r = 0.973*) at 21 days and Ca-P (r = 0.901*) at 60 days, at 5 % level of significance. Organic amendments showed significant increase of soil EC (dSm^{-1}) and organic carbon (%) but significant decrease of soil pH.

Fly ash and farmyard manure influenced the soil biochemical activity and performance of rice variety Tellahamsa grown on inceptisol soils (Reddy et al. 2015b). There was an increase in grain and straw yield of rice with fly ash, FYM, and their interactions. The highest grain (5.84 t ha^{-1}) and straw yield (7.87 t ha^{-1}) were recorded by combined application of fly ash at 10 t ha^{-1} and FYM at 10 t ha^{-1}. Application of fly ash at 15 t ha^{-1} along with FYM at 10 t ha^{-1}has also resulted in highest urease, dehydrogenase, and cellulase activity at 30 DAT (4.48 mg of NH_4^+ released g^{-1}soil h^{-1}, 5.37 mg of TPF produced g^{-1} soil d^{-1}, and 3.50 mg of glucose released g^{-1} soil d^{-1}), 60 DAT (4.80 mg of NH_4^+ released g^{-1} soil h^{-1}, 5.47 mg of TPF produced g^{-1} soil d^{-1}, and 3.32 mg of glucose released g^{-1} soil d^{-1}) and at harvest (2.53 mg of NH_4^+ released g^{-1} soil h^{-1}, 3.07 mg of TPF produced g^{-1} soil d^{-1}, and 2.16 mg of glucose released g^{-1} soil d^{-1}), respectively. The acid and alkaline phosphatase activity was not influenced by fly ash levels at all the stages viz., 30, 60 DAT, and at harvest. However, it was significantly influenced by FYM application and their interactions.

4.11.11 Residual Effects of Organics and Nitrogen

In order to arrive at a profitable, sustainable, and ecofriendly nutrient management, it is necessary to consider cropping sequence as a whole rather than an individual crop. Studies (Veeranna et al. 2015) on residual effect of organics and nitrogen on maize yield and soil properties in rice-no till maize cropping system in Warangal, Telangana State, during *kharif* and *rabi* seasons of 2010–2011 and 2011–2012 on a vertisol soil indicated that among the three organic sources (FYM, Rice straw, Green manure (*Sesbania*)) applied to rice crop, the residual effect was higher with rice straw on kernel yield of maize (6997 kg ha^{-1}). Among the different doses of nitrogen, higher kernel yield was obtained at 300 kg N ha^{-1} (7108 kg ha^{-1}). The pH of the soil was moderated in all the organic treatments and bulk density was reduced under rice straw combined with 300 kg N ha^{-1} (1.28 g cc^{-1}). Organic carbon content of the soil after the harvest of maize crop was improved in all the three organic sources compared to initial status (0.44 %) both at higher and lower nitrogen application rates. FYM and rice straw improved the available P and K status of the soil respectively, while green manuring resulted in higher available status of both P and K in the soil.

4.11.12 Organic and Inorganic Mulching

Moisture conservation in soils is possible by mulching. In addition to moisture conservation, mulching also reduces weed growth, moderates the crop micro-climate, and improves fertility status of the soil. Moisture management in wheat under different agro-ecologies studied through Pusa hydrogel, aqua sowing, and organic mulching during winter seasons of 2010–2011 to 2012–2013 (Bana et al. 2015) in sandy loam having 195, 14.0, and 230 kg ha^{-1} NPK, respectively, and 0.34 % organic carbon with pH of 7.8 indicated that wheat grain and straw yield increased where Pusa hydrogel, organic mulching, and aqua fertilization were applied over the control both under limited irrigation and assured irrigation conditions. However, the highest yield was obtained where organic mulch was applied both under limited as well as assured irrigation conditions.

Pusa hydrogel application at 2.5 kg ha^{-1} under limited irrigation resulted in the highest grain yield of 3.30 t ha^{-1}, while organic mulching and aqua fertilization resulted in yields of 2.79, 3.53, and 3.06 t ha^{-1} respectively during the years of study. Under assured irrigation conditions the grain yield in the hydrogel treatment was 4.37 t ha^{-1} as compared to control, organic mulching, and aqua fertilization where the yields were 4.04, 4.57, and 4.30 t ha^{-1}, respectively (Bana et al. 2015).

Use of various moisture conservation practices gave better output under limited availability of soil moisture than under assured supply of soil moisture. Performance of various moisture management practices including Pusa hydrogel and organic mulch were remained comparatively poor under normal water availability, because the hydrogel and mulching are meant for reducing the water stress under limited water supply.

The study concluded that the use of various moisture conservation practices including Pusa hydrogel has favorable effect on wheat productivity and water use efficiency and their effect is more visible under moisture stress condition.

Nalayini et al. (2015) made evaluation of organic and inorganic mulches for rainfed BT cotton with RCH 20 Bt as test cultivar at Central Institute for Cotton Research, Regional Station, and Coimbatore. Five organic mulch treatments viz., maize stover, sugarcane trash at 50 t ha^{-1}, gunny sheet spread between two cropped rows, sub surface coir pith, and surface coir pith application at 20 t ha^{-1} were compared with two inorganic mulches viz., polyethylene mulching and biodegradable polyethylene mulching (Polyethylene + 4 % d2w additive) and evaluated against no mulch control. In both years, the crop suffered severe moisture stress during boll development and the desired yield could not be obtained. All the mulch treatments except surface coir pith application enhanced the seed cotton yield of RCH 20 Bt significantly under rainfed condition. The yield increase due to various mulching ranged from 6.2 to 92.8 %. Polyethylene mulching recorded the highest (1816 kg ha^{-1} as pooled mean) seed cotton yield and was on par with all other mulch treatments except surface coir pith and no mulch control. Though poly ethylene and biodegradable polyethylene mulching recorded significant increase in yield over no mulching, considering the cost of poly film and the rate of yield increase, it may not be economical under rainfed condition. The sub-soil coir pith application was as good as other crop residue mulches and this waste product of coir industry could be used efficiently as mulch for rainfed cotton.

4.12 Resource Management

Summer rice occupies a sizeable area in Chattisgarh and also in India. Its area is further increasing gradually with irrigated area. Drip irrigation is a relatively new but revolutionary concept in applying water. Drip irrigation may save huge quantity of water in summer rice that may help in enhancement of area of summer rice crop even under available water resources and good plant health.

Sonit et al. (2015) investigated resource management in summer rice varieties MTU 1010 and IR 64 through drip irrigation treatments of (0.6, 0.8, 1.0, 1.2, and 1.4 IW: CPE ratio) micro sprinkler irrigation once in 3 day and flooding as subplots. Panicle length, panicle weight, sound grains $panicle^{-1}$, and numbers of grains $panicle^{-1}$ were statistically similar in drip irrigation at 1.4, 1.2, and 1.0I W: CPE ratio and traditional flooding. The maximum seed yield was obtained in drip irrigation at 1.4 IW: CPE ratio.

The water use was maximum in flooding (3340 mm) followed by micro sprinkler irrigation (1417 mm) and drip irrigation at 1.4 (1300 mm):1.2 (1152 mm) and 1.0 (948 mm) IW: CPE ratio. Drip irrigation at1.4 and 1.2 IW: CPE ratio saved 63 and 59 % of irrigation water, respectively overflooding; indicating that cropped area under summer rice can be doubled with same quantity of available water by using drip irrigation without sacrificing grain yield.

4.13 Effects of Rootstocks on Grape Production

Root stocks have an influence on yield and quality of grapes. Manjuvani et al. (2015) studied the influence of rootstocks (1103 P, SO_4, and Dog Ridge) on yield and quality of table varieties (Thompson Seedless, Flame Seedless and Kishmish Chorni) of grapes. High vigour in terms of pruning weight was observed with varieties grafted on rootstocks when compared to those raised on own roots. Own rooted varieties performed well compared to grafted varieties with respect to number of fruitful canes per vine and number of bunch per vine and resulted in higher yield per vine. Scions on Dog ridge gave high bunch weight, 100 berry weights, and berry diameter. Among the varieties Thompson Seedless was more vigorous. Kishmish Chorni produced more number of fruitful canes, bunches per vine, and high yields. Highest bunch weight and berry diameter was obtained with Flame Seedless. High TSS and low acidity was recorded with scions on 1103 P and Dog Ridge rootstocks. Brix: acid ratio was high in scions on 1103 P. Flame Seedless and Kishmish Chorni recorded high TSS.

4.14 Extending the Vase Life of Flower

Locally available floral preservatives are also effective in extending the vas life of cut flowers. Cut gerberas of variety cv. Savannah were precooled at 5 °C for 6 h, followed by pulsing with sucrose at 20 % + sodium hypochlorite at 50 ppm for 12 h and then kept in locally available preservative floral solutions i.e. sugar, vinegar, lime juice, commercial bleach, and neem extract (Bhanumurthy et al. 2015). Under ambient conditions, 1 % neem extract maintained better water relations, recorded lowest scape bending curvature (4.16°), highest total soluble solids in stalk (10.08 °Brix), anthocyanin content (5.83 mg Congo Red g^{-1} fruit weight) in ligules, with lowest optical density of (0.041). Flowers held in distilled water (control) registered highest scape bending curvature (29.47°) and recorded highest optical density (0.088) which was attributed with highest microbial count (9.11 × 106 cfu ml^{-1}) than neem extract at 1 % (4.26 × 105 cfu ml). These cut gerberas which were held in 1 % neem extract had highest vase life (11.76 days) compared to control (4.53 days).

4.15 Effects of Coal Dust on Nutrient Availability

Coal dust emitted from open cast mining of coal accumulates at different levels up to a distance of 2 km from mining site. The soil samples collected within 2 km from coal mining and beyond 2 km open cast coal mining sites from four places of opencast i.e., Srirampur, Medipalli, Bhupalpally, and Yellandu in Telangana State (Madhavi et al. 2015) indicated that open cast coal mining had an impact on soil available macronutrients.

4.16 Efficacy of Biological Resources Against Rust

White rust caused by *Albugo candida* is one of the important diseases in rapeseed-mustard. Biocontrol agents and botanicals evaluated against white rust incidence in rapeseed-mustard during 2010–2012 (Gopi et al. 2015) indicated that *Trichoderma viride* was effective in reducing the disease incidence of white rust in leaf phase.

4.17 Use of Mycoflora on *Rhizoctonia solani*

The isolates of mycoflora and florescent *Pseudomonas* isolated from rhizosphere soil of tomato fields in different mandals of Ranga Reddy district (Sumalatha et al. 2015) were evaluated against stem blight disease caused by *Rhizoctonia solani*. Among the screened isolates against

Rhizoctonia solani, isolate M10 (*Trichoderma viride*) was highly effective with 79.17 % inhibition of test pathogen under *in vitro* conditions. Further, the *in vitro* studies with eight commercial fungicides indicated that propiconazole, hexaconazole, tebuconazole, captan + hexaconazole, and trifloxystrobin + tebuconazole showed 100 % inhibition of the radial growth of the pathogen at four dosages (recommended, 75 % of recommended, 50 % of recommended, and 25 % of recommended dosage). Potential bioagent *Trichoderma* showed least sensitivity or mostly compatible with most effective fungicide pyroclostrobin + metiram.

4.18 Production of Bioresource

Silk worm cocoon size, shape, and quality play vital role in sustainability of the sericulture industry. For a successful breeding strategy in silkworm, one has to give proper importance on cocoon characters of parental lines. Khan (2015) studied cocoon characters based on multivariate clustering of different types of silkworm breeds, *Bombyx mori* L. (Lepidoptera: Bombycidae). According to them, variability measured in terms of co-efficient of variation was highest for geographic breeds compared to mutant and evolved breeds. Excepting the cocoon weight and length-breadth ratio in male cocoon of nine evolved breeds, wide differences among the breeds for all three types of silkworms' breeds were observed. Hence the better performance could be exploited in future breeding program. Clustering of breeds separately for males and females based on cocoon characters indicated that by and large four groups could be framed and the silkworms from different groups i.e. geographic, evolved and mutant breeds do not follow the same grouping, rather a tendency of mixing among the breeds of different groups were noticed. Among the breeds Raj, Sarupat (W), and Tamilnadu White are significantly different, having the tendency to picture their respective identities separately in dendrogram. The study urged the selection of better breeds for future breeding improvement programmes in silkworm.

4.19 Adjuvants for UV-C Rays Protectability

The effect of UVC rays on biomass development of *Beauveria bassiana* (Balsamo) Vuillemin combination with chemicals, vegetable oils, and other substrates as adjuvants were evaluated by exposing the combination for 10–50 min and 2, 3, and 5 h under laboratory conditions (Jadhav et al. 2015). The biomass produced by *B. bassiana* with or without adjuvants in culture medium after exposure to UVC rays decreased with increase in the exposure periods. After 5 h UVC rays exposure, the treatment with adjuvant sunflower oil 1.0 % maintained its superiority and developed highest (6.13 g) biomass 40 ml^{-1} medium of *B. bassiana* fungal mat. It was at par with sunflower oil 0.5 % (6.07 g). The next effective treatments for the UVC protectability reflecting in fungal biomass production were groundnut oil 1.0 % (5.70 g) and 0.5 % (5.63 g). Treatment with Tween 80 0.5 % (2.63 g), molasses 2.0 % (3.13 g), and Tween 80 1.0 % (zero g) consistently proved to be inefficient for UVC rays protectability for *B. bassiana*. As adjuvants vegetable oils, ghee and indigo 0.5 % and 1 % gave considerable protection to *B. bassiana* from the UVC rays producing the biomass 4.47–6.13 g against 6.20 and 2.92 g biomass development in the unexposed and exposed control, respectively.

4.20 Seed Ball Technology

In Bundelkhand region, Jhansi, Uttar Pradesh, since its birth, the unfertile land has threatened the food grain production, whereas the scarcity of rangelands and pastures for feed and fodder supply has restricted the milk availability to the local people. In this situation, the identification and rejuvenation of the natural grasslands through sowing of grass seeds in this region has drawn attention even at national level since long back. Geographic information system (GIS) based mapping has been done previously by Indian Grassland and Fodder Research Institute (IGFRI), Jhansi, especially for unexplored areas in this region to identify degraded rangelands. Maity et al. (2015) adopted seed ball technology: way forward to revitalizing rangelands in Bundelkhand region, Jhansi, Uttar Pradesh.

Grass seeds are very small and covered with fluff or appendages which lead to blowing up of seeds and hinder its mechanical sowing. To rejuvenate these rangelands through grasses, grass seed ball technology or seed pelleting has been tried in a new form to provide a better microenvironment to the seeds and to address above problems. The experiment was conducted to pellet Dinanath (*Pennisetum pedicellatum*) and guinea grass (*Panicum maximum*) seeds. Dinanath grass produces highly fluffed seed and at harvest stage a substantial portion of seed remains blank due to shedding of caryopsis at maturity leading to poor germination in field. Seeds were defluffed through 'common cotton batting machine' (used for quilt making) to get assured germinable naked seeds out of the fluffs. Naked seeds were pelleted by mixing with soil, required nutrient mixture and water at right proportion through a rotating tyre. The modified pellets were of 10–15 mm diameter and 2.0–4.0 g in weight and could contain 5–10 seeds depending on the purpose, of which >90 % seed could germinate. This technology in which soil balls acted as harbor for defluffed seeds during pelleting solved the earlier problem of non-germination of seed pellets due to compactness and/or hardness of those pellets and uncertainty in presence or absence of real seed in the fluffs (Maity et al. 2015). These balls may be broadcasted or thrown to the barren rangelands identified in Bundelkhand region before monsoon and these would germinate consequently. At small scale, the technology was in practice in IGFRI and holds a greater potential in large scale also.

4.21 Resource Use Efficiency for Fish Production

In India, of the total population, 68.87 % live in rural areas and in Manipur; 66.57 % of the population lives in rural areas; and 57 % of the workforce depends on agriculture, forestry, and fishing (Census 2011) in Devi and Singh 2015. This revealed that in India and in Manipur, rural economy was dominant and large numbers of people depend on agriculture and allied activities. Fishery, both in the country and the state, has been a traditional practice dating back to several centuries especially for the livelihood of the rural people. An attempt made to analyze the resource use efficiency of inland fish production under different categories of farm size in Imphal West district of Manipur during 2010–2011 (Devi and Singh 2015). The farmers in the district utilized available water resources for fish farming. Villages nurtured a variety of fish viz., Katla, Rohu, Mrigal, Common carps, Grass carps, and Silver carps in the ponds. Multistage random sampling procedure was adopted for the selection of the ultimate unit of samples. The data was collected from 100 selected fish farmers of Imphal-West district of Manipur. The fish farmers were classified into two categories i.e. category I large ($\geq$1 ha) and category II small (<1 ha) based on the area under fish farming. The data were analyzed by using Cobb Douglas production function to analyze the expenditures of inputs used. In category I, the ratio of marginal value productivity (MVP) to their factor cost for the variable pond area was found negative indicating over utilization hence the expenditure on this input should be reduced which tends to decrease the cost of production and increase of net returns of the fish farmers, whereas the variables like lime and fingerlings were found positive indicating under utilization hence their further use will increase the net returns. In category II, the expenditures on fingerlings and feed were found positive indicating underutilization hence their further use in the production will increase the net returns and that of human labor was negative indicating over utilization hence the expenditure on this input should be reduced to increase the net returns of the fish farmers. For the overall average farm, the ratio of marginal value productivity (MVP) to their marginal factor cost (MFC) of the significant variables of fertilizers, lime, fingerlings, feed, and human labor were −6.78, 4.98, 42.99, 22.07, and −14.29, respectively. So, the expenditure on fertilizers and human labor is over utilized and it should be reduced. But the expenditure on lime, fingerlings, and feed is underutilized so there is scope for more use of them. By doing so, the gross returns of fish production could be increased in the state.

4.22 Bioresource Management in Society

Hulagur and Swamy (2015) studied convergence and collaboration for bioresource management in society under present climate change. Convergence of bioresource management is the goal of work towards establishing a single set of accounting standards. Convergence is the tendency for different technological systems to evolve towards performing similar tasks. There is a need of the combination of multiple services through lines of status having the same knowledge of the surrounding. The notions, functions, and sequences approach a limit.

Collaboration of bioresource management is more than the intersection of common goals seen in cooperative ventures, but a deep collective determination to reach an identical objective. Collaboration is working with each other to do a task and to achieve shared goals. The institution or organization has objectives of bioresource management to work towards, and that all staff members are made aware of these objectives. Unless the objectives are known to and shared by all staff members, resources cannot be pooled in one direction and the institution or organization bound to waste resources. To organize and mobilize resources for bioresource management in the best possible and most cost-effective way, there is a need for job description for each job function of bioresource management. It is important to distinguish between a particular job function required in an institution or organization and individual staff members. One should have a clear picture of what each staff member of bioresource management should be able to do as per her or his job chart. Probably the training needs will be greater than the resources one has at hand meeting these needs of bioresource management. Once it is known what the different staff members need to be trained on bioresource management and that these training needs have been prioritized.

It is now time to implement the training plan that has been developed for bioresource management. An important issue facing is commitment to training. This is required and it is seldom well realized. Hence, there is a need for knowledge of training models to train professionals. However, the training must start by questioning the basic assumption which has governed our training approach. So an analysis of SWOT, i.e., strength, weakness, opportunity, and threats, called for to enable as to have new conceptualization.

4.23 Alleviation of Heat Stress in Animals

Heat stress is a constant challenge to goats under tropical climatic conditions. Menacing global warming created the need for research over mitigation of heat stress and climate smart animal agriculture. The cytoprotective and immunity enhancive effects of melatonin during heat stress were studied (Iqbal et al. 2015) in goats. The total protein level increased in melatonin treated group with the globulin component contributing to the increase than the albumin, while a decline was seen at 40 °C in melatonin treated group. The cortisol level rose significantly ($P > 0.05$) with increase in exposure temperature in control animals. Significantly ($p < 0.05$) low-level of cortisol was observed in melatonin in treated group at all exposure temperature. HSP60, HSP70, leptin, and ubiquitin showed a significant ($p < 0.05$) upregulation as a protective effects under heat stress in both the groups. The relative expression of HSP60 and leptin was increased many folds in melatonin treated group at 40 °C exposure temperature. The study clearly supported the immune enhancive effects and cytoprotective role of melatonin with elevation in serum globulin level and upregulation of genes like HSP-60 and leptinin PBMCs.

References

Bana RS, Rana KS, Anupama (2015) Studies on pusa hydrogel, aqua sowing and organic mulching for moisture management in wheat under different agro-ecologies. In: Compendium of abstracts of the 2nd international conference on bio-resource and stress management, ANGRAU & PJTSAU, Hyderabad, India, 7–10 January 2015

Bhanumurthy KC, Chandrasekhar R, Prashanth P (2015) Extending the vase life of cut gerbera (Gerbera jamesonii Bolus ex. Hook) cv. Savannah by using locally available floral preservatives under ambient storage. In: Compendium of abstracts of the 2nd international conference on bio-resource and stress management, ANGRAU & PJTSAU, Hyderabad, India, 7–10 January 2015

Bisen Y, Tripathi MP, Dwivedi K et al (2015) Water management studies on finger millet (Eleusine coracona) in alfisol of mid-land situation in Chhattisgarh plains. In: Compendium of abstracts of the 2nd international conference on bio-resource and stress management, ANGRAU & PJTSAU, Hyderabad, India, 7–10 January 2015

Chauhan A, Verma SC, Bharadwaj SK (2015) Impactof land uses and seasons on physico-chemical characteristics of surface water in Solan District of Himachal Pradesh, India. In: Compendium of abstracts of the 2nd international conference on bio-resource and stress management, ANGRAU & PJTSAU, Hyderabad, India, 7–10 January 2015

Devi SZ, Singh NA (2015) Resource use efficiency in Inland fish production under different farm sizes in Imphal-West district, Manipur. In: Compendium of abstracts of the 2nd international conference on bio-resource and stress management, ANGRAU & PJTSAU, Hyderabad, India, 7–10 January 2015

Dipali K, Jadhav P, Mishra PK et al (2015) Effect of different forms of organic manures and systems of planting on growth of scented rice (Oryza sativa L.) cv. Pusa Basmati. In: Compendium of abstracts of the 2nd international conference on bio-resource and stress management, ANGRAU & PJTSAU, Hyderabad, India, 7–10 January 2015

Dutta P, Das K, Roy D et al (2015) Effect of bio-fertilizer on organic fruit production of guava grown in new alluvial zones of West Bengal. In: Compendium of abstracts of the 2nd international conference on bio-resource and stress management, ANGRAU & PJTSAU, Hyderabad, India, 7–10 January 2015

Gopi R, Avasthe RK, Kapoor C (2015) Evaluation of efficacy of some biological resources for the management of white rust of mustard in sikkim himalaya. In: Compendium of abstracts of the 2nd international conference on bio-resource and stress management, ANGRAU & PJTSAU, Hyderabad, India, 7–10 January 2015

Hulagur B, Swamy BKN (2015) Convergence and collaboration for bio-resource management in society under present climate change. In: Compendium of abstracts of the 2nd international conference on bio-resource and stress management, ANGRAU & PJTSAU, Hyderabad, India, 7–10 January 2015

Iqbal H, Sharma S, Ramesh K et al (2015) Cytoprotective and immunity enhancive effects of melatonin during heat stress. In: Compendium of abstracts of the 2nd international conference on bio-resource and stress management, ANGRAU & PJTSAU, Hyderabad, India, 7–10 January 2015

Jadhav RS, Ukey NS, Patil SD et al (2015) Evaluation of some adjuvants for UVC rays protectability for Beauveria bassiana (Balsamo) Vuillemin. In: Compendium of abstracts of the 2nd international conference on bio-resource and stress management, ANGRAU & PJTSAU, Hyderabad, India, 7–10 January 2015

Kashyap R, Verma KS, Brahmi MS et al (2015) Waterquality assessment and suggested restoration measures for a National Lake of India: case study of Rewalsar Lake. In: Compendium of abstracts of the 2nd international conference on bio-resource and stress management, ANGRAU & PJTSAU, Hyderabad, India, 7–10 January 2015

Khan MZ (2015) Cocoon characters based multivariate clustering of different types of silkworm breeds, Bombyx mori L. (Lepidoptera: Bombycidae). In: Compendium of abstracts of the 2nd international conference on bio-resource and stress management, ANGRAU & PJTSAU, Hyderabad, India, 7–10 January 2015

Khan VM, Verma HP, Shivran AC et al (2015) Effect of vermicompost and bio fertilizers on growth and yield of cowpea Vigna unguiculata (L.) Walpon sandy loam soil. In: Compendium of abstracts of the 2nd international conference on bio-resource and stress management, ANGRAU & PJTSAU, Hyderabad, India, 7–10 January 2015

Kharutso A, Singh AP, Longkumer LT (2015) Response of maize (Zea mays L.) to organic manures and Azospirillum in foothills of north-eastern India. In: Compendium of abstracts of the 2nd international conference on bio-resource and stress management, ANGRAU & PJTSAU, Hyderabad, India, 7–10 January 2015

Kumar S, Gupta A, Srivastava A et al (2015) Effect of organic amendments on availability of different chemical fractions of phosphorus. In: Compendium of abstracts of the 2nd international conference on bio-resource and stress management, ANGRAU & PJTSAU, Hyderabad, India, 7–10 January 2015

Madhavi B, Jayasree G, Raj GB (2015) Effect of Opencast coal mining on soil fertility in Telangana state. In: Compendium of abstracts of the 2nd international conference on bio-resource and stress management, ANGRAU & PJTSAU, Hyderabad, India, 7–10 January 2015

Maity A, Vijay D, Malaviya DR et al (2015) Seed ball technology: way forward to revitalizing rangelands in Bundelkhand region, Jhansi, Uttar Pradesh. In: Compendium of abstracts of the 2nd international conference on bio-resource and stress management, ANGRAU & PJTSAU, Hyderabad, India, 7–10 January 2015

Manjuvani B, Srinivas Rao B, Vijaya D (2015) Influence of root stocks on yield and quality of table varieties of grapes. In: Compendium of abstracts of the 2nd international conference on bio-resource and stress management, ANGRAU & PJTSAU, Hyderabad, India, 7–10 January 2015

Mundiyara R, Jakhra ML, Bhardwaj R (2015) Sorghum: a potential source of renewable energy in India. In: Compendium of abstracts of the 2nd international conference on bio-resource and stress management, ANGRAU & PJTSAU, Hyderabad, India, 7–10 January 2015

Nalayini P, Sankaranarayanan K, Velmourougane K (2015) Evaluation of organic and inorganic mulches for rain fed Bt cotton. In: Compendium of abstracts of the 2nd international conference on bio-resource and stress management, ANGRAU & PJTSAU, Hyderabad, India, 7–10 January 2015

Nandapure SP, Wankhade SG, Imade SR (2015) Influence of nutrient management on macro and micronutrients availability of inceptisols under Java citronella (Cymbopogon winterianus). In: Compendium of abstracts of the 2nd international conference on bio-resource and stress management, ANGRAU & PJTSAU, Hyderabad, India, 7–10 January 2015

Pariari A, Khan S, Das S et al (2015) Effect of different source and combination of nitrogen on chilli (Capsicum annuum L.). In: Compendium of abstracts of the 2nd international conference on bio-resource and stress management, ANGRAU & PJTSAU, Hyderabad, India, 7–10 January 2015

Praveen T, Swamy N (2015) Retrospect and prospects of organic fertilizer for enhancing the quality of life. In: Compendium of abstracts of the 2nd international conference on bio-resource and stress management, ANGRAU & PJTSAU, Hyderabad, India, 7–10 January 2015

Reddy GS, Susila T (2015) Effect of spacing and integrated nutrient management on herbage yield of Geranium (Pelargonium graveolens). In: Compendium of abstracts of the 2nd international conference on bio-resource and stress management, ANGRAU & PJTSAU, Hyderabad, India, 7–10 January 2015

Reddy MC, Prakash KV, Swamy BKN (2015a) Performance evaluation of animal operated spreader for donkey manure to conserve biodiversity. In: Compendium of abstracts of the 2nd international conference on bio-resource and stress management, ANGRAU & PJTSAU, Hyderabad, India, 7–10 January 2015

Reddy TP, Umadevi M, Chandrasekhar Rao P (2015b) Effect of fly ash and farm yard manure on soil biochemical activity and performance of rice grown on inceptisol. In: Compendium of abstracts of the 2nd international conference on bio-resource and stress management, ANGRAU & PJTSAU, Hyderabad, India, 7–10 January 2015

Shanti M, Balazzi Naaik R, Susheela R et al (2015) Forage production potential of various maize cultivars for baby corn. In: Compendium of abstracts of the 2nd international conference on bio-resource and stress management, ANGRAU & PJTSAU, Hyderabad, India, 7–10 January 2015

Shinde RB, Deshmukh PW, Shirale AO (2015) Effect of integrated use of inorganic and organic manures on uptake of nutrients and Soybean productivity in Inceptisol. In: Compendium of abstracts of the 2nd international conference on bio-resource and stress management, ANGRAU & PJTSAU, Hyderabad, India, 7–10 January 2015

Sonit A, Hemlata, Rathore AL (2015) Resource management in summer rice through drip irrigation. In: Compendium of abstracts of the 2nd international conference on bio-resource and stress management, ANGRAU & PJTSAU, Hyderabad, India, 7–10 January 2015

Sreenivas C, Rani TU, Sireesha A et al (2015) Spatio temporal water quality monitoring of Kolleru Lake and Upputeru wet land ecosystem, Andhra Pradesh, India. In: Compendium of abstracts of the 2nd international conference on bio-resource and stress management, ANGRAU & PJTSAU, Hyderabad, India, 7–10 January 2015

Sumalatha N, Pushpavathi B, Jagadeeshwar R et al (2015) In vitro evaluation of rhizosphere isolates and fungicides against Rhizoctonia solaniin tomato. In: Compendium of abstracts of the 2nd international conference on bio-resource and stress management, ANGRAU & PJTSAU, Hyderabad, India, 7–10 January 2015

Swamy K, Baleeshwar Reddy A, Hanumanth Rao V et al (2015) Bio-diversity at Professor Jayashankar Telangana State Agricultural University and surrounding environments. In: Compendium of abstracts of the 2nd international conference on bio-resource and stress management, ANGRAU & PJTSAU, Hyderabad, India, 7–10 January 2015

Veeranna G, Raghu Rami Reddy P, Padmaja B et al (2015) Residual effect of organics and nitrogen on maize yield and soil properties in rice-no till maize cropping system. In: Compendium of abstracts of the 2nd international conference on bio-resource and stress management, ANGRAU & PJTSAU, Hyderabad, India, 7–10 January 2015

Venkateswarulu P, Vasudeva Rao V, Ramalingeshwar Rao AVLN (2015) Status and spatial distribution of fruit bats using GIS and GPS Tools at Ranga Reddy District of Telangana State, India. In: Compendium of abstracts of the 2nd international conference on bio-resource and stress management, ANGRAU & PJTSAU, Hyderabad, India, 7–10 January 2015

Vineethaa, N, Sujathammaa R, Ramesh B et al (2015) Performance of rice to different organic manures farming. In: Compendium of abstracts of the 2nd international conference on bio-resource and stress management, ANGRAU & PJTSAU, Hyderabad, India, 7–10 January 2015

Vinutha SN, Narayana Swamy BK (2015) Medicinal and aromatic plants contract farming for sustainable horticulture development under the present climate change. In: Compendium of abstracts of the 2nd international conference on bio-resource and stress management, ANGRAU & PJTSAU, Hyderabad, India, 7–10 January 2015

Yogi R, Sharma OP, Verma HP (2015) Performance of varieties to fertility levels and agro-chemicals on growth and yield of cluster bean Cyamopsis tetragonoloba (L.). In: Compendium of abstracts of the 2nd international conference on bio-resource and stress, management, ANGRAU & PJTSAU, Hyderabad, India, 7–10 January 2015

Agrotechnology and Crop Diversification

5

A. V. Ramanjaneyulu, Narayan Chandra Sarkar, R. V. T. Balazzii Naaiik, and Ratikanta Maiti

Abstract

As an aftermath of green revolution, agriculture today is facing multiple problems of shrinking cultivable area, declining water table, global climate change, loss of soil productivity, pollution of land and water resources at macro level and decline in factor productivity and soil fertility, development of herbicide resistance and pest resurgence, thus unsustainability in productivity and farm income at micro level. This chapter has highlighted the need for adoption of nonmonetary or low monetary inputs and sustainable practices such as optimum sowing window; crop geometry; demand-driven planting methods; crop diversification; water-saving micro-irrigation methods; use of nanoparticles for enhancing nutrient use efficiency; integrated nutrient, pest, disease and weed management options; and location-specific organic farming strategies to make agriculture a profitable enterprise.

5.1 Introduction

Green revolution, a mega success story in the history of Indian agriculture, helped the country to reach to exporting status from begging bowl status. It has contributed to food security through substantial enhancement in food production and productivity. However, this situation supported by input-intensive conventional agriculture production system led to declining or stagnant factor productivity, deterioration in soil fertility and loss of organic matter, emergence of micronutrient deficiencies, deterioration in soil physical environment, declining ground water table and so on, thus stagnating farm incomes. In the light of these facts, there is a need to reverse the trend through

A.V. Ramanjaneyulu (✉)
Regional Agricultural Research Station, Professor Jayashankar Telangana State Agricultural University, Palem, Mahabubnagar District, Telangana 509 215, India
e-mail: avr_agron@rediffmail.com

N.C. Sarkar
Department of ASEPAN (Agronomy), Institute of Agricultural Sciences, Visva-Bharati, Sriniketan, West Bengal, India
e-mail: ncsiari@gmail.com

R.V.T. B. Naaiik
Electronic Media Wing, ARI, PJTSAU, Rajendranagar, Hyderabad, Telangana, India

R.K. Maiti
Facultad de Ciencias Forestales, Universidad Autónoma de Nuevo León, Carr. No 85, km 145, NL 67700 Linares, Mexico
e-mail: ratikanta.maiti@gmail.com

R.K. Maiti et al. (eds.), *Bioresource and Stress Management*, DOI 10.1007/978-981-10-0995-2_5

appropriate location-specific, demand-driven and problem-oriented agro-techniques and crop diversification strategies.

5.2 Crop Geometry and Planting Methods

Nileshsadar et al. (2015) reported that, of the three spacings (45 × 8 cm, 45 × 15 cm, 60 × 15 cm) tested under high-density planting of arboretum cotton, significantly higher kappa's yield (1656 kg ha^{-1}) was obtained besides more leaf area (61.35 dm^2 $plant^{-1}$), chlorophyll content index (35.52 %), total dry matter production (77.74 g $plant^{-1}$) and no. of sympodia (8.12 $plant^{-1}$) at harvest, when the plants were sown at a spacing of 45 × 8 cm as compared to 45 × 15 cm and 60 × 15 cm.

The pooled data (statistical) analysis of 4 years of experimentation revealed that drum seeder planting was found to be significantly superior to line planting, zigzag planting and Mechanized System of Rice Intensification (M-SRI), with regard to plant height, number of productive plant $tillers^{-1}$, number of seeds $panicle^{-1}$, dry matter production m^{-2} and thus significantly higher grain yield (6942 kg ha^{-1}) (net returns (INR 72,376 ha^{-1}) and B:C ratio (2.70)), besides less unfilled grains $panicle^{-1}$. However, different planting techniques didn't differ significantly from each other in respect of panicle length and 1000 seed weight. The grain yield obtained under drum seeder technique was 15.8 %, 25.4 % and 25.9 % higher than that of line planting, zigzag planting and M-SRI, respectively. Further, zigzag planting and M-SRI techniques performed similarly with regard to grain yield and B:C ratio (Neelima et al. 2015). According to Devy et al. (2015), rice cultivation in time is becoming a problem to the farmers due to erratic climatic conditions. Delay in crop season affects the sowing of second crop also. In order to overcome this problem, the farmers are actively taking up direct sowing of rice, sowing with drum seeder and machine planting which helped the farmers to grow the crop within the season besides saving water. Documentation of success cases of direct sown rice and drum seeder revealed that these technologies helped in getting an additional yield of four bags with INR 6426.00 reduction in cost of cultivation. Farmers also get an opportunity to sow black gram or green gram or maize in time. The straw yield was also increased compared to transplanted rice. The major constraints experienced by the farmers in direct sown rice are reduced crop growth during long intervals of rains and resowing of crop if continuous heavy rains occur during early stages of crop. The advantages of direct sown rice are less labour requirement followed by timely sowing. The advantages of machine planting of rice are plantings can be taken up at a stretch during peak season and maintenance of recommended plant population. The constraints of machine planting of rice are technicality in rising of nurseries and difficulty to plant seedlings aged more than 18 days.

Singh and Singh (2015) studied the performance of sugarcane and wheat crops on furrow-irrigated raised bed (FIRB) system in relation to date of planting or sowing. The results revealed that wheat grain yield was the highest (46.6 q ha^{-1}) in November sown wheat. Wheat yielded almost the same in flat as well as FIRB method. However, wheat sown in the month of November yielded higher than wheat sown in the month of December due to higher number of ear head per running metre, number of grains per ear head and test weight. Under FIRB method wheat yield (November: 44.1 q ha^{-1}) is higher over sugarcane + wheat (February to March) in 1:3 row ratio (40.2 q ha^{-1}) and 1:2 row ratio (33.5 q ha^{-1}). Tiller population recorded at different stages indicated that under FIRB method, tiller count was higher in autumn in sole sugarcane and in sugarcane planted with wheat on the third week of February, compared with sugarcane planted with wheat on the third week of March. The lowest tiller population was observed in sugarcane planted with wheat (1:3) under flat method followed by wheat-sugarcane system. The highest tiller count (231.8 thousands ha^{-1}) was recorded in the month of July in sugarcane planted with wheat on the third week of February under FIRB system and the lowest (86.4 thousands ha^{-1}) in sugarcane + wheat (1:3). The cane yield planted with sole sugarcane was highest (89.0 t ha^{-1}) in autumn. Under FIRB method, sugarcane planted with wheat on the third week of February (82.5 t ha^{-1}) was significantly higher than sugarcane planted with wheat on the third week of March and sugarcane + wheat (1:2) due to higher NMC, cane length, cane weight and number of internodes. The lowest cane yield was recorded in wheat – sugarcane system (59.3 t ha^{-1}) and sugarcane + wheat in 1:3 row ratio (60.3 t ha^{-1}).

5.3 Cropping Systems and Crop Diversification

Adoption of monocropping which is in vogue in several farmers' fields can't be advocated in the present scenario in view of the possibility of decline in soil fertility, flare up of pest and disease complex and emergence of super weeds. Crop diversification is the need of the hour as it is required to meet increasing demand for food, fodder and fibre, increase income of the farmers, restore and sustain long-term soil fertility, reduce dependence on off-farm inputs and control the pest or disease or weed. However, crop diversification changes depending on the agro-ecological regions, problems and demand.

Ramana et al. (2015) reported that among 18 cropping systems tested for their suitability in terms of productivity and profitability in semiarid tropics, maize-groundnut cropping system was found to be the best, followed by maize-sunflower, soybean-maize, soybean-wheat and soybean-castor. Resource use efficiency in terms of per day productivity and per day returns was superior with soybean-maize followed by maize-groundnut and maize-sunflower systems. Maize and then groundnut or sunflower or castor were identified as most remunerative cropping systems for semiarid tropics. However, Madhu and Ramana (2015) reported that among different castor-based cropping systems, maize-castor with application of 120 kg N ha^{-1} to castor gave greater system productivity in terms of crop equivalent yield, system gross and net returns and benefit-cost ratio, thus higher system productivity day^{-1} and system profitability. However, it did not differ statistically with green gram-castor at the same level of 'N'.

Ganvir et al. (2015) studied cotton-based intercropping systems in different row proportions under dry land conditions in Maharashtra and found significantly higher seed cotton equivalent yield (963 kg ha^{-1}), gross (INR 39,445 ha^{-1}) and net (INR 22,571 ha^{-1}) returns, and B:C ratio (2.37) with higher sustainability yield index (0.62) and sustainable value index (0.46) in cotton + sorghum + pigeon pea + sorghum (6:1:2:1) intercropping system, followed by cotton + sorghum + pigeon pea + sorghum (3:1:1:1) intercropping system. Though cotton + sorghum + pigeon pea + sorghum (6:1:2:1) intercropping system is more profitable as compared to cotton + sorghum + pigeon pea + sorghum (3:1:1:1) intercropping system, farmers are facing problems during sowing for maintaining 6:1:2:1 row proportion. Hence, it can be concluded that cotton + sorghum + pigeon pea + sorghum (3:1:1:1) intercropping system is the most suitable cropping system under dry land condition in Maharashtra.

Chaturvedi et al. (2015) reported that total rice equivalent yield of *rabi* and summer crops was 9 %, which is 20 % higher in reduced tillage and mulched plot, respectively, than conventional tillage and un-mulched field. Among the cropping systems, rice-potato-maize (CS3) had a significant increased yield of 25–30 % and 54–56 % higher total REY than CS2 and CS1 during 2011–2012 and 2012–2013, respectively. Combined application of RCT practices consumed 6–10 % less energy, while generating 99–143 % and 25–52 % higher output energy in CS3 and CS2, respectively, over conventional practices in rice-wheat system (CS1). Moreover, combined application of RCT practices improved soil physical characteristics such as bulk density, soil temperature and soil moisture conservation and chemical properties such as soil organic carbon %, available N, P and K. Conservation tillage system sequestered three times higher carbon than conventional tillage and mulching four times higher C sequestration than un-mulched condition in agricultural soils.

Higher system returns, B:C ratio and profitability resulted due to combined application of RCT practices, and this was 2.5 and 3 times higher when followed in CS2 and CS3, respectively, compared to rice-wheat (CS1) system under conventional practices. Thus, these holistic approaches of combined RCT practices are technically sound, agronomically efficient, economically attractive, practically feasible and ecologically safe and may be a win-win technology when they are adopted for the long term and are advocated for the Terai regions of Indo-Gangetic plains of India.

5.4 Optimization of Sowing Window

Sowing time is the most important nonmonetary input in agriculture and is the basis for successful crop production. Taking a decision regarding sowing in the presence of optimum moisture either due to rainfall or irrigation is very critical. Sowing should be taken during rainy season only after receiving 60–70 mm of cumulative rainfall. It requires fine tilth for small-seeded crops and fine to rough tilth for large-seeded crops. Sowing crops early during *rabi* season during October helps to establish the optimum plant population and escapes low-temperature stress during November and December and high-temperature stress during summer months, March and April.

Bankar et al. (2015) reported that sowing during the 41st meteorological week resulted in significantly higher yield, attributing it to the size of seed yield (21 q ha^{-1}) and straw yield (46 q ha^{-1}), quality parameters (viz. oil and protein yield) and uptake of N and P by the plants due to favourable weather conditions. According to Shivran (2015), plant height, tillers $plant^{-1}$ and dry matter produced $plant^{-1}$ at 60 DAS and at harvest, spikes $plant^{-1}$, seeds $spike^{-1}$, seed yield $plant^{-1}$, seed (12.69 q ha^{-1}): straw and biological yields and net returns (INR 29555 ha^{-1}) were significantly increased by sowing of isabgol on 10th November as compared to first November sowing, whereas, growth parameters remained at par but yield attributes, yields and net returns were reduced significantly when the sowing of crop was delayed to 20th November. On 10 November sown crop produced 26.7 % higher net returns over sowing on the first of November, but delayed sowing to 20 November reduced net returns by 7.2 %.

5.5 Water- and Nutrient-Saving Technologies

Rapidly depleting water resources, aberrations in monsoon rainfall and changing climate are threatening the sustainability of the agriculture, as well as food security and livelihood of producers and consumers. The challenge is to develop novel technologies and production systems that would allow the farmers to produce more per drop. Alternate furrow irrigation, micro-irrigation, pulse irrigation and low water requiring alternate rice systems and irrigated dry cops are the strategies to be adopted.

5.6 Furrow Irrigation

In the areas where there is no provision for micro-irrigation methods and also the small and marginal farmers who can't afford for the adoption of these methods, the best way to save irrigation water without jeopardizing the productivity is alternate furrow irrigation. As concluded by Sankaranarayanan and Nalayini (2015), in cotton, alternate furrow irrigation used less quantity of irrigation water (18.4 cm^{-1}) and recorded the highest water use efficiency of 69.2 kg $ha\text{-}cm^{-1}$ and water productivity of INR 32.2 m^{-3} and required less amount of water (1444 l kg^{-1}) to produce unit kilogramme of seed cotton yield.

5.7 Micro-Irrigation

Due to depletion of underground water in scarce rainfall areas, farmers have been facing many problems in irrigating the crops by traditional methods. At the same time, even in irrigated areas, water availability over a longer period of time is becoming a problem. In both the situations, enhancing factor productivity is critical. In such situations, drip method of irrigation proved efficient. Enough empirical evidence is available from different parts of the country to suggest that drip technology saves water in comparison to surface method of irrigation from 18.7 % to 47.7 % in orchards and fruits; 2.1–42.9 % in field crops; 11.9–38 % in vegetable crops; 14.3–51.3 % in root, bulb and tuber crops; 12–56 % in plantation crops; 36.7–46.7 % in spice crops; and 41.4 % in flowers in a properly designed and managed drip irrigation system. Likewise, in various experiments fertigation registered significant fertilizer savings (20–60 %) through drip fertigation (Rao and Anitha 2015).

The research on drip irrigation in groundnut is meagre. Padmalatha et al. (2015) have conducted the experiment to identify optimum irrigation depth and suitable irrigation interval under drip method of irrigation to *rabi* groundnut, and the results revealed that significantly higher pod yield was obtained with 10 mm depth of irrigation, which was at par with 15 mm depth given on the third day and 15 mm depth given on the second day. The lowest groundnut pod yield was with check basin method at IW/CPE 1.0, which was 187 % and 23 % lower than that of 10 mm depth scheduled on the third day during 2006–2007 and 2007–2008, respectively. Higher field water use efficiency was with 10 mm depth of irrigation scheduled on the third day during 2006–2007 (7.7 kg ha mm^{-1}) and 10 mm depth scheduled on the fourth day during 2007–2008 (4.2 kg ha mm^{-1}). When the quantity of irrigation water available is equal to that of check basin method (700 mm^{-1}), irrigations could be scheduled with 10 mm depth of water on the fourth day (300 mm^{-1}) such that 136 times additional area could be brought under cultivation with the available irrigation water.

The results of experiments conducted by Saileela et al. (2015) revealed that higher green pod yield (3556 kg ha^{-1}) was realized by scheduling irrigation at 100 % Epan which was 9.5 % and 22 % higher than the irrigation scheduling at 80 % and 60 % Epan. However, the increase in the yield with 100 kg ha^{-1} N and K was to the tune of 14 % and 21 % than application of 50 kg N + 50 kg K_2O ha^{-1} and no fertilizer application, respectively. Total nutrient uptake of N, P and K was higher with drip irrigation scheduling at 100 % Epan when compared to other irrigation levels. However, application of optimum dose of fertilizers 100:100 kg ha^{-1} N and K through drip resulted in maximum uptake of nutrients at all stages compared to other fertigation levels.

Kombali et al. (2015) emphasized the importance of drip fertigation in improving the productivity of aerobic rice, and their results revealed that the higher plant height, number of tillers per hill, leaf area per hill and total dry matter production per hill were recorded under drip fertigation at 1.5 Epan up to maturity with 100 % RDF through water soluble fertilizers (WSF) (53.57 cm, 48.13, 3529.8 cm^2 and 118.4 g $plant^{-1}$, respectively) which was superior over surface irrigation and soil application of RDF and was on par with drip fertigation at 1.5 Epan up to maturity with 75 % RDF through WSF. This is mainly because of WSF through fertigation that resulted in continuous supply of nutrients besides maintaining optimum water availability which leads to higher uptake of nutrients and in turn recorded higher growth attributes. Rekha et al. (2015) studied the effect of drip fertigation with the combination of water soluble fertilizers and normal fertilizers or alone on economics of aerobic rice. Through drip fertigation with water soluble fertilizer, application of 100 % RDF recorded higher gross returns of INR121222 ha^{-1} followed by 100 % and 75 % RDF in which 50 % was applied as basal with normal fertilizer and 50 % top dress (INR 114,364 and 105,093 ha^{-1}, respectively). This was mainly due to an increase in grain and straw yield. The pooled data revealed that the application of 100 % recommended dose of fertilizer (100:50:50 kg NPK ha^{-1}) with micronutrients (4 kg Zn ha^{-1} + 5 kg Fe ha^{-1}) in the form of water soluble fertilizer (WSF) through drip had significantly higher seed cotton yield (18.19 q ha^{-1}) which was found at par with 75 % recommended dose of fertilizer (RDF) with micronutrients (3 kg Zn ha^{-1} + 3.75 kg Fe ha^{-1}) in the form of WSF through fertigation.

Concentration and uptake of nutrients recorded significantly highest values with the application of 100 % RD along with micronutrients through drip and the efficiency of major nutrients and nutrient use, besides improving the soil nutrient status after the harvest of the cotton. The B:C ratio was found highest (1.82) in the treatment of 75 % RDF along with micronutrients (3 kg Zn ha^{-1} + 3.75 kg Fe ha^{-1}) followed by 100 % RDF (100:50:50 kg NPK ha^{-1}) + 4 kg Zn ha^{-1} + 5 kg Fe ha^{-1} (1.75) through water soluble fertilizer (Magare et al. 2015). The fertilizer application at 125 % of RD in the form of WSF with 0.75–1.5 m × 0.75 m spacing showed the best effect for yield and quality. Drip method of irrigation resulted into 59.12 % and 55.18 % more seed cotton and cotton seed yield than the conventional method of irrigation. Similarly, the yield of seed cotton and cotton seed was 17.17 % and 16.53 % more in WSF than CF (Jawale and Jawale 2015).

Application of 50 % RDF (250, 250 and 500 kg ha^{-1} of $N:P_2O_5:K_2O$, respectively) through fertigation produced the yield and TSS equivalent to 80 % RDF through fertigation and 100 % RDF through soil application. Variation in yield was attributed to a bunch number and 100 berries weight rather than the bunch weight. The treatments with higher doses of RDF (80 % and 100 %) had higher nutrient content in petioles at both bud differentiation and full-bloom stages. Significantly higher Ca content in petioles was recorded with fertigation treatments (Vijaya et al. 2015).

5.8 Pulse Irrigation

Precision water management is the process of determining and controlling the volume, frequency and application rate of irrigation water in a planned and efficient manner. Maize is both water-responsive and drought-tolerant crop. It requires 500–600 mm of water and demand eight to ten irrigations. Seedling, knee height, tasselling, silking and grain filling are critical stages for irrigation. Peak consumption of water occurs during tasselling and silking period. Water shortage for 2 days at this stage reduces the yield by 20 %. Pulse drip irrigation is a recent concept where small frequent irrigation applications are applied to saturate the soil and meet the plant water requirements while reducing the leaching and run-off losses. The higher the irrigation frequency, the smaller the soil wetting volume. Higher soil water content in a small range can be maintained by pulse drip irrigation. It has increased the maize grain yield (4443.8 kg ha^{-1}) by 11.8 % compared with continuous drip irrigation (3974.2 kg ha^{-1}). The total applied water was saved by 2.01 % for pulse drip irrigation than that of continuous drip. Water use efficiency under pulse drip irrigation was 2.34 kg m^{-3} compared with 2.06 kg m^{-3} for continuous drip irrigation recording an increase of 13.55 % (Madhu et al. 2015).

5.9 Nutrient Management

With vast geographical land area and diverse agro-ecological regions across the country, there is a considerable variation in soil health and fertility. Further, indiscriminate use of chemical fertilizers threatened soil fertility. Long-term application of inorganic fertilizers along with organic manures in integrated manner plays a vital role in not only obtaining higher crop yields but also sustaining soil fertility and sequestering high amounts of organic carbon in the long run under scarce, medium and high rainfall regions. The results also suggest that cropping systems like rice-maize if continued for long term may deplete soil organic carbon. Recommended management practices (RMPs) like conservation tillage, erosion control and residue recycling are suggested. In rice-rice cropping system, prior rising of sun hemp or green gram is found to be the best option to sequester organic carbon and to bring sustainability. Since significant reduction is observed in seasonal CH_4 emission, scientific irrigation practices like SRI cultivation and aerobic rice are better options in tank-fed and bore or well-irrigated areas. In Nalgonda district, most of the area comes under carbon density ranges of 3.5–5.5 kg m^{-2}. The highest

SOC is observed in agri-silviculture system followed by silvi-pasture and agri-silvi-horticulture system. Regular additions of nutrients through fertilizers along with organic manures are found necessary for carbon sequestration, particularly in soil with nutrient deficiencies. With the introduction of carbon trading, agroforestry systems may become more attractive. In crop husbandry as well as agroforestry, research addressing both biophysical and socio-economic issues and identifying, developing and bringing out best management practices (BMPs) with reference to carbon sequestration and sustainable production needs to be intensified (Rao and Padmaja 2015).

5.10 STCR-Based Fertilizer Recommendations to Save Fertilizer and Increase FUE

The shortage and high cost of fertilizers necessitate that every unit of fertilizer be used judiciously. In view of high location-specific nutrient needs and soil test-based fertilizer recommendation for specific crop and crop sequences, soil testing has updated its methods and techniques to remove soil fertility constraints. According to Shanwad et al. (2015), STCR approach leads to precise use of fertilizer dose compared to RDF method, and ultimately it can reduce the cost of cultivation, increase yield and act as a basic tool for sustainable agriculture production as experimented and proved in sunflower crop.

5.11 Integrated Nutrient Management

Continuous use of inorganic fertilizers may affect the soil fertility and productivity. Effect of conjunctive use of organic and inorganic manures as against their independent application on soil fertility and crop productivity studied revealed that among the different combinations tried, application of 75 % RDFN + 25 % N-VC recorded the highest nutrient uptake, dry matter yield, grain and straw yields at flowering and maturity. The maize crop grown during *rabi* responded favourably to the residual effect of treatments applied to rice as well as fertilizer N levels applied to it. Among rice treatments, the residual effects on maize were highest with the application of 50 % RDFN in combination with 50 % RDN as vermicompost. Among the fertilizer N levels applied to maize, application of 100 % RDFN resulted in highest dry matter production, grain and stover yields as well as uptake of nutrients. The fertilizer needs of maize crop can be reduced to the extent of 50 % by substituting 25 % fertilizer N through poultry manure or vermicompost or FYM+BGA to rice in rice-maize cropping system. The organic carbon, N, P and K status and activities of enzymes were not reduced after the harvest of the following maize crop compared to initial values due to integrated nitrogen supply to preceding rice. The soil health was best maintained by the integrated nitrogen management strategy in rice due to the residual effect of the slow mineralization of organic sources of nitrogen.

Vesicular-arbuscular mycorrhizal (VAM) fungi are the most important ubiquitous beneficial below-ground symbiotic community for increasing the sustainability of agricultural system. They are obligate biotrophs, active, diverse and ubiquitous in distribution and can multiply successfully only in the presence of living plant roots. The role of AMF in agriculture and horticulture is well understood. Mycorrhizae played a key role in mobilizing phosphorus efficiently and increased the absorptive area of root, thus improved the plant growth and offered many other benefits. The mycorrhizal fungi can protect the associated plant against various soil-borne pathogens, heavy metals and toxic compounds and thereby reduce the indiscriminate usage of chemicals in the field. Experimental studies have shown that many photosynthesizing plants fail to develop into strong individuals, if deprived of their mycorrhizal partner; besides these it has the potentiality of mobilizing different nutrients, particularly nitrogen and phosphorus, and can also produce hormones in plants under stress conditions. Though some studies have been done in mycorrhizae, its research did not reach the expected peak. Still mycorrhizae will find application as biofertilizers by farmers in the near future (Ngomle 2015).

Application of Zn 20 kg ha^{-1}+*Rhizobium* inoculation) was found to be significantly superior to other treatments in yield attributes, viz. a number of pods per plant and biological and grain yield. However, the yield attributing to its parameters such as seed index and number of seed pods^{-1} did not show any significant difference. The economic analysis revealed that the net returns and benefit-cost ratio were highest in the same treatment (Kamble et al. 2015).

Tamgadge et al. (2015) reported significantly higher seed cotton yield (14.15 q ha^{-1}) as compared to recommended dose of fertilizer (RDF) + 12.5 kg K_2O ha^{-1} and at par with RDF + 37.5 kg K_2O ha^{-1} besides uptake of nutrients (NPK): better fibre quality and improvement in soil fertility.

Ramanjaneyulu et al. (2015) based on 3 years of experimentation concluded that application of 20 kg P_2O_5 ha^{-1} besides seed treatment with Biophos has resulted in significantly higher seed yield (1439 kg ha^{-1}) than that of 60 kg P_2O_5 ha^{-1} with or without seed treatment with Biophos and control (no phosphorus application). However, it was found at par with that of 20 kg P_2O_5 ha^{-1} without seed treatment with Biophos (1373 kg ha^{-1}) and 40 kg P_2O_5 ha^{-1} (1421 kg P_2O_5 ha^{-1}) with or without seed treatment with Biophos (1416 kg P_2O_5 ha^{-1}). Higher gross (INR 46,555 ha^{-1}) and net returns (INR 25,675 ha^{-1}) and B:C ratio (2.21) were observed when 20 kg P_2O_5 ha^{-1} was applied besides seed treatment with Biophos in castor. The value of alkaline phosphatase enzyme before sowing was 230.9 μg g^{-1} dwt soil ha^{-1}. Its activity increased due to addition of Biophos irrespective of phosphorus doses. The maximum increase was found in the treatment of 40 kg P_2O_5 ha^{-1} with Biophos. Seed treatment with Biophos + application of 20 kg P_2O_5 ha^{-1} can be recommended for castor under rain-fed conditions on red sandy loam soils with medium phosphorus level as it has resulted in higher seed yield and net monetary benefits.

FYM at 10 t ha^{-1} recorded significantly greater seed cotton yield (13.75 q ha^{-1}) and gross monetary returns ha^{-1} (INR 26,204 ha^{-1}). Treatment seed inoculation with *Azotobacter* + PSB recorded significantly maximum value of net monetary returns and B:C ratio in both the years and pooled analysis (INR 11,535 ha^{-1} and 2.31) (Solunke et al. 2015).

Gabhane et al. (2015) recommended that conjunctive use of FYM and *Gliricidia* leaf manuring (50 % recommended nitrogen) along with 50 % RDF under low tillage (50 % CT + hand weeding) was found beneficial in improving soil fertility, higher nutrient uptake and productivity of rain-fed sorghum grown in inceptisol under semiarid climatic conditions. Fifty percent RDN to intercropped pearl millet was sufficient for optimum growth and productivity of pearl millet when it is grown as intercrop with soybean in 1:2 additive series (Layek et al. 2015).

The productivity of rice is low due to cultivation of low yielding local varieties, poor agronomic practices, inadequate nutrient supply and deterioration of soil quality due to removal or burning of residues, excessive tillage, etc.

The results of five cropping cycles revealed that highest grain yield of rice obtained under NT (4.65 t ha^{-1}) followed by MT (4.26 t ha^{-1}) and CT (4.20 t ha^{-1}). Among the nutrient management practices, the highest grain yield was recorded fewer than 50 % NPK + weed biomass (4.64 t ha^{-1}) followed by 50 % NPK + green leaf manuring (4.55 t ha^{-1}). The soil penetration resistance (PR) recorded after the harvest of rice at different depths (0–5 cm, 5–10 cm and 10–15 cm) increased with the soil depth. The PR was minimum under NT and maximum under CT at all the depths. The productivity (green pod) of pea (Arkel) was significantly higher under residual effect of CT while Prakash under residual effect of MT as compared to NT. The bulk density of soil at different depths was highest in CT followed by MT and NT. The soil organic carbon and soil microbial biomass carbon content at 0–15 cm soil depth were significantly higher under NT as compared to MT and CT. The soil available N, P, K, Ca, Mg and S status increased over the years under NT. Hence, NT plays an important role in improving crop productivity, soil fertility and carbon sequestration. Among different nutrient management options, application of 50 % NPK through fertilizer and 50 % NPK either through weed biomass or green leaf manure is viable option for improving crop and soil productivity.

The grain, shank and stover yield, dry matter production and yield attributes of maize were significantly higher with integration of 25 % RDN through FYM or poultry manure or vermicompost with 125 % RDN (150 kg N ha^{-1}) through urea than that of 75 or 100 (120:60:30 kg N, P_2O_5, K_2O ha^{-1}) or 125 % RDN through urea alone (Kumar et al. 2015b).

As reported by Bhargavi et al. (2015), number of pods plant-1 (42): seeds pod^{-1} (10): fresh and dry herbage yield of *Andrographis paniculata* was significantly higher with 50% RDF (37.5:25:25 kg N, P_2O_5, K_2O ha^{-1}) + 50% neem cake (0.5 t ha^{-1}) over control followed by 50 % RDF + 50 % vermicompost (1.0 t ha^{-1}). Fresh and dry herbage yield (2968.9 and 2078.2 kg ha^{-1}) was recorded in 50 % RDF +50 % neem cake compared to 1386.6 and 970.6 kg ha^{-1}in control.

According to Angha and Longkumer (2015), the plant height, number of plant $leaves^{-1}$, total dry matter accumulation, leaf area index, size of head and seed and stover yields, besides the highest values of N, P, K uptake, highest soil available N, K (kg ha^{-1}) and the highest B:C ratio, were recorded when FYM at 10 t ha^{-1} was applied along with 50 % RDF (20:30:20 kg ha^{-1}).

In an experiment conducted by Patel et al. (2015), the economics of various treatments showed that fertigation at 60 % of RDN gave the highest net return of INR 176,759 ha^{-1} with B:C ratio of 4.03, thus saving 40 % applied N in sweetcorn during *rabi* season in Gujarat. Srinivas et al. (2015) reported positive effects of combined application of organic FYM or vermicompost + *Rhizobium* and PSB as seed treatment at 25 g kg^{-1} of seed on yield and quality of fenugreek. However, total chlorophyll content, leaf moisture, leaf protein content and fibre content were higher due to application of FYM + *Rhizobium* and PSB seed treatment.

Application of poultry manure at 6 t ha^{-1} + *Rhizobium* at 20 g kg^{-1} seed + phosphatica at 20 g kg^{-1} seed showed maximum number of leaves $plant^{-1}$, number of primary branches, number of root nodules and fresh weight of root nodules $plant^{-1}$ which was at par with pig manure at 7 t ha^{-1} + *Rhizobium* at 20 g kg^{-1} seed + phosphatica at 20 g kg^{-1} seed. A maximum number of plant $pods^{-1}$, pod $seeds^{-1}$, grain weight, grain yield and stover yield were observed with poultry manure at 6 t ha^{-1} + *Rhizobium* at 20 g kg^{-1} seed + phosphatica at 20 g kg^{-1} seed. The highest harvest index was also observed with poultry manure at 6 t ha^{-1} + *Rhizobium* at 20 g kg^{-1} seed + phosphatica at 20 g kg^{-1} seed. This treatment was found significantly superior over the rest of the treatments including control. Hence, these findings revealed that poultry manure with recommended dose of biofertilizers can play a significant role in soybean crop production (Changkija and Longkumer 2015).

Vikram et al. (2015) from Spiti, a cold desert region of Himachal Pradesh, reported that combined application of 50 % RD NPK+ sheep manure at 10 tonnes ha^{-1} + seed treatment with *Rhizobium* and PSM was the best and resulted in 50 % saving of NPK fertilizer over recommended fertilizer dosage, giving the highest yields. The study also revealed the existence of negative correlation between nodule no. of $plant^{-1}$ and inorganic fertilizers.

Katkar et al. (2015) studied the long-term effect of nutrient management on vertical distribution of sulphur fractions under the sorghum-wheat crop sequence after 22nd cycle in vertisols. The continuous application of 100 % NPK along with sulphur at 37.5 kg ha^{-1} showed increase in total sulphur (361 mg kg^{-1}), organic sulphur (219 mg kg^{-1}), sulphate sulphur (26 mg kg^{-1}), heat soluble sulphur (19 mg kg^{-1}) and adsorbed sulphur (39 mg kg^{-1}) fractions in the soil at 0–15 cm depth. These fractions were decreased in the similar trend with the increase in soil depth (15–30 and 30–45 cm). The organic S, sulphate S, water-soluble S, adsorbed S and non-sulphate S are correlated with each other. Correlation studies of pH, EC, organic carbon, $CaCO_3$ and different forms of sulphur indicated that significant correlation of total S ($r = 0.438^{**}$), organic S ($r = 0.525^{**}$), sulphate S ($r = 0.464^{**}$), heat soluble S ($r = 0.347^{*}$) and adsorbed S ($r = 0.362^{*}$)

was observed with organic carbon except non-sulphate sulphur. The different forms of sulphur did not show correlation with pH.

Chauhan et al. (2015) reported that the seed yield and all yield contributing factors such as the number of pods per cluster (4.60), diameter of pods (0.7 cm) and length of pods (20.66 cm) were significantly higher in the treatment that received 75 % RDF + seed treatment with *Rhizobium* + PSB compared to control and RDF alone. Nitrogen status in soil after harvest was found significant in treatment, i.e. 75 % RDF + biofertilizers (*Rhizobium* + PSB), while phosphorus and potassium status in soil was found significant in treatment, i.e. 50 % RDF + vermicompost + biofertilizers (*Rhizobium* + PSB).

Highest rice yield was obtained with 75 % RDF + 5 t ha^{-1} FYM and was significantly superior over rest of nutrient management practices. Among intercropping treatments, highest rice seed yield was obtained with sole rice crop, and the rest of the treatments were comparable with each other, except rice + black gram (4:2 replacement series) which was superior to the remaining intercropping treatments with soybean or black gram. Seventy-five percent RDF + 5 % t ha^{-1} FYM gave significantly higher soybean equivalent yield than rest of the nutrient management treatments under study. Among intercropping treatments rice + soybean (3:2 and 4:2 in replacement series) gave comparable soybean equivalent yield and was significantly superior to the rest of the intercropping treatments (Kumar et al. 2015a).

According to Harne et al. (2015), seed inoculation with *Bradyrhizobium japonicum* + *Bacillus megaterium* var. *phosphaticum* + *Trichoderma* spp. treatment recorded maximum number of nodules per plant, i.e. 16.60, 31.66 and 60.23 at 30, 45 and 60 DAS, respectively, and nodule dry weight, i.e. 34.20, 95.36 and 155.56 mg $plant^{-1}$ at 30, 45 and 60 DAS.

Long-term fertilizer experiments carried out by Meshram and Ismail (2015) revealed the fact that the yield of soybean and safflower besides micronutrient availability (Zn, Fe, Mn, Cu and B) was significantly improved with the application of 100 % NPK+FYM at 5 t ha^{-1} compared to other treatment combinations.

Highest dry matter production at panicle initiation (5867 kg ha^{-1}): grain (5547 kg ha^{-1}) and straw (7630 kg ha^{-1}) yields were recorded by the integrated application of FYM at 10 t ha^{-1} + $ZnSO_4$ at 50 kg ha^{-1} along with RDF followed by the treatment supplied with FYM at 10 t ha^{-1} + $ZnSO_4$ at 25 kg ha^{-1} along with RDF. The macro- and micronutrient (Zn, Fe, Mn and Cu) contents and their uptake by rice at panicle initiation and harvest were markedly influenced by the treatments with highest values recorded supplied with FYM at 10 t ha^{-1} + $ZnSO_4$ at 50 kg ha^{-1} along with RDF. Increased level of zinc application along with RDF significantly increased the yield, contents and uptake of nutrients by rice. The results of the investigation indicated that the integrated use of RDF along with FYM at 10 t ha^{-1} and $ZnSO_4$ at 50 kg ha^{-1} was proved to be beneficial in improving performance of rice (Gopi et al. 2015).

The optimization of the mineral nutrition is the key to optimize the production of groundnut, as it has very high nutrient requirement, and the recently released high yielding groundnut varieties remove still more nutrients from the soil. On contrary groundnut farmers' use very less nutrient fertilizer and sometimes only one or two nutrients resulting in severe mineral nutrient deficiencies. Inadequate and imbalance use of nutrients is one of the major factors responsible for low yields in groundnut. India is the world's largest producer of groundnut where nutritional disorders cause yield reduction from 30 % to 70 % depending upon the soil types. Thus, it is time to look into the multi-micronutrient mixture (MMM) aspects of groundnut for achieving high yield. Application of mixture of zinc, ferrous iron, boron, calcium and magnesium (Multinelf 16) at 60 kg ha^{-1} resulted in significantly higher pod yield (3284 kg ha^{-1}) compared to the remaining treatments excluding easy mix (3145 kg ha^{-1}), zinc high (3059 kg ha^{-1}), vasundhara (3018 kg ha^{-1}) and boracal (2803 kg ha^{-1}) at 60 kg ha^{-1} over its higher and lower level at the time of basal application. The higher yield with Multinelf 16

(MMM) at 60 kg ha^{-1} was mainly due to higher plant biomass, plant $branches^{-1}$, plant $pods^{-1}$, dry pod weight and harvest index. However, 100-kernel weight was influenced significantly due to application of different multi-micronutrient fertilizer mixtures and, similarly, fetched significantly higher net return (INR 96,492 ha^{-1}) and B:C ratio of 3.18, followed by easy mix at 60 kg ha^{-1} (INR 92,377 ha^{-1}, 3.12), zinc high at 60 kg ha^{-1} (INR 88,259 ha^{-1}, 3.00), vasundhara at 60 kg ha^{-1} (INR 85,233 ha^{-1}, 2.95) and boracal 12 at 60 kg ha^{-1} (INR 77,883 ha^{-1}, 2.81) over other treatments. Hence, to achiever a maximum economic return from summer groundnut, basal application Multinelf 16, i.e. mixture of zinc, ferrous iron, boron, calcium and magnesium at 60 kg ha^{-1}, is found to be profitable through improvement of productivity of groundnut (Mahatale et al. 2015).

Keeping in view the ever-increasing negative impacts on human health and ecology due to present commercial farming systems following intensive usage of synthetic inputs, researches and policy makers are forced to find out an alternative way of farming (Ramanjaneyulu et al. 2013). Organic farming is one such system which provides healthy and safe food without ecological harm. Application of 75 % vermicompost (on N-equivalent basis of RDF) + *Trichoderma harzianum* (30–40 ml $plant^{-1}$) + azadirachtin (1 % at 3–4 ml lit^{-1}) + *Pseudomonas fluorescens* (30–40 ml $plant^{-1}$) was effective in improving growth, yield and quality parameters of Nagpur mandarin and recorded significantly more plant height (4.81 m); canopy volume (57.40 m3); fruit yield (92.76 kg $plant^{-1}$, i.e. 25.70 t ha^{-1} having 670.25 number of plant $fruits^{-1}$); juice content (49.15 %); and TSS (11.20 °B) with minimum acidity (0.72 %) compared to the remaining treatments. The leaf nutrient status was also improved with this treatment (Paithankar et al. 2015).

Dhage et al. (2015) reported that grain and straw yield, uptake of phosphorus and sulphur increased with increase in rate of application of P and S individually as well as in various combinations. Applied S and P increased the number of plant $nodules^{-1}$, fresh weight of nodules, root length and dry root weight of soybean. The synergistic effect of phosphorus and sulphur was observed on root length and dry root weight of soybean at 60 DAS, grain, straw and total biological yield of soybean. Similar result was observed in total nutrient uptake of nitrogen, phosphorous and sulphur by soybean.

Katore et al. (2015) gave conclusive evidence that legumes such as black gram, green gram and soybean save 15.45, 9.47 and 8.17 kg ha^{-1} nitrogen compared to sunflower and fallow.

Prasad (2015) stated that among the much advancement in sciences, nanotechnology is being visualized as a rapidly evolving field that has potential to revolutionize agriculture and food systems and improve the condition of the farmer. Technical convergence across the fields of physics, engineering, chemistry, biology, agriculture and food sciences is the essential core of development of nanotechnology. According to the US Environmental Protection Agency, nanotechnology is defined as the science of understanding and control of matter at dimensions of roughly 1–100 nm, where unique physical and chemical properties make novel applications possible. The Asia Pacific Economic Cooperation (APEC) Center for Technology Foresight has predicted that nanotechnology will revolutionize all aspects of our economy and society, with associated large-scale social upheaval. According to Singh et al. (2015), intensive agriculture practice due to the green revolution in India has resulted in progressive depletion of soil nutrients and nutrient demand. In Indian agriculture, these are going to increase tremendously in coming years. Nanotechnology is seen as the important technology for future agriculture production. Smaller size, higher specific surface area and reactivity of nano-fertilizers compared to bulk one may increase the solubility, diffusion and hence availability to plants and enhance crop productivity. Nanoparticles and nano capsules provide an efficient means to distribute fertilizers and chemicals in a controlled fashion with high site specificity. Nanoparticles proved to be beneficial in seed germination and plant growth. Nanosensors can be useful in precision farming and effective water management in agriculture. Nanohydrogels may be used effectively in water-deficit desert areas. In one of the experiments at DIHAR, carboxy methyl cellulose (CMC)-based nanohydrogels were developed in the laboratory. Field experiments were undertaken in high-altitude cold desert Ladakh region to observe effects of different doses of developed nanohydrogels and irrigation interval on growth and yield of capsicum (var. California Wonder) in trenches. Results showed that in nanohydrogel-treated plots at 75 kg ha^{-1}, significant increase (24.2 %) in capsicum yield (42.6 t ha^{-1}) was observed compared to control plots (34.3 t ha^{-1}). Significant increase in leaf chlorophyll content and relative water content (RWC) and membrane stability index (MSI) were observed in treated plots when compared with untreated plots. In the present study, high amount of irrigation water savings was also achieved with improved water use efficiency.

5.12 Integrated Weed Management

Weeds are ubiquitous in nature and difficult to be eradicated. No single weed control method was found to be successful in all situations. Hence, integrated approach involving cultural, mechanical, chemical and biological methods is the need of the hour based on the agro-ecological situation.

It is basically an integration of different weed management strategies that can be used economically by the producers as a part of sound farm management system. Integrated weed management system is not meant for replacing selective, safe and efficient herbicides but is a sound strategy to encourage judicious use of herbicides along with other safe, effective, economical and eco-friendly control measures. The use of clean crop seeds and seeders and field sanitation (weed-free irrigation canals and bunds) should be integrated for effective weed management. Combining good agronomic practices, timeliness of operations, fertilizer and water management and retaining crop residues on the soil surface improve the weed control efficiency of applied herbicides and competitiveness against weeds. Approaches such as stale seedbed practice, uniform and dense crop establishment, use of cover crops and crop residues as mulch, crop rotations and practices or enhanced crop competitiveness with a combination of pre- and post-emergence herbicides could be integrated to develop sustainable and effective weed management strategies.

Mandal et al. (2015) reported that weed management treatments involving pre-emergence application of herbicides such as Targasuper at 15 g a.i. ha^{-1}+hand weeding at 45 days after sowing resulted in higher bulb yield of onion (22.73 and 21.63 t ha^{-1}, respectively) due to maintenance of weed-free condition during the initial stages with pre-emergence application of herbicides and control of late emerging weeds by one-hand weeding at 45 days after sowing. The weed control efficiency was more than 77.34 and 76.46 % at 50 DAA. The weed index values ranged from 6.23 to 9.24 % in sole herbicide and integrated treatments as compared to 68.11 % with weedy check indicating least crop weed competition. Weed-free check fetched the higher cost of cultivation (INR 68,192 ha^{-1}), gross monetary returns (INR193920 ha^{-1}) and net return (INR 125,728 ha^{-1}) over all the other treatments, but it had less benefit-cost ratio (2.84).

As reported by Lalita et al. (2015) based on the results of four seasons of experimentation, the mean pod yield of groundnut was significantly higher with 150 % RDF (1027 kg ha^{-1}) followed by 100 % RDF+FYM at 10 t ha^{-1} (1008 kg ha^{-1}), and both the treatments were on par. Application of 100 % RDF+FYM at 10 t ha^{-1} also maintained highest post harvest available N status revealing the importance of organic nutrient sources in nutrient management of groundnut.

According to Saras et al. (2015), weed dry weight of monocot and dicot weeds was lowest in intercultural operation followed by hand weeding at 20 and 40 DAS with maximum weed control efficiency (100 %) and also recorded higher yield attributes (seed and stover yield of 16.55 and 49.36 q ha^{-1}, respectively) which gave maximum net monetary returns and B:C ratio (INR 59,404 ha^{-1} and 2.97, respectively). In the case of chemical treatment, application of herbicide combination (pendimethalin and imazethapyr (premixed) at 800 g ha^{-1} pre-emergence) followed by hand weeding at 30 DAS was found statistically at par with intercultural operation followed by hand weeding at 20 and 40 DAS. It led to minimum dry weight of monocot and dicot weeds and higher weed control efficiency and yield attributes and gave second maximum net monetary returns and B:C ratio (INR 45,872 ha^{-1} and 2.20, respectively).

As reported by Das et al. (2015a), it was found that post-emergence application of bispyribac sodium 10 % SC at 30 g a.i. ha^{-1} gave significantly lower total weed density, weed dry weight and higher weed control efficiency at all the stages. Application of bispyribac sodium 10 % SC at 20 g a.i. ha^{-1} kept the weed density and dry weight below the economic threshold level and increased the grain yield in rice. Succeeding crop black gram sown immediately after the harvest of rice was not affected by the residue of bispyribac sodium at all different doses. Sreedevi et al. (2015) reported that among the weed management practices, pre-emergence application of pendimethalin+rice: dhaincha (1:1) and 2,4-D Na application at 25–30 DAS, and pendimethalin+rice: dhaincha (1:1) and one-hand weeding at 60 DAS were effective in reducing weed population and weed biomass and in increasing grain yields under aerobic unpuddled conditions. As concluded and reported by Kumar et al. (2015a), N at 200 kg N ha^{-1} has to be applied for achieving significantly higher grain yields (5974 kg ha^{-1} and 4480 kg ha^{-1}) during *kharif* season under mechanized transplanting system. Black polythene mulch of 50 μ and black polythene mulch of 50 μ plus 3 % $CaNO_3$ spray treatments were superior over other treatments in most of the parameters studied. Black polythene mulch of 50 μ plus 3 % $CaNO_3$ spray showed superiority in plant height, number of branches, plant spread, flower diameter, mean flower weight, number of flowers per plant, yield per plant, yield per plot, yield per hectare and vase life. Weed control efficiency and soil temperature was also recorded maximum under this treatment. It also resulted in higher contents of major and secondary nutrients (Vimala et al. 2015).

Sharma and Singh (2015) opined that shifts in weed populations from annuals to perennials have become common in conservation agriculture systems. Perennial weeds are known to thrive better in reduced or no-tillage systems. Most perennial weeds have the ability to reproduce from several structural organs other than seeds. For example, Bermuda grass (*Cynodon dactylon*), nut sedge (*Cyperus rotundus*) and Johnson grass (*Sorghum halepense*) generally reproduce from underground plant storage structures: stolons, tubers or nuts and rhizomes, respectively. Crop yield losses in conservation agriculture due to weeds may vary depending on weed dynamics and weed intensity. However, the presence of post-emergence broad-spectrum herbicides provides an opportunity to control weeds in CA. Crop yields can be similar for conventional and conservation tillage systems if weeds are controlled and crop stands are uniform. Results of on-farm trials at several locations in Haryana, India, revealed that population density of *Phalaris minor* was considerably lower and grain yield of wheat was comparatively higher under zero tillage than conventional tillage.

In rainy season, when the weed problem is generally more, growing crops with zero tillage require additional measures for effective weed control, including use of nonselective herbicides like paraquat and glyphosate. Zero-till sowing in standing crop residues along with application of herbicides in proper combination, sequence or in rotation led to lower weed population and higher yield than conventional planting. However, changing from tillage-based farming to no-till farming is not easy. No till incurs a greater risk of crop failure or lower net returns than conventional agriculture, and this perception has seriously hindered its adoption in countries outside North and South America. Yields of no-till crops may be lower by 5–10 % in the initial years, especially on fine-textured and poorly drained soils. No-till farming demands use of extra N fertilizer and heavy reliance on herbicides. The continued practice of no-till is, therefore, highly dependent on development of new herbicide formulations and integrated weed management options (Sharma and Singh 2015).

5.13 Integrated Crop Management

Rice (*Oryza sativa* L.) – wheat (*Triticum aestivum* L.) is a predominant system in India. Long-term experiments across Indo-Gangetic plains (IGP) revealed that productivity and sustainability of this system are threatened as yields of both rice and wheat are either stagnant or declining. Moreover, there has been enormous damage to natural resources. Hence, the development of new integrated resource management strategies is need of hour for sustainable crop production in the region.

Murthy et al. (2015) investigated the effect of varieties, irrigation regimes, weed management and N levels on productivity of rice under aerobic conditions and concluded that Vasundhara and then Samba Mahsuri were the best high-yielding varieties under high rainfall conditions. Application of 120 kg N ha^{-1} (4991 and 4730 kg ha^{-1}) gave significantly higher grain yield of rice under aerobic conditions. Scheduling of irrigation at IW:CPE ratio of 1:2 produced significantly higher yield attributes and grain yield. The lowest of density of weeds and their corresponding dry weight was recorded with hand weeding at 20 and 40 DAS, which were comparable with oxadiargyl 0.07 kg a.i. ha^{-1} supplemented with HW at 40 DAS for density of grasses and produced the higher grain yield (5561 and 5595 kg ha^{-1}).

5.14 Conclusions

To address problems like climate change and decline in available cultivated area, soil fertility and emergence of multi-micronutrient deficiencies, complex weed problems, and pest and disease flare-up, adoption of modern agro-techniques and crop diversification aside from organic farming and integrated crop management strategies to make agriculture sustainable and profitable in the long run is the need of the hour.

References

Angha M, Longkumer T (2015) Effect of integrated nutrient management in Sunflower (*Helianthus annuus* L.). In: Compendium of abstracts of the 2nd international conference on bio-resource and stress management, ANGRAU & PJTSAU, Hyderabad, India, 7–10 January 2015

Bankar SL, Kubde KJ, Prachand SS et al (2015) Response of safflower varieties on yield, quality and uptake of nutrient. In: Compendium of abstracts of the 2nd international conference on bio-resource and stress management, ANGRAU & PJTSAU, Hyderabad, India, 7–10 January 2015

Bhargavi DD, Madhavi LA, Srinivas A et al (2015) Yield and yield parameters of kalmegh (*Andrographis Paniculata*) as influenced by integrated nutrient management in pongamia based agri-silvi system. In: Compendium of abstracts of the 2nd international conference on bio-resource and stress management, ANGRAU & PJTSAU, Hyderabad, India, 7–10 January 2015

Changkija S, Longkumer LT (2015) Effect of organic nutrient sources on the productivity of soybean (*Glycine max*, L., Merrill.). In: Compendium of abstracts of the 2nd international conference on bio-resource and stress management, ANGRAU & PJTSAU, Hyderabad, India, 7–10 January 2015

Chaturvedi S, Pandiaraj T, Singh R et al (2015) Agricultural management practices to sustain yield, improve soil and conserve resources in rice based cropping systems of Indo-Gangetic Plains. In: Compendium of abstracts of the 2nd international conference on bio-resource and stress management, ANGRAU & PJTSAU, Hyderabad, India, 7–10 January 2015

Chauhan J, Paithankar DH, Khichi P et al (2015) Studies on integrated nutrient management in Cowpea. In: Compendium of abstracts of the 2nd international conference on bio-resource and stress management, ANGRAU & PJTSAU, Hyderabad, India, 7–10 January 2015

Das R, Bera S, Ghosh RK (2015a) Weed management in transplanted rice through bispyribac sodium 10% SC and its effect on soil micro flora and succeeding crop-Black gram. In: Compendium of abstracts of the 2nd international conference on bio-resource and stress management, ANGRAU & PJTSAU, Hyderabad, India, 7–10 January 2015

Devy MR, Rao GR, Venkateswarlu E (2015) Water saving rice production technologies and constraints analysis. In: Compendium of abstracts of the 2nd international conference on bio-resource and stress management, ANGRAU & PJTSAU, Hyderabad, India, 7–10 January 2015

Dhage SJ, Patil VD, Patange MJ (2015) Interaction effect of P and S nutrition on nodulation, root parameters, yield and nutrient uptake of soybean (*Glycine max* (L.) Merrill.). In: Compendium of abstracts of the 2nd international conference on bio-resource and stress management, ANGRAU & PJTSAU, Hyderabad, India, 7–10 January 2015

Gabhane VV, Gite AD, Ganvir MM et al (2015) Effect of tillage and nutrient management on soil fertility and productivity of sorghum in inceptisols under semi arid conditions in Maharashtra. In: Compendium of abstracts of the 2nd international conference on bio-resource and stress management, ANGRAU & PJTSAU, Hyderabad, India, 7–10 January 2015

Ganvir MM, Nagdeve MB, Karunakar AP et al (2015) Productivity and sustainability of cotton based intercropping system in different row proportions under dry land Conditions in Maharashtra. In: Compendium of abstracts of the 2nd international conference on bio-resource and stress management, ANGRAU & PJTSAU, Hyderabad, India, 7–10 January 2015

Gopi Y, Madhu Vani P, Prasuna Rani P (2015) Effect of paddy straw, FYM and zinc on performance of rice (*Oryza sativa* L.)." In: Compendium of abstracts of the 2nd international conference on bio-resource and stress management, ANGRAU & PJTSAU, Hyderabad, India, 7–10 January 2015

Harne AD, Gupta VR, Rakhonde PN et al (2015) Effect of combine inoculations of biofertilizers and bioagent on soybean. In: Compendium of abstracts of the 2nd international conference on bio-resource and stress management, ANGRAU & PJTSAU, Hyderabad, India, 7–10 January 2015

Jawale S, Jawale S (2015) Impact of soil soluble fertilizer on productivity of Bt cotton. In: Compendium of abstracts of the 2nd international conference on bio-resource and stress management, ANGRAU & PJTSAU, Hyderabad, India, 7–10 January 2015

Kamble D, Jadhav P, Singh BJ et al (2015) Effect of Zinc, Rhizobium and PSB on the yield and economics of Chickpea (*Cicer arietinum* L.). In: Compendium of abstracts of the 2nd international conference on bio-resource and stress management, ANGRAU & PJTSAU, Hyderabad, India, 7–10 January 2015

Katkar RN, Jadhao SD, Kharche VK et al (2015) Effect of long term nutrient management on vertical distribution of sulphur fractions under sorghum-wheat sequence on Vertisols. In: Compendium of abstracts of the 2nd international conference on bio-resource and stress management, ANGRAU & PJTSAU, Hyderabad, India, 7–10 January 2015

Katore JR, Ingole PG, Paslawar AN (2015) Effect of preceding legumes, crop residue and nitrogen levels on nitrogen economy in sunflower. In: Compendium of abstracts of the 2nd international conference on bio-resource and stress management, ANGRAU & PJTSAU, Hyderabad, India, 7–10 January 2015

Kombali G, Nagaraju, Rekha MB et al (2015) Growth attributes and dry matter production of aerobic rice at varied levels of water and nutrients under drip fertigation. In: Compendium of abstracts of the 2nd international conference on bio-resource and stress management, ANGRAU & PJTSAU, Hyderabad, India, 7–10 January 2015

Kumar KA, Reddy BB, Reddy MD et al (2015a) Yield and dry matter *kharif* maize (*Zea mays*) under integrated nitrogen management. In: Compendium of abstracts of the 2nd international conference on bio-resource and stress management, ANGRAU & PJTSAU, Hyderabad, India, 7–10 January 2015

Kumar MP, Lateef PM, Bhadru D et al (2015b) Performance of rice in mechanized transplanting system for different nitrogen levels and schedules under NSP left canal command area. In: Compendium of abstracts of the 2nd international conference on bio-resource and stress management, ANGRAU & PJTSAU, Hyderabad, India, 7–10 January 2015

Lalita KA, Kishore BG, Sandhya Rani P (2015) Long term soil fertility management in groundnut – groundnut cropping system in Alfisols. In: Compendium of abstracts of the 2nd international conference on bio-resource and stress management, ANGRAU & PJTSAU, Hyderabad, India, 7–10 January 2015

Layek J, Shivakumar BG, Das A et al (2015) Effect of levels of nitrogen on growth, yield and quality of pearl millet (*Pennisetum typhoides*) intercropped with soybean. In: Compendium of abstracts of the 2nd international conference on bio-resource and stress management, ANGRAU & PJTSAU, Hyderabad, India, 7–10 January 2015

Madhu M, Ramana MV (2015) Nitrogen management for zero-till *rabi* castor (*Ricinus communis* L.) under the influence of different preceding crops. In: Compendium of abstracts of the 2nd international conference on bio-resource and stress management, ANGRAU & PJTSAU, Hyderabad, India, 7–10 January 2015

Madhu GK, Sanju HR, Bharath R et al (2015) Influence of precision water management practices on sustained maize production. In: Compendium of abstracts of the 2nd international conference on bio-resource and stress management, ANGRAU & PJTSAU, Hyderabad, India, 7–10 January 2015

Magare PN, Katkar RN, Kharche VK et al (2015) Effect of nutrient application through drip irrigation on yield and quality of cotton in vertisol. In: Compendium of abstracts of the 2nd international conference on bio-resource and stress management, ANGRAU & PJTSAU, Hyderabad, India, 7–10 January 2015

Mahatale PV, Chaudhari BN, Mahatale YV et al (2015) Soil application Multi micro nutrient fertilizer mixtures (MMM) influenced in groundnut yield. In: Compendium of abstracts of the 2nd international conference on bio-resource and stress management, ANGRAU & PJTSAU, Hyderabad, India, 7–10 January 2015

Mandal MK, Pramanick M, Bandopadhyay P et al (2015) Comparative efficacy of herbicides and hand weeding for control of weeds in onion. In: Compendium of abstracts of the 2nd international conference on bio-resource and stress management, ANGRAU & PJTSAU, Hyderabad, India, 7–10 January 2015

Meshram NA, Ismail S (2015) Yield, dynamics of micronutrients in soil and their uptake by plants as influenced by long-term use of organic manure and fertilizers under soybean-safflower cropping sequence in vertisol. In: Compendium of abstracts of the 2nd international conference on bio-resource and stress management, ANGRAU & PJTSAU, Hyderabad, India, 7–10 January 2015

Murthy KV, Prabhakara Reddy G, Srinivasulu Reddy D (2015) Standardisation of agro techniques for aerobic rice in north coastal zone of Andhra Pradesh. In: Compendium of abstracts of the 2nd international conference on bio-resource and stress management, ANGRAU & PJTSAU, Hyderabad, India, 7–10 January 2015

Neelima TL, Jemimha N, Anitha V et al (2015) Study of comparative performance of various rice planting techniques. In: Compendium of abstracts of the 2nd international conference on bio-resource and stress management, ANGRAU & PJTSAU, Hyderabad, India, 7–10 January 2015

Ngomle S (2015) Mycorrhizae: a component of sustainable agro-ecosystems. In: Compendium of abstracts of the 2nd international conference on bio-resource and stress management, ANGRAU & PJTSAU, Hyderabad, India, 7–10 January 2015

Nileshsadar, Amarshettiwar SB, Parihar LB (2015) Morphophysiological study of arboreum cotton under high density planting system in rainfed condition. In: Compendium of abstracts of the 2nd international conference on bio-resource and stress management, ANGRAU & PJTSAU, Hyderabad, India, 7–10 January 2015

Padmalatha Y, Bhargavi K, Narayanaswamy G (2015) Influence of irrigation depth and interval under drip irrigation method on productivity of *rabi* groundnut in water scarce regions. In: Compendium of abstracts of the 2nd international conference on bio-resource and stress management, ANGRAU & PJTSAU, Hyderabad, India, 7–10 January 2015

Paithankar DH, Nagre P, Ekta Bagde (2015) Organic production of mandarin. In: Compendium of abstracts of the 2nd international conference on bio-resource and stress Management, ANGRAU & PJTSAU, Hyderabad, India, 7–10 January 2015

Patel AS, Kacha RP, Patel CJ et al (2015) Effect of nitrogen and phosphorus fertigation on sweet corn in goradu soil of middle Gujarat conditions. In: Compendium of abstracts of the 2nd international conference on bio-resource and stress management, ANGRAU & PJTSAU, Hyderabad, India, 7–10 January 2015

Prasad (2015) Agri-nanotechnology – a prosperous approach to Indian agriculture. In: Compendium of abstracts of the 2nd international conference on bio-resource and stress management, ANGRAU & PJTSAU, Hyderabad, India, 7–10 January 2015

Ramana MV, Sridevi S, Goverdhan M (2015) Resource efficient diversified cropping systems for Southern telangana region of telangana state. In: Compendium of abstracts of the 2nd international conference on bio-resource and stress management, ANGRAU & PJTSAU, Hyderabad, India, 7–10 January 2015

Ramanjaneyulu AV, Sarkar NC, Thakur AK et al (2013) Organic farming- a perspective. Int J Bioresour Stress Manag 4(3):i–ii

Ramanjaneyulu AV, Neelima TL, Madhavi A et al (2015) Effect of biophos on the performance of castor under rain fed conditions. In: Compendium of abstracts of the 2nd international conference on bio-resource and stress management, ANGRAU & PJTSAU, Hyderabad, India, 7–10 January 2015

Rao VP, Anitha (2015) Micro irrigation technologies for water conservation and sustainable crop production. In: Compendium of abstracts of the 2nd international conference on bio-resource and stress management, ANGRAU & PJTSAU, Hyderabad, India, 7–10 January 2015

Rao PC, Padmaja G (2015) Soil quality and enhanced productivity through soil organic matter. In: Compendium of abstracts of the 2nd international conference on bio-resource and stress management, ANGRAU & PJTSAU, Hyderabad, India, 7–10 January 2015

Rekha MB, Jayadeva HM, Kombali G et al (2015) Economic evaluation of drip fertigation in aerobic rice. In: Compendium of abstracts of the 2nd international conference on bio-resource and stress management, ANGRAU & PJTSAU, Hyderabad, India, 7–10 January 2015

Saileela K, Raji Reddy D, Devender Reddy M et al (2015) Drip fertigation effects on yield and nutrient uptake of bush bean (*Lablab purpurium* var. *typical* L.). In: Compendium of abstracts of the 2nd international conference on bio-resource and stress management, ANGRAU & PJTSAU, Hyderabad, India, 7–10 January 2015

Sankaranarayanan K, Nalayini P (2015) Performance of surface irrigation methods on productivity and water use efficiency of ELS Cotton. In: Compendium of abstracts of the 2nd international conference on bio-resource and stress management, ANGRAU & PJTSAU, Hyderabad, India, 7–10 January 2015

Saras PK, Parmar SK, Patel BD et al (2015) Weed Management in *kharif* cluster bean (*Cyamopsis tetragonoloba* (L.) Taub) under semi-arid conditions. In: Compendium of abstracts of the 2nd international conference on bio-resource and stress management, ANGRAU & PJTSAU, Hyderabad, India, 7–10 January 2015

Shanwad UK, Veeresh H, Bhatt SN et al (2015) Soil test crop response (STCR) approach for sustainable yield of sunflower (*Helianthus annuus* L.): a case study from North-eastern dry zone of Karnataka. In: Compendium of abstracts of the 2nd international conference on bio-resource and stress management, ANGRAU & PJTSAU, Hyderabad, India, 7–10 January 2015

Sharma AR, Singh (2015) Weed management in conservation agriculture systems. In: Compendium of abstracts of the 2nd international conference on bio-resource and stress management, ANGRAU & PJTSAU, Hyderabad, India, 7–10 January 2015.

Shivran AC (2015) Influence of sowing time and N levels on Isabgol (*Plantago ovata* forsk) varieties. In: Compendium of abstracts of the 2nd international conference on bio-resource and stress management, ANGRAU & PJTSAU, Hyderabad, India, 7–10 January 2015

Singh I, Singh SN (2015) Enhancing sugarcane productivity and profitability under wheat–sugarcane cropping through furrow irrigated raised bed (FIRB) system. In: Compendium of abstracts of the 2nd international conference on bio-resource and stress management, ANGRAU & PJTSAU, Hyderabad, India, 7–10 January 2015

Singh R, Acharya S, Hota SK et al (2015) Applications of nanotechnology in plant nutrition and water savings in agriculture. In: Compendium of abstracts of the 2nd international conference on bio-resource and stress management, ANGRAU & PJTSAU, Hyderabad, India, 7–10 January 2015

Solunke PS, Gite BD, Patil SV et al (2015) Study of Various Sources of Nutrients with Different plant Protection Measures on seed Cotton Yield and Monetary Returns and B:C ratio of desi Cotton (*Gossypium arboreum* L.). In: Compendium of abstracts of the 2nd international conference on bio-resource and stress management, ANGRAU & PJTSAU, Hyderabad, India, 7–10 January 2015

Sreedevi B, Latha PC, Sandhyarani A et al (2015) Nitrogen doses and weed management practices on grain yield of aerobic rice. In: Compendium of abstracts of the 2nd international conference on bio-resource and stress management, ANGRAU & PJTSAU, Hyderabad, India, 7–10 January 2015

Srinivas J, Kale VS, Nagre PK et al (2015) Effect of different organic sources of nutrient on growth and yield of Fenugreek. In: Compendium of abstracts of the 2nd international conference on bio-resource and stress management, ANGRAU & PJTSAU, Hyderabad, India, 7–10 January 2015

Tamgadge DB, Magare PN, Katkar RN et al (2015) Effect of potassium on growth, nutrient uptake and yield of cotton in vertisol. In: Compendium of abstracts of the 2nd international conference on bio-resource and stress management, ANGRAU & PJTSAU, Hyderabad, India, 7–10 January 2015

Vijaya D, Ram Reddy G, Joshi V (2015) Influence of fertigation schedule on yield, petiole mineral concentration and soil nutrients in grape (cv. Thompson Seedless grafted Dogridge). In: Compendium of abstracts of the 2nd international conference on bio-resource and stress management, ANGRAU & PJTSAU, Hyderabad, India, 7–10 January 2015

Vikram A, Kumar M, Thakur AK (2015) Integrated nutrient management studies in peas (*Pisum sativum* L.) under cold desert conditions of HP. In: Compendium of abstracts of the 2nd international conference on bio-resource and stress management, ANGRAU & PJTSAU, Hyderabad, India, 7–10 January 2015

Vimala B, Indira B, Pratap M (2015) Studies on the influence of inorganic mulches and foliar spray of calcium nitrate on growth, flower yield and vase life of chrysanthemum. In: Compendium of abstracts of the 2nd international conference on bio-resource and stress management, ANGRAU & PJTSAU, Hyderabad, India, 7–10 January 2015

Physiological Basis of Crop Productivity

6

Ratikanta Maiti and Aruna Kumari

Abstract

This chapter reviews various aspects of the physiological basis of crop production under climatic changes and the use of different crop geometry planting techniques and their influence on yield. This includes diurnal variations in canopy temperature and stress degree, and various abiotic and biotic stress factors such as chlorophyll, nitrogen and potassium content under post-flowering moisture stress in sorghum, sodium and potassium accumulation under salt stress, chlorophyll-scavenging enzymes under drought stress, physiological responses to foliar spray of organic manure, zinc, integrated nutrient management, application of nutrients for shelf life, sustainable land water and crop management, nutrient cycling and distribution, foliar application of zinc, boron, GA, plant growth regulators, and application of low-cost technologies such as *Rhizobium* and PSB for soil fertility improvement. It also discusses the role of biotechnology and genetic engineering, green technology, nanotechnology, bio-fortification in bio-resource and stress management, with a special emphasis on pest management.

6.1 Introduction

The productivity of crops is influenced by various physiological functions of the plants among which may be mentioned absorption of water and nutrients from soils by roots, photosynthesis, respiration, transpiration, water relations and so on. Besides the environmental conditions of precipitation, temperature, humidity, heat and cold stress, other abiotic stresses such as salinity and drought have direct impacts on the physiological functions of the plants and their productivity. Significant advances have been made on the

R.K. Maiti (✉)
Facultad de Ciencias Forestales, Universidad Autónoma de Nuevo León, Carr. No 85, km 145, NL 67700 Linares, Mexico
e-mail: ratikanta.maiti@gmail.com

A. Kumari
Department of Crop Physiology, Agricultural College, Professor Jayashankar Telangana State Agricultural University, Polasa, Jagtial, Karimnagar 505 529, Telangana, India
e-mail: arunasujanagcjgl@gmail.com

R.K. Maiti et al. (eds.), *Bioresource and Stress Management*, DOI 10.1007/978-981-10-0995-2_6

physiology of plants and mechanisms of resistance to various abiotic stresses. Several papers were presented on these various aspects at the Second International Conference on Bio-resource and Stress Management in Hyderabad in January 2015.

6.2 Impact of Climate Change on Physiological and Yield Parameters

Climate change has an impact on various physiological and yield parameters. In sorghum the physiological parameters are analysed in leaves at the boot stage, 50 % flowering and dough stages. The physiological parameters (viz chlorophyll a and b, total chlorophyll, RWC, photosynthetic rate, PAR, stomatal conductance and leaf thickness) and biochemical parameters (viz reducing sugars, non-reducing sugars, total sugars and soluble protein) were found to be higher in the Phule Revati variety of sorghum when sowings were taken in the month of October for the *rabi* season, indicating that the *rabi* sorghum crop sown in late October favours obtaining higher grain and fodder yields. Thus the *rabi* sorghum variety Phule Revati, owing to its physiological and yield-contributing characters, was found to sustain climatic changes and could be exploited in breeding programmes (Baviskar 2015).

6.3 Diurnal Variations in Canopy Temperature and Stress Degree Day

Diurnal variations in canopy temperature and stress degree day are seen in different crops at different dates of sowing. Four mung bean varieties (V1: Pant Mung-5; V2: Bireswar; V3: RMG-62; V4: Sukumar) sown on three different dates (D1: 15 February; D2: 1 March; D3: 15 March) in a sub-humid tropical environment revealed increases in canopy temperatures at 7:30 and 11:30 h. The mean canopy temperature under D2 was lower than under D1 and D3 in both the years, which indicated no moisture stress suffered by the crop during that period. The SDDI was higher at 11.30 h, which indicated moisture stress at that time. Overall, the moisture adequacy was better under D2 sowing, giving the probability of good yield under this sowing, and was closely followed by D3 sown crops (Tzudir et al. 2015).

6.4 Abiotic and Biotic Stress Management

6.4.1 Post-flowering Moisture Stress in Sorghum

The chlorophyll, nitrogen and potassium content in the fully expanded third leaf from the top was also positively and significantly correlated with grain yield in sorghum genotypes CRS 4, CRS 19, CRS 20, PEC 17, CSV 18, M 35–1, Phule Chitra, Phule Moulee, EP 57 and CRS 1 under post-flowering moisture stress (Devkumar et al. 2015).

6.4.2 Sodium and Potassium Accumulation Under Salt Stress

The tolerance type of plant response to salinity and drought stress is caused by cellular and molecular modifications, which lend themselves to biotechnological manipulation. Salt tolerance is related to the efficiency of a tissue to absorb, deposit and transport the levels of inorganic solutes in response to salt stress. The plants that regenerated from the salt-tolerant calli of sugarcane varieties, namely Co 99004, were selected when grown in a pot culture system under salinity stress showing that an increasing supply of NaCl more adversely affected root growth than shoot growth. The chlorophyll content, photosynthesis rate, stomatal conductance and dry matter showed a decreasing trend but at a slower rate in tolerant mutants than in normal plants. Further, it was observed that at high salt concentrations, Na^+ content and K^+ content in shoots of tolerant mutants were higher compared to normal plants. The results indicated that leaf area, number of leaves, CCI, photosynthesis rate, stomatal conductance, shoot length, root length, fresh weight, dry weight and some mineral solutes, i.e. K^+, had a positive role to play in the tolerance of salinity by the generated plant (Gadakh et al. 2015).

6.4.3 Mannitol Stress and Chlorophyll-Scavenging Enzymes for Drought Tolerance

Biochemical markers can be utilized for identifying genotypes for drought tolerance. Changes in chlorophyll content and activities of scavenging enzymes were determined in 14th-day seedlings of 20 rice genotypes, each with a different genetic background. Water-deficit conditions were induced by treatment with mannitol at 20 % concentration added to the supporting micronutrient medium (Yoshida 1981) in Deogirkar 2015. The estimation of chlorophyll content and activity of scavenging enzymes in the leaf tissue carried out after 24 h exposure to stress indicated an increase in the activity of SOD and POD in rice genotypes exposed to the water-limiting situation. The increased antioxidant activity was in correspondence with raised levels of free radicals. The increase in SOD and POD of the antioxidant system indicated that the increase in oxidative stress caused by drought might have been overwhelmed by this enzymatic system. The leaf SOD activity of Rajendra (34.39): Vandana, IR 64 (31.48), Anjali (30.38) and Varalu (30.23) at 20 % mannitol were higher than that of check Vandana at no mannitol stress. Genotype Rajendra was superior to check Vandana at 20 % mannitol stress. The results clearly indicated the free radical scavenging ability of these varieties under the influence of drought stress by correspondingly enhancing the production levels of SOD. The POD activity in Vijetha (138.60), MTU 1010 (133.40), Vandana (136.50), BPT 5204 (127.77), IR 64 (112.90), Prasanna (114.30), B 133 (114.80) and Azucena (102.50) was found to be desirable and reflected an increased ROS scavenging capacity of these genotypes. A significant reduction in leaf chlorophyll content was also observed under mannitol stress. This study indicated that chlorophyll b was more sensitive than chlorophyll a for water-stress conditions. The data showed that the status of scavenging enzymes and chlorophyll could provide a meaningful tool for depicting drought tolerance of a rice genotype (Vanisri et al. 2015).

6.4.4 Physiological Responses to Foliar Sprays

In today's changing climatic conditions, use of organic manures is required for sustainable agriculture rather than use of fertilizers alone in the field for producing higher yields. This was evident in chickpea foliar sprayed with 6 % cow urine, which recorded the highest seed protein content (21.93), number of pods per plant (47.60), 100-seed weight (24.00 g), seed yield per plant (6.77 g), seed yield per plot (1.35 kg) and seed yield per hectare (22.58 q) rather than only fertilizer use (Deogirkar 2015).

6.5 Role of Society in Bio-resource Management

6.5.1 Integrated Nutrient Management and Biofertilizers

An efficient means of managing resources is the adoption of usage of integrated nutrient management and biofertilizers. Fertilizer or manure application alone or in combination significantly increased the N content in seed and straw and total uptake of NPK in the Isabgol (*Plantago ovate*) cultivar. There were also increases in the levels of N content in seed and straw and NPK uptake, and higher seed and straw yields with vermicompost at 1.5 t ha^{-1} + 50 % RDF. Though the application of *Azotobacter* and *Azospirillum* inoculants marginally increased NPK content, nutrient uptake in seed and straw was found to be higher with biofertilizer. Application of fertilizers or manures in an integrated manner improves the protein content and swelling capacity and results in maximum net returns. Though the fertilizers or manures improve organic carbon and available NPK in soil, a marginal improvement in soil fertility is also achieved by biofertilizers (Shivran et al. 2015).

6.5.2 Impact of Foliar Spray of Nanoformulated Zinc Sulphate on Growth

Nanoformulated zinc fertilizer with biopolymers affects growth and development. The nanoformulated zinc fertilizer with or without a zinc coating is synthesized through ionic gelation process. The cationic biopolymer chitosan and anionic polymer sodium tripolyphosphate (TPP) are used to encapsulate the zinc ions by producing chitosan nanoparticles. The newly synthesized nanoformulated zinc fertilizer is then characterized for confirmation of size, shape, zeta potential, crystalline nature and elemental composition using a particle size analyser, scanning electron microscopy with EDAX, X-ray diffraction and FTIR. The characterized nanoformulated zinc fertilizer with or without a zinc coating when foliar sprayed at 100 ppm concentration on 35-day-old cotton (var. Suraj) plants increased plant height, number of leaves, number of squares, root length, and shoot and root weight at 8 weeks (20 days after spraying) (Kanjana 2015).

6.5.3 Impact of Supplemental Pollinators on Flowering and Fruit Set Under Changing Climatic Scenario

Pollination is an important ecosystem service considered by the Millennium Ecosystem Assessment. It is fundamental to the reproduction of flowering plants and is essential for production of about one third of the food consumed by humans (Klein et al. 2007). Pollinators are considered keystone species in many situations because they not only support humanity but also maintain diversity in an ecosystem. There is mounting evidence of a decline in pollinators, which would threaten the reproductive cycle of many plants and may reduce the quality of fruit and seeds, many of which are of nutritional and medicinal importance to humans. A growing number of studies suggest that climate change may be one of the biggest disturbance factors imposed on ecosystems today (Walther et al. 2002). A recent review has emphasized that plant–pollinator interactions can be affected by changes in climatic conditions in subtle ways. However, data on the impacts of climate change on crop pollination are still limited. An investigation was conducted to study the impact of climate change on pollination in apples in the Kullu Valley with the pollinator's diversity, wherein honey bee colonies (*Apis mellifera*) were also supplemented to apple orchards. A shift was observed in the numbers of days for the onset, peak and end of bloom for the apple cultivar, which further might be related to climate change. The diversity and density of different pollinators, viz honey bees, bumble bees, other wild bees, syrphids, drone flies and others insects recorded by the use of scan sampling and the sweep net method revealed that syrphids were the most abundant pollinators (2.24), followed by honey bees (1.55). The fruit set (on a spur basis) at 60–200 m from bee colonies varied between 34.13–49.88 % and 79.5–71.76 % during 2 years of study. No specific varied trend was seen in the total soluble solids (TSS) of fruits or the number of seeds per fruit (Gupta et al. 2015a).

6.5.4 Impacts of Denavelling and Stalk End Application of Nutrients on Yield and Shelf Life

Denavelling and stalk end application of nutrients influence yield and quality of banana. In banana cv. Dwarf Cavendish, denavelling and stalk end application of nutrients improved bunch characters and yield. A denavelled bunch treated with a combination of 500 g of fresh cow dung slurry with 20 g of sulphate of potash resulted in maximum bunch yield (73.10 t ha^{-1}), bunch weight (20.57 kg), weight of the hand (2.60 kg) and weight of the finger (127.66 g), and finger parameters of the bunch like finger length (18.43 cm), finger circumference (12.90 cm), weight of the pulp (109.16 g) and weight of the peel (42.31 g) also increased. It also resulted in the longest shelf life (7.56 days) (Priyanka et al. 2015).

6.6 Sustainable Land, Water and Crop Management, Conservation Agriculture and Homestead Farming

6.6.1 Nutrient Cycling and Distribution

The soil nutrient distribution varies under plantations in comparison to cultivable land. The soil nutrient status of bamboo plantations was found to vary with cultivable land. The soil pH was lower in bamboo followed by other plantations and cultivable land. Bamboo plantations had higher OC, IOC, available nutrients N, P, K and S, and exchangeable Ca and Mg concentrations than cultivable land. The organic carbon and total nutrients added through leaf biomass (t ha^{-1}) to soils was higher in bamboo, followed by other plantations. There were no total returns of nutrients and OC to cultivable land (Kumar et al. 2015b).

6.6.2 Distribution of Zinc and Zinc Fractions in Paddy Soils

In the upper Krishna project command area the distribution of zinc and zinc fractions in paddy soils at two depths (0–15 and 15–30) from paddy growing areas of Surpur (Shorapur) and Shahapur taluk, when analysed for physicochemical properties, revealed that the soils of the command area were neutral to slightly alkaline in reaction with varying electrical conductivity. The organic and free calcium carbonate in soils increased with increases in depth. The soils were calcareous in nature with varying cation exchange capacity (CEC) values of 34.30–59.00 cmol (p+) kg^{-1} and 36.20–61.20 cmol (p+) kg^{-1} in surface and subsurface soils, respectively. The order of zinc distribution in paddy soils were water soluble plus exchangeable zinc < organically bound zinc < crystalline bound zinc < amorphous sesquioxide bound zinc < manganese oxide bound zinc < residual zinc (Rajini et al. 2015).

6.6.3 Rooting Responses Under Shade Net Conditions

Vegetative propagation of pomegranate (*Punica granatum* L.) through cuttings is the most convenient and cheap method of obtaining true-to-the-type, fully developed plants in considerably less time with reduced mortality of rooted cuttings. Pomegranate (*Punica granatum* L.) is an ancient and favourite fruit of many tropical and subtropical areas because of its refreshing and medicinal qualities. The pomegranate is native from Iran to the Himalayas in Northern India. An investigation carried out to evaluate the effect of substrate and IBA on root characteristics of pomegranate cuttings indicated that different substrates respond differently to IBA concentrations. The cuttings treated with 4000 ppm IBA and planted in coir pith + FYM had the highest percentage of rooted cuttings, survival percentage of rooted cuttings, number and length of roots, fresh and dry weight of roots, root-to-shoot ratio and percentage of field establishment of rooted cuttings (Ratnakumari et al. 2015).

6.6.4 Impacts of Potting Media on Growth and Quality of Ornamental Foliage Plants

Potting media influence the growth and quality of ornamental plants. In the ornamental plant Aglaonema cv. Ernesto's Favourite a potting medium containing cocopeat + sand + FYM + vermicompost in a 2:1:1:0.5 ratio v/v was found to be better in increasing the plant height, number of leaves, plant growth index, fresh and dry weight of root, visual plant grade, colour grade, root grade and K content (Swetha et al. 2015).

6.6.5 Foliar Application of Zinc and Boron

Micronutrients like zinc and boron improve the growth and yield. They also have an influence on the quality of fruits. In a tomato (Azad T-6) variety, the foliar application of these nutrients greatly influenced the vegetative growth in terms of plant height and number of branches at various stages (30, 60 and 90 days after transplanting) and resulted in early flowering with a maximum number of flowers and fruit yield. Further, these micronutrients improved the physicochemical qualities of tomato fruits and especially improved the TSS:acid ratio. Thus, application of boron and zinc either solely or in combination is quite beneficial for vegetative growth, flowering and fruiting as well as quality improvement of tomato fruits grown under high-pH soil (pH 8.2) (Meena et al. 2015a).

6.6.6 Impact of Boron in Improving Spikelet Fertility

Under the changing climatic conditions of prevailing high or low temperatures there is a decrease in the spikelet fertility and subsequently a reduction in grain density. Four levels of boron (B) sprays in seven genotypes of rice have shown differential variations in grain filling high-density grains. The grain filling percentage was higher in Rasi at 0.4 ppm B spray and a high-density grain weight was seen in IET 21106 (17.73 g) with less chaff weight in IET 20979 (0.62 g) and Rasi (0.68 g). Grain yield was maximum in IET 20979. B resulted in an increase in spikelet fertility, grain density and yield of rice genotypes IET 20979, IET 21106 and IET 21114, which responded more positively to B application (Guru et al. 2015).

6.6.7 Impacts of Foliar Spray of Gibberellic Acid on Physicochemical Properties

Foliar sprays of gibberellic acid improved fruit quality. Cape gooseberry (*Physalis peruviana* L.) is an herbaceous or soft-woody, perennial plant and is grown for its edible fruits, which have high nutritional value and potential health benefits. GA_3 when foliar sprayed increased the fruit size, fruit weight, TSS and ascorbic acid levels. An increase in the GA_3 concentration beyond 250 ppm decreased the fruit acidity. A still higher concentration of GA_3 (300 ppm) had a detrimental effect on the fruit quality of cape gooseberry. Thus an optimum level of foliar application of GA_3 is only effective in improving the physicochemical properties of cape gooseberry (Kumar et al. 2015a, b).

6.6.8 Impacts of Salicylic Acid on Growth and Bulb Yield

Exogenous application of salicylic acid was found to increase plant growth and alter several physiological processes. In onion var. Agrifound Light Red and garlic, exogenous application of salicylic acid (SA) during the winter season influenced the growth and yield. It increased vegetative growth in terms of plant height (68.18–71.08 cm) and collar thickness (16.90–18.51 mm) with a higher chlorophyll content of leaves (31.53–33.01 SPAD). Foliar spray of SA at 30 days after sowing (DAS), 30 and 45 or 60 days after transplanting (DAT) was more effective in producing maximum vegetative parameters, bulb yield and yield-contributing parameters. Spraying of SA three times had better efficacy than spraying two times in terms of bulb diameter (polar: 65.60–67.94 mm and equatorial: 49.10–49.80 mm), bulb weight (59.50–69.25 g), marketable bulb yield (180.91–183.10 q ha^{-1}) and total bulb yield (266.99–290.91 q ha^{-1}). Application of SA at 30 DAS, 30 DAT and/or 45 or 60 DAT not only increases the vegetative growth but also increases bulb yield in onion variety ALR (Pradhan et al. 2015).

6.6.9 Impacts of Plant Growth Regulators and Antioxidants on In Vitro Multiplication

Over the past few years, plant growth regulators such as IAA, BAP and antioxidants like adenine sulphate and ascorbic acid have been used for in vitro multiplication of crops. In in vitro multiplication of banana cv. Karpurachekkarakeli (AAB) there was a higher primordial emergence coupled with maximum shoot length and number of leaves by the addition of these to the MS medium (Usha et al. 2015).

6.6.10 Impacts of Pinching and Plant Growth Regulators on Flowering and Yield

Pinching and plant growth regulators (GA_3 or NAA) influence flowering and seed yield. In fenugreek cv. APHU Methi-1, single pinching at 25 DAS (P1) resulted in a maximum number of pods, length of pods, number of seeds per pod and seed yield. Among the plant growth regulators, foliar spraying of GA_3 50 ppm (G1) three times (at 25, 45 and 65 DAS) resulted in the best performance of yield parameters like the number of pods, length of pods, number of seeds per pod and seed yield of fenugreek. Early flower initiation and early maturity were observed with application of GA_3. Pinching in combination with growth regulator (GA_3) resulted in maximum yield attributes like the number of pods, length of pods, seed yield and test weight (Krishnaveni et al. 2015).

6.6.11 Impacts of Bio-regulators on Growth and Flowering

Plant bio-regulators influence growth, flowering and seed yield. Various plant growth regulators, viz GA_3, SA, maleic hydrazide, alar and paclobutrazol were used as foliar sprays twice at 20 and 35 days after transplanting in China aster cv. Kamini, affecting plant growth, flowering and seed yield. SA (200 ppm) spray achieved a higher plant height, number of primary branches per plant and number of secondary branches per plant at 90 DAT, number of flowers per plant, flower yield per plant, flower yield per hectare, seed yield per plant, seed yield per hectare and 1000-seed weight, and GA_3 200 ppm foliar spray was found to be on par with SA 200 ppm. SA foliar spray could be recommended for increased growth, flower and seed yield in China aster cv. Kamini, as it is less expensive than GA_3 (Pavankumar et al. 2015).

In *Gladiolus* cv. Darshan, foliar sprays of SA and calcium nitrate induced early flowering in plants that were raised from cormels and resulted in maximum vegetative growth, number of spikes per plant, number of florets per spike and spike longevity (Padmalatha et al. 2015).

6.6.12 Impacts of Growth Regulators on Hastabahar Regulation

Regulation of hastabahar or flowering in most fruit crops is brought about by sprays of growth regulators. Use of plant growth regulators, viz GA_3 50 ppm in June + cycocel 1000 and 2000 ppm and paclobutrazol 2.5 g a.i. and 3.5 a.i. per tree and 1000 and 2000 ppm in September nutrients, viz KNO_3, zinc (0.3 %) and boron (0.1 %), used for the regulation of hastabahar in acid lime, has revealed better performance of acid lime on treatment. There were increases in fruit yield (32.29 kg per plant, i.e. 8.90 t ha^{-1}) fruit juice (48.95 %), TSS (8.13 °B), acidity (6.94 %), ascorbic acid content (30.62 mg 100 ml^{-1}) and weight of fruit (42.49 g) in acid lime (Deshmukh et al. 2015).

Shoots of mango cv. Kesar were selected and pruned back by 10 cm and sprayed with 2 % urea spray during the last week of May, i.e. after harvest. Spraying of 6-BA at 200 ppm on the post-harvest vegetative flush as well as old shoots showed a profound effect in early panicle emergence with a maximum length of panicle (41.83 cm) without cool inductive temperature for flower bud differentiation. Spraying of 6-BA at 200 ppm is effective towards early induction of flowering in mango cv. Kesar (Shankaraswamy et al. 2015).

Foliar application of GA_3 (200 ppm) + KNO_3 (1 %) in marigold improved the days of opening of the flower from flower bud emergence, 50 % flowering and first harvesting from transplanting and maximum flowering span, and increased flower yield. Similarly, there was also an increase in the diameter of the fully open flower, length of flower stalk, weight of flower and vase life of flowers under foliar application of GA_3 400 ppm + KNO_3 2 % (Taksande et al. 2015).

6.6.13 Impacts of Plant Growth Retardants on Growth of Plants

Growth retardants in some ornamental plants are used to retard the growth of foliage of ornamental plants. Two growth retardants, viz paclobutrazol and ancymidol, in ornamental foliage plants, i.e. Aglonema cv. Ernesto's Favourite and *Schefflera arboricola*, when applied as soil drench at monthly intervals from 30 DAT, drastically reduced the plant height, number of leaves produced, canopy width, leaf length and width, leaf area, petiole length and stem diameter (Kameswari et al. 2015).

Growth characters like the number of leaves per plant, number of branches per plant and dry matter accumulation per plant are influenced by the time of application and the concentration of growth retardants. In soybean, foliar application of CCC 2000 ppm at 40 DAS reduced plant height and led to an increase in seed and straw yield (Badkhal et al. 2015).

6.6.14 Impacts of Growth Regulators on Shelf Life

Post-harvest treatment of fruits with growth regulators, fungicides and wax help in the reduction of losses due to wastage. At present there is high domestic market potential for sweet oranges. The effect of growth regulators 2–4 D, benzyl adenine and GA_3 with wax influenced the shelf life of sweet oranges. Of the various physicochemical parameters like physiological loss in weight (PLW), juice content, peel content, firmness, spoilage, colour index, shelf life, TSS, acidity, sugars and ascorbic acid analysed at an interval of 5 days, fruits treated with BA 50 ppm + wax 6 % recorded the lowest PLW, spoilage and colour index, thereby having an improved shelf life, with the highest juice content, peel content and firmness. Further, this treatment also resulted in the lowest TSS, total sugars and highest ascorbic acid content. The analysis of various physicochemical parameters indicated that BA 50 ppm + wax 6 % preserved the quality of sweet orange with a 45 % concomitant increase in shelf life (Hemalatha et al. 2015).

6.6.15 Impact of Gibberellic Acid in Modified Atmospheric Packaging

Sweet orange (*Citrus sinensis* cv. Nucellar) fruit is harvested at a physiological stage of maturity at the light green stage. Most physiological and biochemical changes are affected during storage and ripening. When sweet orange fruits were harvested at physiological maturity and treated with GA_3 and paraffin wax in a dose-dependent manner, stored in 100-gauge polythene bags with 5 % ventilation and without ventilation, and kept in corrugated fibre board boxes (CFB) at room temperature (27 + 2 °C) and humidity of 60–70 %, most physiological and biochemical changes during storage and ripening were affected by GA_3 and paraffin wax in a dose-dependent manner. There was a gradual increase in TSS content and a reduction in the titratable acidity and ascorbic acid content. Further, the fruit ripening was retarded and a significant reduction in the decay incidence of GA_3 and paraffin wax-treated fruit was observed under the storage condition. The fruits treated with GA_3 100 ppm in combination with 6 % wax, wrapped in 100-gauge polyethylene bags stored at ambient temperature, had a prolonged storage life up to the 49th day wherein they retained desirable quality, colour, flavour and taste, and overall acceptability (Pimpalpalle et al. 2015).

6.6.16 Bio-stimulants of Growth

Certain bio-stimulants such as seaweed extracts significantly enhance growth, chlorophyll content and yield. In soybean, foliar spray of seaweed extracts (*Kappaphycus* and *Gracilaria*) enhanced growth in terms of plant height, number of branches, root length, dry matter accumulation, chlorophyll content index and yield (Mahajan et al. 2015).

6.7 Low-Cost Technologies for Bio-resource Productivity Enhancement

6.7.1 Integrated Nutrient Management in Growth Improvement

Biofertilizers and organic manures play a major role in supplementing nutrients and thereby help in reducing the quantities of application of inorganic fertilizers. Integration of organic manures with fertilizers has been traditionally important in production, for the maintenance of soil fertility and yield stability in crops. Continuous use of organic manures could undoubtedly meet nutrient demand in less intensive production systems though it is inadequate in the present exhaustive agriculture. Nevertheless, integrated use of fertilizer, organic manures and biofertilizer helps in maintaining yield stability and correcting marginal deficiencies of secondary and micronutrients, enhancing the efficiency of applied nutrients and providing favourable soil physical and chemical conditions. There is a need to optimize INM in the present exhaustive intensive agricultural production systems.

Gladiolus hybridus is an important cut flower grown under open field conditions in the Peri urban areas of Hyderabad. It is much valued for its attractive long-lasting spikes. *Gladiolus* plants respond well to an ample supply of fertilizers. An integrated supply of organic manures (75 % RDF, FYM, vermicompost) and biofertilizers (*Azospirillum* and PSB) improved the growth and flowering in *Gladiolus* variety 'Arka Amar'. There was an early emergence of the spike, with maximum spike length, number of florets per spike and individual floret size (Girwani et al. 2015).

In green gram, integrated nutrient management produced maximum seed yield and enhanced the uptake of micronutrients like Fe, Zn, Mn and Cu (Aware et al. 2015). The interactive effect of fertility levels and biofertilizers influenced the seed yield of cowpea. Maximum seed yields are obtained with 75 % RDF + VC at 2 t ha^{-1} and a *Rhizobium* + PSB combination. The increase in seed yield is positively correlated with plant height, number of branches per plant, number of pods per plant and number of seeds per pod. Further, seed inoculation with *Rhizobium* + PSB led to an increase in plant height, number of branches at harvest, leaf area index, chlorophyll content, number of total and effective root nodules, fresh and dry weight of root nodules, leg haemoglobin content in root nodules, number of pods per plant, number of seeds per pod, and seed, straw and biological yield (Meena et al. 2015b).

6.7.2 Drip Irrigation Improves Physiological Performance

Irrigation of several crops influences the physiological performance by improving the water use efficiency. In healthy runners of cv. Chandler of strawberry, drip irrigation led to an increase in the net photosynthetic rate and stomatal conductance with a concomitant improvement in fruit yield. It resulted in an improvement in its physiological performance with an increase of 16–30 % in fruit yield (Kachwaya et al. 2015).

6.7.3 Low-Cost Technology for Higher Phosphorus Use Efficiency in Rice

The availability of phosphorus in soils is less, resulting in yield losses. A low-cost method of phosphorous application for higher use efficiency and yield is required in the current changing conditions of intensified agricultural systems. In rice, root dipping of 25-day-old rice seedlings before transplanting for 12 h in diammonium phosphate (DAP) solution at 20 % was effective in increasing the number of effective tillers, yield and the benefit:cost ratio. Phosphorous agronomic efficiency (PAE) was the highest in foliar application. Though root dipping treatments exhibited seedling mortality, it was reduced in soil application of P, while foliar application led to slight scorching of leaves (Mallareddy et al. 2015).

6.7.4 Impacts of Micronutrients on Growth and Yield

In red and laterite soils and those that are acidic in nature, soil availability of micronutrients is a common problem. It results in reductions of vigour, yield and fruit quality. Micronutrient sprays ($ZnSO_4$ 0.1 %, 0.2 %; Borax 0.2 %, 0.3 %; $FeSO_4$ 0.1 %, 0.15 %; $CuSO_4$ 0.1 %, 0.15 %; and $MnSO_4$ 0.1 %, 0.15 %) in the grape Pusa Navrang cultivar decreased the mortality percentage and increased the vine growth of the shoots in terms of the diameter and vigour of the shoots. Foliar application of these micronutrients (Borax at 0.3 %) at 7 days after fruit set and 21 days after the first spray increased the juice percentage and TSS. Among the different micronutrients, borax promoted production and improved the vine growth, fruit yield and quality of fruits (Pal et al. 2015).

6.7.5 Nutrient Uptake

Cane thrash and mycorrhizae in the presence of phosphorous levels in soils influence the nutrient uptake in crops. Phosphorus application in a ratoon crop influences the uptake of nitrogen and potassium. In sugarcane, the uptake of NPK was not influenced by phosphorus application at the formative stage in both plant and ratoon crops. At the grand growth stage application of mycorrhizae or cane trash plus mycorrhizae improved P uptake even when inorganic P was not applied. Though the uptake of nitrogen and potassium was not influenced by phosphorus application, higher P uptake was recorded when inorganic P at 50 kg P_2O_5 ha^{-1} was applied in the presence of both cane trash and mycorrhizae. Higher uptake of nitrogen, phosphorus and potassium in the first and second plant crops was also observed at the grand growth stage with cane trash plus mycorrhizae application in both plant and ratoon crops, and at the maturity stage, there was higher uptake of P with 150 kg P_2O_5 ha^{-1} (Usharani et al. 2015).

6.7.6 Bio-fortification with Zinc

Bio-fortification with zinc in speciality corn (popcorn, sweetcorn, quality protein maize) in sandy soils by soil application or as a foliar spray at the tasselling and milking stage influenced growth parameters like plant height, LAI and dry matter production, and yield attributes like cob number, number of grains per cob, cob girth and yield. Maximum values were registered with 25 kg $ZnSO_4$ ha^{-1} along with two foliar sprays, i.e. at the tasselling and milking stage. Sweetcorn had a higher number of cobs, longer cobs and greater maximum cob girth. Further, the number of grain rows per cob of popcorn were greater. Zinc fortification in sweetcorn resulted in a higher yield. Bio-fortification also led to maximum green fodder yield and nutrient uptake (N, P and K) and zinc uptake in sweetcorn. Zinc fortification was maximum (zinc content and zinc uptake) with 25 kg $ZnSO_4$ ha^{-1} along with two foliar sprays at the tasselling and milking stage ($ZnSO_4$) in both plant (leaves) and grain (Debnath et al. 2015).

6.7.7 High P Alfisols

Organics like bio-char influence phosphorus availability and recycling in soils that have inherently low exchange capacities by influencing the CEC over long time scales. By reducing the presence of free Al^{3+} and Fe^{3+} near root surfaces, this may promote the formation and recycling of labile P fractions. Humic substances resulting from organic matter decomposition play a significant role in mobilizing insoluble phosphates. Bio-char combined with 30 kg P_2O_5 ha^{-1} resulted in higher seed and haulm yields and also high P uptake in soybean (Kalyani and Sailaja 2015).

6.8 Organic Agriculture

6.8.1 Impacts of Organic Liquid Formulations on Growth and Yield

Organic liquid formulations such as *jeevamrutha*, cow urine and *Panchagavya* are equally efficient in maintaining the soil fertility and growth of plants. In *Capsicum*, application of *jeevamrutha* resulted in higher plant height, number of branches per plant, number of fruits per plant and fruit yield (Boraiah et al. 2015).

6.8.2 Organic Nutrients and Their Influence on Enzyme Activity

Different organic nutrient sources influence soil enzymatic activities. In a subtemperate hilly agro-ecosystem, addition of poultry manure (Pm), farmyard manure (FM) and vermicompost (VC) in a tomato pea cropping system increased soil organic carbon (SOC) by 35 %, 57 % and 53 %, respectively. Farmyard manure resulted in a 40–198 % increase in dehydrogenase activity, while poultry manure resulted in higher β-glucosidase activity. All three organic manures led to increased activities of protease and cellulose enzymes and different hydrolase activities. Alkaline phosphatase and urease activity were more influenced by application of FYM. The increased phosphatase activity is attributed to microbial stimulation by organic C and soil pH (Kumar et al. 2015a).

6.8.3 Indigenous Plant Growth-Promoting Bacteria

The indiscriminate use of chemical inputs in modern agriculture is severely modifying our environment. The application of eco-friendly, beneficial soil organisms that could enhance plant growth by various modes of action could be a potential alternative to increase agricultural productivity in a sustainable manner. Antagonistic indigenous plant growth-promoting rhizobacteria are effective against fungal pathogens causing diseases. In *Capsicum*, four rhizospheric isolates (RS11, RS15, RS14 and RS17) possessing two or more PGP traits showed antagonism against *Ralstonia solanacearum* and enhanced growth of *Capsicum*. The isolates were applied as seed and soil treatments. Seed bacterization with the RS15 isolate not only showed maximum seed germination (83.8 %) and vigour index (914.83) but also resulted in maximum increases in plant parameters such as shoot length, shoot biomass, root length and root biomass. The isolate RS15 can act as a potential biofertilizer agent for *Capsicum* (Gupta et al. 2015b).

6.8.4 Impacts of Phosphorus Levels and PSB Strains on Soil Microbial and Enzymatic Activities Under Acidic Soils

Soil microorganisms are supportive in the transformation of soil phosphorus (P) and are thus an important component of the soil P cycle. These are effective in releasing P from both inorganic and organic pools of total soil P through their respective solubilization and also enhancement of soil microbial and enzymatic activities. In a local high-yielding variety of rice (cv. Shahsarang), phosphorus levels and PSB strains influenced the soil microbial and enzymatic activities. The soil microbial carbon (C), nitrogen (N), phosphorus (P) and dehydrogenase activity (DHA) and phosphatase activity were found to be higher (5.61 μg TPF g^{-1} h^{-1}) at a 60 kg P_2O_5 ha^{-1} dose and with inoculation of *Pseudomonas* species (Raghuveer et al. 2015).

6.8.5 Impacts of PGPRs on Growth and Yield

Plant growth-promoting rhizobacteria such as *Pseudomonas* or *Azospirillium* present in the soil rhizosphere have an influence in increasing growth parameters due to their ability to synthesize growth regulators such as IAA, nitrogen fixation, p-solubilization and biocontrol potential. Six efficient strains of TBP Karnataka were identified on the basis of their IAA production, nitrogen fixation, siderophore production, P-solubilization and biocontrol potential in improving the chilli growth. The chilli seedlings treated with efficient strains of PGPR showed increases in plant height, number of branches per plant, shoot dry weight, root dry weight, shoot N and P content, and number of fruits per plant. The PS-5 strain of *Pseudomonas* increased plant height (42.96 cm), number of branches per plant (6.67), shoot dry weight (12.08 g per plant), root dry weight (0.98 g per plant), shoot N content (with 2.23 % of nitrogen) and P content (with 0.21 % of phosphorus) per plant, and number of fruits per plant (52 fruits per plant). Similarly, AZP-26 of *Azospirillum* resulted in greater plant height (39.34 cm per plant), number of branches per plant (6.00), shoot dry weight (11.83 g per plant),: root dry weight (0.80 g per plant), shoot N content (2.70 % N per plant), shoot P content (0.18 % P per plant) and number of fruits per plant (48 fruits per plant) (Sabnis et al. 2015).

6.8.6 Rhizobium Symbiosis in Soybean

Symbiotic native *Rhizobium* among the soybean advanced genotypes, i.e. RKS-113, NRC-93, MACS-1407, KDS-705, JS-2071, DSb-19, Bragg, RKS-18 and JS-9752, increased nodulation (92.62 per plant), N uptake in nodules (5.4 mg per plant^{-1}), dehydrogenase activity of soil (96.2 μg.TPF gram^{-1} h^{-1}. soil), rhizobial population (9.81 × 104 g^{-1} soil) and grain yield (29.72 q ha^{-1}) in soybean variety NRC-93 because of its effective symbiosis with native *Rhizobium* (Deshmukh and Singh 2015).

6.9 Roles of Biotechnology and Genetic Engineering, Green Technology, Nanotechnology, Bio-fortification in Bio-resource and Stress Management, with Special Emphasis on Pest Management

6.9.1 Molecular Characterization and Identification of Phyllody Isolates

Phyllody of sesame is an important disease caused by a pleomorphic phytoplasma and is transmitted by a leaf hopper called *Orosius albicinctus*. An incidence of 13–68 % was seen from 94 locations in 10 districts of three states in India (Andhra Pradesh, Karnataka and Maharashtra). Ninety-five phyllody samples were collected from surveyed areas and their DNA was extracted. The representative 18 samples, representing different geographic regions, were used for amplification with nested PCR. Upon amplification at 1200 base pairs and sequencing, the nucleotide sequences of phytoplasma 16S rDNA shared the highest nucleotide identity (95 %) with sequences of members of the 16SrI phytoplasmal group. Seed treatment of imidacloprid at 5 g kg^{-1} followed by spraying of clothionidin at 0.5 g l^{-1} recorded only 21 % of phyllody incidence and subsequently gave a yield of 310 kg ha^{-1} (Bharathreddy et al. 2015).

References

Aware S, Parmar KB, Gupta A (2015) Yield potential of green gram (*Vigna radiata*) under integrated nutrient management and micro nutrient uptake. In: Compendium of abstracts of the 2nd international conference on bio-resource and stress management. ANGRAU & PJTSAU, Hyderabad, 7–10 Jan 2015

Badkhal AM, Chouhan PP, Mahajan RV et al. (2015) Effect of growth retardant on growth and yield of soybean. In: Compendium of abstracts of the 2nd international conference on bio-resource and stress management. ANGRAU & PJTSAU, Hyderabad, 7–10 Jan 2015

Baviskar SB (2015) Effect of climate change on physiological and yield parameters of *rabi* sorghum. In: Compendium of abstracts of the 2nd international conference on bio-resource and stress management. ANGRAU & PJTSAU, Hyderabad, 7–10 January 2015

Bharathreddy NS, Naik MK, Savitha AS et al (2015) Molecular characterisation and identification of phyllody of sesame isolates and its management. In: Compendium of abstracts of the 2nd international conference on bio-resource and stress management. ANGRAU & PJTSAU, Hyderabad, 7–10 January 2015

Boraiah B, Devakumar N, Palanna KB et al (2015) Growth and yield of capsicum (*Capsicum annuum* L. var. grossum) as influenced by organic liquid formulations. In: Compendium of abstracts of the 2nd international conference on bio-resource and stress management. ANGRAU & PJTSAU, Hyderabad, 7–10 January 2015

Debnath P, Hemalatha S, Sreelatha D et al (2015) Biofortification of speciality corn with zinc. In: Compendium of abstracts of the 2nd international conference on bio-resource and stress management. ANGRAU & PJTSAU, Hyderabad, 7–10 January 2015

Deogirkar MS (2015) Physiological responses of chickpea to foliar sprays of cow urine and NAA. In: Compendium of abstracts of the 2nd international conference on bio-resource and stress management. ANGRAU & PJTSAU, Hyderabad, 7–10 January 2015

Deshmukh SL, Singh D (2015) Rhizobium symbiosis on soybean varieties. In: Compendium of abstracts of the 2nd international conference on bio-resource and stress management. ANGRAU & PJTSAU, Hyderabad, 7–10 January 2015

Deshmukh HK, Paithankar DH, Dewangan RK et al (2015) Studies on regulation of Hasta bahar in acid lime. In: Compendium of abstracts of the 2nd international conference on bio-resource and stress management. ANGRAU & PJTSAU, Hyderabad, 7–10 January 2015

Devkumar D, Padma V, Talwar HS et al (2015) Effect of bio chemical aspects of chlorophyll content, nitrogen content and potassium content with grain yield in sorghum genotypes under post flowering moisture stress conditions. In: Compendium of abstracts of the 2nd international conference on bio-resource and stress management. ANGRAU & PJTSAU, Hyderabad, 7–10 January 2015

Gadakh SS, Patel DU, Singh C et al (2015) Sodium, potassium accumulation, photosynthesis and growth of sugarcane mutant's lines in response to salt stress. In: Compendium of abstracts of the 2nd international conference on bio-resource and stress management. ANGRAU & PJTSAU, Hyderabad, 7–10 January 2015

Girwani A, Kameswari L, Padmavathamma AS (2015) Effect of Integrated nutrient management practices on the growth, flowering and corm production in *Gladiolus hybridus* var. Arka Amar. In: Compendium of abstracts of the 2nd international conference on bio-resource and stress management. ANGRAU & PJTSAU, Hyderabad, 7–10 January 2015

Gupta JK, Sharma HK, Suman V et al (2015a) Studies on flowering phenology, supplemented pollination, pollinators diversity and their impact on fruit set in apple under changing climatic scenario in Kullu Valley of Himachal Pradesh. In: Compendium of abstracts of the 2nd international conference on bio-resource and stress management. ANGRAU & PJTSAU, Hyderabad, 7–10 January 2015

Gupta S, Kaushal R, Sood G et al (2015b) Studies on indigenous plant growth promoting rhizobacteria on suppressing *Ralstonia solanacearum* and promoting capsicum plant growth. In: Compendium of abstracts of the 2nd international conference on bio-resource and stress management. ANGRAU & PJTSAU, Hyderabad, 7–10 January 2015

Guru T, Veronica N, Thatikunta R et al (2015) Boron application as a physiological approach to improve Spikelet fertility and Grain density of Rice genotypes. In: Compendium of abstracts of the 2nd international conference on bio-resource and stress management. ANGRAU & PJTSAU, Hyderabad, 7–10 January 2015

Hemalatha V, Dilip Babu J, Siva Sankar A (2015) Influence of growth regulators on the shelf life of sweet orange cv. Sathgudi. In: Compendium of abstracts of the 2nd international conference on bio-resource and stress management. ANGRAU & PJTSAU, Hyderabad, 7–10 January 2015

Kachwaya DS, Chandel JS, Sarvade S et al (2015) Physiological performance and yield of strawberry (*Fragaria xananassa* duch.) cv. chandler as influenced by drip and furrow irrigation. In: Compendium of abstracts of the 2nd international conference on bio-resource and stress management. ANGRAU & PJTSAU, Hyderabad, 7–10 January 2015

Kalyani K, Sailaja V (2015) Effect of organics and inorganic Phosphorus on P uptake by soybean (*Glycine max* L.) on a high P Alfisol. In: Compendium of abstracts of the 2nd international conference on bio-resource and stress management. ANGRAU & PJTSAU, Hyderabad, 7–10 January 2015

Kameswari PL, Dhanumjaya Rao K, Girwani A (2015) Studies on effectiveness of plant growth retardants in retarding the growth of ornamental foliage plants. In: Compendium of abstracts of the 2nd international conference on bio-resource and stress management. ANGRAU & PJTSAU, Hyderabad, 7–10 January 2015

Kanjana D (2015) Effect of nanoformulated zinc sulphate with biopolymers on growth of cotton by foliar spray method. In: Compendium of abstracts of the 2nd international conference on bio-resource and stress management. ANGRAU & PJTSAU, Hyderabad, 7–10 January 2015

Klein AM, Vaissière BE, Cane JH et al (2007) Importance of pollinators in changing landscapes for world crops. Proc R Soc B Biol Sci 274:303–313

Krishnaveni V, Padmalatha T, Vijaya Padma SS et al (2015) Influence of Pinching and Plant Growth Regulators on Flowering, Yield and Economics of

Fenugreek (*Trigonella foenum-graecum* L.). In: Compendium of abstracts of the 2nd international conference on bio-resource and stress management. ANGRAU & PJTSAU, Hyderabad, 7–10 January 2015
Kumar R, Singh SP, Maji S et al (2015a) Influence of foliar spray of gibberellic acid (GA_3) on physic-chemical properties of cape gooseberry (*Phyalis Peruveana* L.). In: Compendium of abstracts of the 2nd international conference on bio-resource and stress management. ANGRAU & PJTSAU, Hyderabad, 7–10 January 2015
Kumar KSB, Guldekar VD, Illorkar VM (2015b) Nutrient cycling and distribution in different aged plantations of bamboo (*Dendrocalamus strictus*) in Vertisols of Nagpur. In: Compendium of abstracts of the 2nd international conference on bio-resource and stress management. ANGRAU & PJTSAU, Hyderabad, 7–10 January 2015
Mahajan RV, Bhale VM, Deshmukh JP et al (2015) Response of bio-stimulant for enhancement of growth, chlorophyll content index and yield of soybean. In: Compendium of abstracts of the 2nd international conference on bio-resource and stress management. ANGRAU & PJTSAU, Hyderabad, 7–10 January 2015
Mallareddy M, Veeranna G, Sri Jaya T et al. (2015) Low cost technology for higher phosphorus use efficiency in rice. In: Compendium of abstracts of the 2nd international conference on bio-resource and stress management. ANGRAU & PJTSAU, Hyderabad, 7–10 January 2015
Meena DC, Maji S, Meena JK et al (2015a) Improvement of growth, yield and quality of tomato (*Solanum lycopersicon* L.) cv. Azad T-6 with foliar application of zinc and boron. In: Compendium of abstracts of the 2nd international conference on bio-resource and stress management. ANGRAU & PJTSAU, Hyderabad, 7–10 January 2015
Meena JS, Verma HP, Shivran AC et al (2015b) Effect ofintegrated nutrient management on growth and yield of cowpea [*Vigna Unguiculata* (L.)Walp.]. In: Compendium of abstracts of the 2nd international conference on bio-resource and stress management. ANGRAU & PJTSAU, Hyderabad, 7–10 January 2015
Padmalatha T, Satyanarayana Reddy G, Chandrasekhar R et al (2015) Effect of foliar sprays of bioregulators on growth and flowering in *Gladiolus* plants raised from cormels. In: Compendium of abstracts of the 2nd international conference on bio-resource and stress management. ANGRAU & PJTSAU, Hyderabad, 7–10 January 2015
Pal R, Roy D, Das S et al (2015) Influence of different micronutrients yield and on growth, quality parameters in grape (*Vitis vinifera* L.). In: Compendium of abstracts of the 2nd international conference on bio-resource and stress management. ANGRAU & PJTSAU, Hyderabad, 7–10 January 2015
Pavankumar K, Padmalatha T, Pratap M et al (2015) Effect of plant bio-regulators on growth, flowering and seed yield in China aster (*Callistephus chinensis* L. Nees) cv. kamini. In: Compendium of abstracts of the 2nd international conference on bio-resource and stress management. ANGRAU & PJTSAU, Hyderabad, 7–10 January 2015
Pimpalpalle LP, Kadam AR, Ghawade PM et al (2015) Combined effects of giberellic acid (GA_3) and paraffin wax with modified atmospheric packaging (MAP) during storage at ambient temperature. In: Compendium of abstracts of the 2nd international conference on bio-resource and stress management. ANGRAU & PJTSAU, Hyderabad, 7–10 January 2015
Pradhan M, Tripathy P, Mandal P et al (2015) Effect of salicylic acid on growth and bulb yield of onion (*Allium cepa* L.). In: Compendium of abstracts of the 2nd international conference on bio-resource and stress management. ANGRAU & PJTSAU, Hyderabad, 7–10 January 2015
Priyanka I, Sudha Vani V, Bhagavan BVK et al (2015) Influence of denavelling and stalk end application of nutrients on yield and shelf life of banana cv. Dwarf Cavendish. In: Compendium of abstracts of the 2nd international conference on bio-resource and stress management. ANGRAU & PJTSAU, Hyderabad, 7–10 January 2015
Raghuveer M, Ram V, Ipsita Kar (2015) Impact of phosphorus levels and psb strains on soil microbial and enzymatic activities under acidic soils of north east India. In: Compendium of abstracts of the 2nd international conference on bio-resource and stress management. ANGRAU & PJTSAU, Hyderabad, 7–10 January 2015
Rajini SR, Narayana Rao K, Bharath Kumar KS (2015) Distribution of zinc forms in paddy soils of upper Krishna project command area of north Karnataka. In: Compendium of abstracts of the 2nd international conference on bio-resource and stress management. ANGRAU & PJTSAU, Hyderabad, 7–10 January 2015
Ratnakumari K, Nageswara Rao MB, Sudha Vani V et al (2015) Rooting response of cuttings of pomegranate (*Punica granatum* L.) cv. Bhagwa as effected by rooting media and IBA under shade net conditions. In: Compendium of abstracts of the 2nd international conference on bio-resource and stress management. ANGRAU & PJTSAU, Hyderabad, 7–10 January 2015
Sabnis PS, Mahadevaswamy, Nagraj SB (2015) Effect of Efficient isolates of pgpr on growth and yield of chilli (*Capsicum Annuum* L.) under pot culture conditions. In: Compendium of abstracts of the 2nd international conference on bio-resource and stress management. ANGRAU & PJTSAU, Hyderabad, 7–10 January 2015
Shankaraswamy J, Neelavathi R, Chovatia RS (2015) Study on exogenous environmental cues mimic of plant growth regulators, nutrients and pruning in early induction of flowering in mango (*Mangiferaindica* L.) cv. Kesar in India. In: Compendium of abstracts of the 2nd international conference on bio-resource and stress management. ANGRAU & PJTSAU, Hyderabad, 7–10 January 2015
Shivran AC, Choudhary I, Puniya MM (2015) Nutrient uptake, yield and soil fertility of isabgol (*Plantago ovata)* cultivation in response to integrated nutrient management and biofertilizers. In: Compendium of abstracts of the 2nd international conference on bio-

resource and stress management. ANGRAU & PJTSAU, Hyderabad, 7–10 January 2015

Swetha S, Padmalatha T, Dhanumjaya Rao K et al (2015) Effect of potting media on growth and quality of *Aglaonema*. In: Compendium of abstracts of the 2nd international conference on bio-resource and stress management. ANGRAU & PJTSAU, Hyderabad, 7–10 January 2015

Taksande S, Panchbhai DM, Parihar L (2015) Effect offoliar application of GA_3 and KNO_3 on flowering and yield quality of african marigold in winter season. In: Compendium of abstracts of the 2nd international conference on bio-resource and stress management. ANGRAU & PJTSAU, Hyderabad, 7–10 January 2015

Tzudir L, Bera PS, Chakraborty PK et al (2015) Diurnal variation in canopy temperature and stress degree day index of mungbean (*Vigna radiata*) varieties under different dates of sowing. In: Compendium of abstracts of the 2nd international conference on bio-resource and stress management. ANGRAU & PJTSAU, Hyderabad, 7–10 January 2015

Usha P, Bhagavan BVK, Sudha Vani V et al (2015) Aninfluence of plant growth regulators and antioxidants on *in vitro* multiplication of banana cv. Karpurachekkarakeli. In: Compendium of abstracts of the 2nd international conference on bio-resource and stress management. ANGRAU & PJTSAU, Hyderabad, 7–10 January 2015

Usharani T, Balaguravaiah D, Bharatha Lakshmi M et al (2015) Nutrient uptake of sugarcane plant and ratoon crops at different stages influenced by cane trash and mycorrhizae in presence of phosphorus levels. In: Compendium of abstracts of the 2nd international conference on bio-resource and stress management. ANGRAU & PJTSAU, Hyderabad, 7–10 January 2015

Vanisri S, Sreedhar M, Jeevan L et al (2015) Evaluation of rice genotypes for chlorophyll content and scavenging enzyme activity under the influence of mannitol stress towards drought tolerance. In: Compendium of abstracts of the 2nd international conference on bio-resource and stress management. ANGRAU & PJTSAU, Hyderabad, 7–10 January 2015

Walther GR, Post E, Convey P et al (2002) Ecological responses to recent climate change. Nature 416: 389–395

Biotic Stress: Diseases 7

Susanta Banik and Ratikanta Maiti

Abstract

The chapter makes a brief review of concerted research activities that have been directed to study the biology of each disease, its epidemiology, control measures, screening of crop cultivars for resistance to disease and mechanisms of resistance. The disease problems and control measures of various crops such as pulse, oil seeds, cereals and other crops and the screening techniques for a few pathogens in these crops are briefly discussed. It also discusses the compatibility of different agrochemicals on reduction of the cost of material application in combating diseases.

7.1 Introduction

The productivity of crops is affected greatly by disease infection in the absence of effective control measures. Concerted research activities have been directed to study the biology of each disease, its epidemiology, control measures, screening of crop cultivars for resistance to disease and mechanisms of resistance. Several papers are presented in Second International Conference on Bio-resource and Stress Management. We are narrating a brief summary of these.

S. Banik (✉)
Department of Plant Pathology, School of Agricultural Sciences and Rural Development, Nagaland University, Medziphema, Nagaland, India
e-mail: susanta.iari@gmail.com

R.K. Maiti
Facultad de Ciencias Forestales, Universidad Autónoma de Nuevo León, Carr. No 85, km 145, NL 67700 Linares, Mexico
e-mail: ratikanta.maiti@gmail.com

Plants synthesize enormous variety of compounds which are essential for its normal growth and provide plant its characteristic property which differentiates it within its own family members or in its class. These compounds are known as primary or secondary metabolites necessary for the normal growth of the plant or its ecological interaction with the environment. Among these compounds are the secondary metabolites, viz. isoprenoids or terpenoids, whose precursors are formed through the mevalonate and non-mevalonate or methylerythritol phosphate (MEP) pathway. The MEP pathway occurs only in bacteria, malarial parasite, green algae and higher plants whereas it is absent in humans and fungi. Various plants such as *Withania somnifera*, *Glycyrrhiza glabra*, *Salvia miltiorrhiza*, etc. are explored to study the secondary metabolic pathways. Furthermore, designing specific inhibitors of MEP pathway enzymes should give in herbicides and drugs with broad-spectrum antimicrobial activity without

R.K. Maiti et al. (eds.), *Bioresource and Stress Management*, DOI 10.1007/978-981-10-0995-2_7

toxicity to humans as these conserved proteins do not show homology with the mammalian proteins. Therefore, inhibiting or altering the structure of proteins involved in such pathways could provide road block for the growth of infectious prokaryotes (Pagoch 2015).

In plant protection schedule, combined applications of different agrochemicals are used due to several considerations. It is very economical to mix agrochemicals together to save on the cost of material application. Keeping the above in view, the compatibility experiment in vitro was conducted through poisoned food technique by mixing different fungicides in nutrient agar inoculated with *Pseudomonas fluorescens*. Among the 12 fungicides tested, triazoles, viz. tebuconazole, propiconazole and hexaconazole, are highly compatible with *P. fluorescens*, whereas the contact fungicides, viz. mancozeb and propineb, are less compatible with *P. fluorescens* by showing less growth (Hanuman et al. 2015).

7.2 Pulse Crops

Wilt in pigeon pea is considered as the most important soil-borne disease caused by *Fusarium oxysporum* f. sp. *udum* leading to major economic loss. *P. fluorescens* strains are widely employed in biological control of *F. udum* (Raghavendra et al. 2015). There are limited management options for this disease occurring in large area as the pathogen is soil-borne. *P. fluorescens* strains isolated from rhizospheric soils of pigeon pea from the farmer's fields of Southern Telangana Zone, Telangana State, India, were found to give promising results in vitro in terms of inhibition of radial growth and inhibition of conidial germination of the pathogen (Raghavendra et al. 2015).

Among other diseases of pigeon pea, *Phytophthora* blight has been gaining importance for the last few years in Gulbarga district in Northern Karnataka, India. The disease becomes severe in low-lying areas and particularly when the early crop growth coincides with the rainy season (June to September). In an experiment with fungicides and bioagents for management of the pathogens, it was found that seed treatment with sprays of Ridomil MZ 72 WP at 4 g kg^{-1} seed and at 4 g lit^{-1} at 30 and 45 days after sowing recorded the lowest mean % of disease incidence of 10.20 and maximum yield of 877.08 kg ha^{-1} (Jayalakshmi et al. 2015).

Chickpea is an important pulse crop in India. Wilt disease is one of the most important chickpea diseases caused by *F. oxysporum* f. sp. *ciceri* (FOC). Twenty-seven chickpea wilted samples from different chickpea-growing states of India were collected, and 18 isolates of FOC were isolated and identified based on morphological and molecular characteristics. The isolates of FOC were tentatively divided into five groups based on their pathogenic reaction as highly, strongly, moderately, weakly and non-pathogenic. The FOC isolates were tested for different reactions against a set of 12 host differentials. Eighteen isolates were classified into five groups on the basis of host differential reaction. Group 1 included isolate Akola, Nagpur, Parbhani, Badnapur, Rahuri, Hyderabad and Dharwad. Group 2 consisted of Kanpur, Allahabad and Varanasi isolates. Group 3 included only one isolate Gurdaspur. Similarly, group 4 comprised of isolates from Jabalpur, Rajnandgaon, Raipur and Delhi and that of group 5 included isolates from Nimboda, Udaipur and Bikaner (Rakhonde et al. 2015).

All 18 isolates of FOC were selected for SSR and ISSR analysis. Out of the 9 SSR primers screened 18 produced reproducible and scorable bands with high degree of polymorphism except MB-13. The SSR primers clustered all 18 FOC isolates into five main clusters according to race type. In the first subgroup race-1, seven isolates Akola, Badnapur, Nagpur, Parbhani, Rahuri, Hyderabad and Dharwad were combined in one cluster. The second subgroup race-4 consisted of virulent isolates, Jabalpur, Rajnandgaon, Raipur and Delhi. The third subgroup race-2 consisted of virulent isolates, Kanpur, Allahabad and Varanasi. The fourth group race-4A consists of Nimboda, Udaipur and Bikaner, whereas Gurdaspur isolate was identified as race-3. The isolates were further analysed by using inter simple sequence repeats (ISSR) primers. The unweighted paired group method with arithmetic average grouped the isolates into five categories same as obtained by SSR analysis at a genetic similarities ranging from 50 % to 94 %. The present race profiling for the Indian population of the pathogen and its distribution pattern is entirely new. The racial pattern obtained during this study was altogether different from what has been reported earlier (Rakhonde et al. 2015).

Collar rot disease of chickpea caused by *Sclerotium rolfsii* Sacc is a soil-borne disease that causes considerable damage to the plant stand. A study was undertaken to test six different antagonists in vitro and in pot culture conditions in an attempt to find out a potential biocontrol agent for *S. rolfsii*. The best inhibition of *S. rolfsii* was obtained with *Trichoderma* spp. (PSJ) followed by *T. harzianum* (Indore). In pot culture, most effective species of *Trichoderma* were *Trichoderma viride* (Indore) and *T. harzianum* (Indore) against *S. rolfsii* as the antagonists increased the percentage of seed germination and minimized the pre- and post-emergence mortality. Leaf extracts of eight medicinal plants were also evaluated under in vitro conditions. The radial growth of *S. rolfsii* was significantly less in the presence of leaf extract of Ashwagandha followed by Van-tulsi. Among the synthetic chemicals or fungicides, complete inhibition of mycelial growth of *S. rolfsii* was observed with Hinosan, Antracol, Mancozeb, Kitazin and Kavach, in vitro. These fungicides also increased seed germination and decreased pre- and postemergence mortality under pot culture conditions (Jayale et al. 2015).

Pea (*Pisum sativum* L.) is one of the most widely grown pulse crop in the world. Pea crop suffers from many diseases, and among them *Fusarium* wilt of pea caused by *F. oxysporum* f. sp. *pisi* is an important disease. Twenty isolates of *F. oxysporum* f. sp. *pisi* were collected from major pea growing tracts of UP (India), and physiological studies of the representative isolates of three groups were made with six different fungicides at 0.1 % concentration. The minimum mean radial growth (2.00–2.99 mm) was recorded in group-C of isolates F8, F11, F12 and F13 and maximum (4.00–4.99 mm) in group-A consisting of F10, F17 and F19 isolates. The maximum micro-conidia were produced by group-A (2.60–2.99 million ml^{-1}) of isolates F10, F17 and F19, while that in group-C of isolates F8, F11, F12, F13 and F15 have minimum micro-conidia (1.00–1.99 million ml^{-1}). The maximum macro-conidia were formed in group-A (0.70–0.79 million ml^{-1}) of F5, F10, F16, F17 and F19. The minimum macro-conidia were produced in group-C (0.40–0.59 million ml^{-1}) of F8, F9, F11, F12, F13, F14 and F15 (Dabbas et al. 2015).

There is considerable loss of yield reported in cowpea caused by bacterial leaf blight (*Xanthomonas campestris* pv. *vignicola*) in India. Due to the low yield of the crop and neglected conditions under which it is grown especially under rain-fed situation, plant protection measures are uneconomical. Therefore, resistant breeding appears to be the most important approach in disease management. For developing high-yielding bacterial leaf blight (BLB)-resistant varieties of cowpea, knowledge of the inheritance of resistance to BLB is of prime importance in cowpea breeding programmes. The inheritance pattern was studied in the cross C-152 (susceptible) × V-16 (resistant) using % disease incidence in F_1 and other segregating generations, viz. F_2, F_3, BC_1P_1 and BC_1P_2. The expression of resistance reaction in F_1 generation is an indication of the role of dominant gene in controlling BLB in cowpea. Based on the classification of plants into resistant and susceptible in F_2 segregating generation of the cross C-152 × V-16, a good fit of 3:1 (242 resistant, 91 susceptible) ratio revealed the role of major dominant gene for field resistance. The F_3 population of this cross showed 1:2:1 ratio (52 resistant, 92 segregating, 36 susceptible) and back cross generations BC_1P_1 and BC_1P_2 segregated in 1:1 (22 resistant, 27 susceptible) and 1:0 (52 resistant, 2 susceptible) test ratio, respectively. The segregation pattern in different generations showed very good fit with the expected ratio confirmed the role of single dominant gene in controlling resistance to cowpea BLB (Khan and Viswanatha 2015).

Lentil, an important pulse crop, is affected by wilt disease which is the most important disease of lentil plant caused by *F. oxysporum* f. sp. *lentis*. The plants are infected from seedling stage to later stages of their development. Several lentil lines were collected from lentil-growing areas for the purpose of resistance screening against the wilt pathogen. The collected lines were screened with a local isolate of the pathogen under greenhouse conditions, and significant differences in disease incidence were observed among them. In the greenhouse, test plants were inoculated by immersing root tips in spore suspension and sowing seeds in pre-infested pot soil. Field screening of the test entries was carried out in a naturally highly infested farm. At all stages, the plant response to the disease was based on the percentage of dead plants. Cluster analyses of the greenhouse and field data led to the selection of three lines that were resistant under greenhouse and field conditions (Puri et al. 2015a).

7.3 Oilseeds

Sunflower, an important oilseed crop in India, is limited by the lack of genetic variability and susceptibility to diseases. Powdery mildew disease caused by *Golovinomyces cichoracearum* is emerging to be a serious constraint in the cultivation of sunflower crop especially in warmer sunflower growing region causing around 20–25 % yield loss. The past decade has evidenced this disease regularly during *rabi*-summer seasons. Under severe conditions infection ranges from the cotyledonary leaves up to ray florets. Host plant resistance in released hybrids and parental lines is limited. Identification of resistance source would be the immediate approach especially from the germplasm lines and promising CMS (cytoplasmic male sterility) lines utilized in pre-breeding. Sunflower germplasm comprising of 120 accessions was screened under artificial inoculation conditions. Among 120 accessions screened, 2 restorer lines were identified to be resistant and 48 accessions were categorized as medium resistant. Of the remaining lines, 63 were susceptible and 7 accessions were highly susceptible. The microscopic observations of conidia stained with lactophenol blue were also in accordance with host reaction. Highly susceptible check 'Morden' showed 468 conidia per microscopic field compared to resistant accessions like R-GM-41 and R-GM-49 recording 52.2 and 56.8 conidia per microscopic field, respectively. The fungal conidial count ranged from 61.1 to 129 in medium-resistant accessions and more than 367 conidia per microscopic field in all highly susceptible accessions. The study indicated the scarcity of resistant sources in cultivated sunflower and the practical utility of these identified sources of resistance for breeding powdery mildew-resistant sunflower hybrids (Kulkarni et al. 2015).

In another study, 38 advanced breeding genotypes were evaluated for powdery mildew disease reaction under natural epiphytotic (field) condition during *rabi* 2013. Each genotype was sown in two rows of 3 m length with two replications. The percentage of leaf area covered by white powdery conidial spores is considered for scoring the genotypes as per the standard rating 0–9 scale at 15-day interval from 30 DAS. Among 38 genotypes screened, 8 genotypes were found to be resistant with less than 10 % PDS and 13 genotypes showed medium-resistant reaction (PDS = 10–25 %) in field screening. Subsequently, resistant and medium-resistant genotypes were screened in greenhouse condition artificially by spraying fungal spore suspension in 1 % sucrose solution at 30 and 45 DAS. Later observations were taken from bottom, middle and top leaves at 15 days interval starting from 30 DAS up to 90 DAS. In artificial screening conditions, only two genotypes (RCR-1947/3-2-9 and RCR-1947/4-1-2) were found to be resistant (<10 % PDS), whereas the remaining six genotypes were categorized as medium resistant (>10 % PDS) revealing the importance of artificial screening conditions. However, 13 genotypes remained in medium-resistant category with less than 25 % PDS. In both field and artificial screening, the susceptible check Morden recorded highly susceptible disease reaction to powdery mildew (>50 % PDS) (Supriya et al. 2015).

Stem or pod rot caused by *S. rolfsii* is an important disease in groundnut. Seven bioagents tested in vitro showed significant mycelial growth inhibition of *S. rolfsii*. However, *T. harzianum* gave significantly higher mycelial growth inhibition (78.37 %): followed by *T. viride* (74.70 %)

and *T. hamatum* (73.96 %). The rest of the bioagents showed 61.22–72.22 % mycelial inhibition of the test pathogen. All seven bioagents tested in vitro were further evaluated (alone and in combination) as pre-sowing seed treatment (protective), post-sowing soil application (curative) and combination of protective + curative methods against *S. rolfsii* (sick soil), sowing susceptible sunflower cv. JL-24 in pot culture under screen house conditions. The three methods of bioagent treatments have been applied: protective + curative was found the most effective, followed by the protective and curative one. Significantly highest seed germination of 93.33, 83.33 and 71.67 % was observed with the bioagent *T. harzianum* + *P. fluorescens* in protective + curative, protective and curative methods, respectively (Kuldhar et al. 2015).

Soybean crop is affected by many fungal, bacterial and viral diseases. Among different fungal diseases, wilt of soybean caused by *F. oxysporum* f. sp. *glycine* is the most destructive disease reported to cause significant yield losses. Seven fungicides, viz. metalaxyl (0.2 %), copper oxychloride (0.3 %), thiram (0.3 %), carbendazim (0.1 %), benomyl (0.2 %), bordeaux mixture (1 %) and a biocontrol agent *T. harzianum*, were evaluated in vitro against *F. oxysporum* f. sp. *glycine*. The fungicide benomyl (0.2 %) recorded maximum growth inhibition (100 %) of the test pathogen and was found significantly superior to the rest of the fungicidal treatments. The next best treatment was found to be the bioagent *T. harzianum* (82.60 %) with mean colony diameter 15.66 mm. It was followed by carbendazim (77.77 %), thiram (77.04) and copper oxychloride (76.3 %), which gave mean colony diameter of 20.00 mm, 20.66 mm and 21.33 mm, respectively, and these treatments were statistically at par with each other. Metalaxyl and bordeaux mixture showed least growth inhibition of 65.55 % and 34.44 % of the test pathogen, respectively (Puri et al. 2015b).

Powdery mildew of linseed (*Linum usitatissimum* L.) caused by *Oidium liniskoric* is one of the major biotic constraints in successful cultivation of linseed. The disease causes dropping of flowers and chaffy seed setting leading to serious loss in the linseed production. In a study, to manage the disease by exploiting cheaper and effective means, the potential of sulphur application in soil and on foliage was examined. It was found that experimental plots with two sprays of wettable sulphur (0.4 %) (sulphex) gave minimum PDI of 13.50 with maximum yield (554.67 kg ha^{-1}) and B:C ratio (2.68) which was significantly superior to all the other treatments. The next best treatment was soil application of sulphur (30 kg ha^{-1}) through gypsum at the time of sowing followed by two sprays of 0.3 % wettable sulphur (sulphex) (PDI, 20.83; yield, 526 kg ha^{-1}; and B:C ratio, 2.44) (Ajithkumar et al. 2015).

Niger is a minor oilseed crop cultivated in India. It constitutes about 3 % of Indian oilseed production. Niger seed contains 35–45 % oil, 20 % protein and 12 % soluble sugars. Both seed and oil are edible and consumed by rural and tribal people. The productivity of niger crop is low owing to its cultivation in unproductive land and lack of high-yielding and fertilizer-responsive, disease- and pest-resistant varieties. Niger is susceptible to diseases like root rot and *Alternaria* leaf spot. In order to develop a resistant variety for *Alternaria* leaf spot, a study was undertaken to evaluate the germplasm comprising of 200 lines for disease resistance during *kharif* 2013. Each germplasm line was sown in a single line. The same were evaluated for seed yield and other ancillary characteristics, viz. days to 50 % flowering, days to maturity, plant height, number of primary branches, number of capitula and seed yield. The 0–5 scale was used to score the disease. Thirty-two lines are found to be immune as these didn't show any infection (i.e. 0 % infection). The 100 lines were scored 1 by showing 1–10 % disease incidence and are found to be resistant. The remaining germplasm (57) lines were scored 2 by showing 11–25 % disease incidence and are moderately resistant. None of the genotypes were found to be susceptible (Suvarna et al. 2015).

Castor (*Ricinus communis* L.) is an important nonedible industrial oilseed crop grown in tropical and subtropical regions across the world. *Fusarium* wilt caused by the fungus *F. oxysporum* f. sp. *ricini* is an economically important disease of castor (*Ricinus communis* L.). Identification of best male combiners coupled with wilt resistance is one of the important objectives in castor hybrid breeding programme. In this study, 42 castor male combiners were screened for 2 years in a permanent wilt sick plot at the Regional Agricultural Research Station, Palem, India, to identify potential sources of wilt resistance. The inoculum load in the wilt sick plot was maintained at 2.1 × 103 colony-forming units per gram of soil, and care was taken to avoid water logging and to suppress nematode population. The susceptible and resistant checks, viz. Kranthi and Haritha, recorded >91 % and <5 % wilt incidence, respectively. Among the tested genotypes, eight castor male lines, viz. PCS 301, 302, 303, 304, 315, 322, 334 and 342, were selected as resistant with <20 % wilt incidence, and eight lines were found moderately resistant with <50 % wilt incidence (Ramya et al. 2015).

Breakdown of *Fusarium* wilt resistance has been observed in castor necessitating the development of durable wilt-resistant cultivars. An understanding on the host-pathogen interaction at microscopic level and how the resistance expression affects *F. o.* f. sp. *ricini* colonization is lacking. Though there is limited knowledge on the genes involved in resistance expression in castor, there is no information on the genes conveying virulence in the pathogen. Marker-assisted selection following the development of reliable markers for tagging of wilt resistant genes will hasten the breeding programme in evolving wilt-resistant cultivars. A review has been made by Kumar et al. (2015) on the progress made along with future outlook towards *Fusarium* wilt resistance in castor.

7.4 Vegetables

A survey was undertaken for two years during September 2011 and 2012 on diseases of tomato grown in semiarid regions of Karnataka, India. The crop was found to be affected by different diseases in both years. While late blight (*Phytophthora infestans*), *Septoria* leaf spot (*Septoria lycopersici*), bacterial spot (*Xanthomonas campestris* pv. *vesicatoria*), sclerotium wilt (*S. rolfsii*) and buck eye rot (*Phytophthora nicotianae* var. *parasitica*) appeared only during 2011, powdery mildew (Oidium stage of *Erysiphe lycopersici*) was prevalent during 2012. Early blight, leaf curl (tomato leaf curl virus) was observed in both years of the surveyed region. Bud blight (peanut bud necrosis virus) was recorded only from Bellary district in both the years. Bacterial spot, buck eye rot and late blight and powdery mildew were recorded from Davangere and Belgaum districts, respectively. Amount of rainfall, its distribution and dry spell appeared to influence disease diversity and its intensity (Bhat et al. 2015).

Tomato is under the constant threat of diseases and about 200 diseases are known to infect tomatoes worldwide. Among these, cucumber mosaic virus (CMV) is the most devastating disease, as it can completely destroy the crop. The characteristic field symptoms of CMV disease include stunting, yellowing, mottling of leaves, extreme filiformity or shoe stringing of leaf blades, depending on virus strain and the host. Host range studies conducted on 29 plants and 18 weed species belonging to 13 families for their susceptibility to CMV through sap inoculation under insect-proof glasshouse conditions at National Bureau of Plant Genetic Resources Regional Station, Rajendranagar, Hyderabad, during rainy season 2013–2014 indicated that 20 plants and 17 weed species expressed visible symptoms. Some of the test plants (crop species or weeds) produced local symptoms, viz. chlorotic spots, chlorotic ring spots, brown necrotic spots, necrotic spots, necrotic ring spots, necrotic local lesions, mottling, mosaic and yellow mosaic, and some exhibited systemic symptoms, viz. necrotic spots, mottling, systemic mosaic and mosaic mottle. The members of the families belonging to Aizoaceae, Chenopodiaceae, Cyperaceae, Gramineae, Malvaceae and Portulacaceae, showed only local symptoms, whereas the members of families Amaranthaceae, Asteraceae, Cucurbitaceae, Euphorbiaceae, Fabaceae and Solanaceae exhibited both local and systemic symptoms. *Datura stramonium*, *Solanum melongena*, *Helianthus annuus*, *Citrullus lana-*

tus, *Arachis hypogaea*, *Cajanus cajan*, *Phaseolus vulgaris*, *Vigna unguiculata*, *Abelmoschus esculentus* and *Zea mays* failed to exhibit both local and systemic symptoms (Jalender et al. 2015).

Among the fungal diseases infecting tomato crops, early blight caused by *Alternaria solani* (Ellis and Martin) Jones and Grout is one of the most catastrophic diseases, causing accountable losses. A study was undertaken on the pathogenic, cultural, morphological and molecular variability among the isolates of *A. solani*, if any. The results revealed all of the eight isolates of *A. solani* as pathogenic to tomato (cv. Pusa Ruby), showing variability among them. The test isolates could grow better on the basic culture medium potato dextrose agar (PDA); however, highest mycelial growth recorded of the isolate AsLt (88.50 mm), followed by AsBd (82.36 mm) and AsHl (78.40 mm), with excellent sporulation. All of the eight test isolates showed a wide range of variability in respect of their mycelial and conidial dimensions and septation also. RAPD-PCR analysis of the four most virulent *A. solani* isolates, using 13 OPA primers, revealed that the isolates AsBd (Beed) and AsLt (Latur) were closely related with 85 % genetic similarity, whereas the isolates, AsHl (Hingoli) and AsJl (Jalna), were closely related with 50 % genetic similarity, but distinct from that of AsLt and AsBd isolates (Nikam et al. 2015a).

In a further study on growth media of early blight pathogen of tomato *A. solani*, 12 culture media were tested. Of them, potato dextrose agar (PDA) supported best mycelial growth of *A. solani* (88.50 mm), followed by tomato leaf extract agar (84.22 mm), tomato fruit extract agar (82.52 mm), Richards's agar (82.42 mm) and Czepek's dox agar (78.22 mm). The excellent (++++) sporulation was observed in PDA, oat meal agar, tomato leaf extract agar and Richards' agar. The maximum mycelial growth and sporulation was detected at temperature 25 °C, followed by 30 °C and 20 °C. Maximum mycelial growth (88.58 mm) of the test fungus was recorded at pH 7.0, followed by pH 6.5 (78.44 mm) and 7.5 (76.52 mm). The excellent (++++) sporulation was observed at pH of 7.00. The best mycelial growth was recorded at alternate dark and light (88.72 mm), followed by continuous dark (82.34 mm). The excellent (++++) sporulation was observed at continuous dark condition. Among the eight carbon sources tested, highest mycelial growth was observed in starch (74.33 mm), followed by sucrose (58.88 mm) and maltose (56.78 mm). Carbon sources, viz. sucrose, maltose and glucose, recorded excellent (++++) sporulation. Of the eight nitrogen sources tested, highest mean mycelial growth and excellent (++++) sporulation were recorded with potassium nitrate (89.00 mm), followed by magnesium nitrate (87.00 mm) and sodium nitrate (86.40 mm) (Nikam et al. 2015b).

The pathogen *A. solani* also causes early blight disease of potato. In a field study on integrated management of early blight disease of potato, fungicides (spray) were found most effective in checking the disease, followed by the bioagents and the botanicals. Of the treatments, fungicide, iprodione caused highest reduction (67.04 %) in the disease incidence. This was followed by mancozeb (59.43), SAFF (53.37 %) and chlorothalonil (51.14 %). Less than 50 % reduction in disease incidence was recorded with carbendazim (47.48 %) followed by hexaconazole (42.29 %), tebuconazole (41.47), the bioagent *T. harzianum* (35.03 %) and botanical *Azadirachta indica* (32.39 %). Botanical *Zingiber officinale* and bioagent *P. fluorescens* were found less effective. Highest tuber yield was obtained on fungicide iprodione with (350.26 q ha^{-1}) and over unsprayed control (yield, 150.20 q ha^{-1}) with significantly less mean *Alternaria* blight incidence (16.16 %) and intensity (21.96 %). The second and third best fungicides found were mancozeb (at 0.2 %) and SAFF (0.25 %) which recorded tuber yield of 300.61 q ha^{-1} and 288.61 q ha^{-1}, respectively (Wagh et al. 2015).

Tomato (*Solanum lycopersicum*) is one of the important vegetables in Sikkim, India. Among various diseases reported for tomato in Sikkim, late blight caused by *Phytophthora infestans* (Mont.) de Bary is one of the most serious diseases causing significant yield and quality losses. To manage the disease organically, field experiment was undertaken during 2013 in two separate plots. The treatments included cow urine 10 %, garlic bulb extract 5 %, copper oxychloride 0.3 %, copper hydroxide 0.3 %, onion bulb extract 5 %, *Artemisia* leaf extract 10 % and *T. viride* 0.2 % and *P. fluorescens* 0.2 % along with metalaxyl 0.1 % and untreated control. The symptom of the disease first appeared 53 days after transplanting. The treatments were given after the initial appearance of the disease. Among the treatments, copper oxychloride was most effective with % disease intensity (PDI) of 27.20 followed by copper hydroxide (33.87 PDI) over all other treatments, and they were at par with metalaxyl 0.1 %. It was found that the yield of tomato was highest in the plots treated with copper oxychloride (38.33 t ha^{-1}) and copper hydroxide (36.00 t ha^{-1}). It was also found that if plant protection measures are not given in time, the yield reduction varied from 97.39 to 100 % (Gopi et al. 2015).

Bacterial wilt caused by *Ralstonia solanacearum* is regarded as one of the world's most important bacterial plant pathogens because of its aggressiveness, large host range, broad geographical distribution and long persistence in soil and water environments. *R. solanacearum* constitutes a serious obstacle to the cultivation of the economically important brinjal among other crops, causing total damage of plantations before as well as after bearing fruits. This soil-borne bacterium enters plant roots, invades the xylem vessels and spreads rapidly to aerial parts of the plant through the vascular system. Typical disease symptoms include browning of the xylem, foliar epinasty and lethal generalized wilting and can finally lead to plant death. Cultural practices and chemicals, if judiciously used, can reduce disease incidence and severity, but alone are expensive and ineffective. The use of resistant varieties is an effective control strategy to manage bacterial wilt. Fifty genotypes of brinjal were screened in a polyhouse for resistance to bacterial wilt caused by *R. solanacearum* by artificial inoculation of bacterial suspension both by soil drenching and axil puncturing method during *rabi* 2011. Among the 50 brinjal genotypes, four genotypes, viz. Arka Keshav, Surya, Arka Neelkanth and Arka Nidhi, were found resistant. The resistant genotypes identified may be used for developing brinjal varieties resistant to bacterial wilt (Sadarunnisa et al. 2015).

Mushroom is a more preferred choice of vegetable owing to its immense health benefits. Though it is grown indoors, it is affected by several biotic factors. A study was undertaken on the impact of disease incidence on yield attributes of oyster mushrooms (*Pleurotus* spp.), and its management through effective fungicides in three different seasons was carried out during the years 2012 and 2013. The results revealed that the association of different fungal species, viz. *Aspergillus niger* (1–5 %), *Aspergillus flavus* (5–7.5 %), *Aspergillus fumigatus* (1–2 %), *Coprinus* spp. (5–19 %), *Cladobotryum* (1–3 %), *Gliocladium virens* (0.5–1 %) and *Trichoderma* spp. (7–10 %), was observed in different mushroom species. The presence of total fungal flora varied from 7.8 to 26.8 % in three different seasons. Irrespective of the seasons, maximum occurrence of fungal flora was observed in pink oyster mushroom, *Pleurotus* (9.15 %). Mean temperature and mean relative humidity ranged from 25 to 28.5 °C and 75–90%, respectively. The differences in the occurrence of fungal flora in different mushroom species may be attributed to the variations in temperature, and relative humidity prevailed during different seasons. Complete elimination of the fungal flora was observed in mushroom beds sprayed with carbendazim at 0.05 % followed by benomyl at 0.01 %. Mean yield of oyster mushrooms ranged from 0.54 to 1.32 kg. Among the seasons, mean yield was higher during winter and rainy seasons (1.32 and 1.0 kg, respectively) with lesser fungal flora (7.8 % and 18.1 %, respectively). During summer season, the mean yield was found less (0.54 kg only) due to maximum occurrence of fungal flora (26.8 %) coupled with high-temperature and low relative humidity. Cultivation of oyster mushrooms was preferred during rainy and winter seasons as compared to summer season (Rajeswari and Devi 2015).

7.5 Cereals

About 400 genotypes of finger millet derived from the cross between a highly blast-susceptible VR 708 and blast-resistant GPU 48 were screened for blast resistance. Data were recorded on all quantitative and qualitative characters along with scoring of leaf blast, neck blast and finger blast and concluded that the leaf blast ranged from 1.00 to 4.00 with a mean of 1.94, the neck blast ranged from 7.20 to 52.10 with a mean of 21.18 and the finger blast ranged from 7.05 to 48.25 with a mean of 21.43. For leaf blast 113 lines, for neck blast nine lines and for finger blast 3 lines showed resistance to blast pathogen. So these identified resistant lines may serve as potential parental genotypes for future breeding programmes to develop desirable stable segregants for finger millet crop improvement programmes (Patro et al. 2015).

Rice genotypes were screened against neck blast during *rabi* 2011–2013 under field conditions by standard protocols of IRRI, Philippines. The test entries were artificially inoculated with highly virulent blast pathogen by spraying spore suspension in order to ensure high disease pressure. Of the 758 entries tested, 18 entries, viz. HRI –173, RP 5130-12-3-5-21-3, RTN 8-4-2-1-2, RH-1531, OR 2172–7, HUR-913, NDR-370135, NPG-209, PAU 3761-26-3-1, NP-3112, WGL-451, 27 P 52, NDR 6311, CR 2656-11-3-4-2, CR 2304-5-3-7-1, NP 5031, TRC 2008–4 and CN 1729, recorded score 1 and were found resistant against neck blast during 2011. Out of 818 genotypes screened, eight entries, viz. RGL 7004, CR 2251-1-1-1-1-3, NLR 20146, NLR 20131, OR 2391–3, CR 2934–35, CR 3607-2-2-1-2 and CRL 67-27-1-1-1, recorded score 0 and found immune; ten entries, viz. MAS 9, UPR 3425-14-3-1, US 315 (Hybrid), Lalat, Savitri, CR 3603-7-2-1-1-3, CRL 67-118-1-3-1, CR 3609-1-2-1-1-1, AD 07250 and OR 2394–3, recorded score 1 and were found resistant against neck blast during 2012. Out of 979 entries tested, three entries, viz. NLR 3091, AD 07260 and NLR 3140, recorded score 0; 14 entries, viz. NP 9381, CRG 1190-2-2, OR 2361–1, NP 9380, OR 2325–12, CR 2713–179, HKR 08–1, WGL 536, HKR 08–62, PR-113, HKR 08–29, WR 37-2-1-1, MCM 104 and MTU 1112, recorded score 1 and were found resistant to neck blast during 2013 (Bhuvaneswari et al. 2015).

A novel compound GALILEO® (picoxystrobin 25SC) which is considered to be an effective tool against the rice blast disease has been tested. Picoxystrobin is a fungicide belonging to the strobilurin group of chemicals. The redistribution properties of picoxystrobin are unique among commercial strobilurin fungicides and those in development, because it shows both surface diffusion activity and systemic action through xylem movement. Picoxystrobin possesses characteristics of rapid absorption, diffusion in the cuticular waxes, systemic and translaminar movement, redistribution by surface diffusion and protection to the new growth. These complimentary properties give a uniform protection of the plant during its growth. Results from across the regions in South Asia showed the superiority of this product in managing the disease when compared with the existing standards. GALILEO® (picoxystrobin 25SC) at 150 g a.i. ha^{-1} has shown superior efficacy for the management of leaf, collar and neck blast with highest grain yield and better grain quality disease when compared with existing market standards (Bhaik et al. 2015).

Turcicum leaf blight (TLB) caused by *Exserohilum turcicum* (Pass.) Leonard and Suggs is a major foliar disease of maize in India. A study was conducted to evaluate different chemicals and biocontrol agent *P. fluorescens* against TLB under field conditions during rainy seasons of 2009 and 2010 with seven treatments, viz. *P. fluorescens* at 5 g l^{-1} spray, propiconazole at 1 ml l^{-1}, salicylic acid at 100 ppm, mono-potassium phosphate (MPP) at 10 g l^{-1}, mancozeb at 2.5 g l^{-1}, carbendazim + mancozeb at 2 g l^{-1} and untreated control with test variety Pioneer 30v92. Three foliar sprays, at 15 days interval, were given starting from the first appearance of the disease. It was found that propiconazole and salicylic acid were on par with % disease intensity (PDI) of 26.6 and 31.33 respectively, followed by *P. fluorescens* spray (34.66 PDI). Mancozeb, carbendazim + mancozeb and MPP treatments recorded PDI of 42, 42.66 and 38.66, respectively. Induction of systemic resistance was evaluated by assaying defence-related peroxidase enzyme activity, which was highest in salicylic acid treatment (340 U ml^{-1}) followed by *P. fluorescens* (325 U ml^{-1}). Propiconazole sprays has recorded significantly higher yield (10,333 kg ha^{-1}) followed by carbendazim + mancozeb (9333 kg ha^{-1}) as against control plots (Madhavi et al. 2015a).

Variability among the population of banded leaf and sheath blight fungus *Rhizoctonia solani* in Andhra Pradesh was assessed from 27 isolates collected from different maize varieties and 1 from rice grown in various regions of Andhra Pradesh for easy breeding task. These isolates were analysed for their morphological and genetic characterization. Morphological characterization revealed that majority of the isolates was fast growing with raised and fluffy colonies. The hyphal width varied from 5.00 (RS 2, RS20) to 7.97 μm (RS10) of rice isolate (7.34 μm). The isolates RS7, RS8, RS9 (Karimnagar), RS10, RS11, RS12 (Khammam), RS16 (Krishna) and RS26 (Kurnool) and one rice isolate RS28 produced barrel-shaped monilioid cells. The clamp connection was present only in RS1 isolate from Pragnapur Mandal of Medak district. The three isolates RS16, RS17, RS18 from Krishna district and rice isolate RS28 recorded maximum colony growth of 90.00 mm within 48 h of incubation. The colour of most isolates varied from light grey to brown, while the rice isolate showed a typical yellowish brown shade. The sclerotial number per 5.00 mm culture disc of test isolates ranged from (7.00) to (17.67). Their size varied between 0.21 μm (RS 24) and 1.57 μm (RS11). The rice isolate RS28 recorded highest (19.00) sclerotial count compared to that of maize isolates. Sclerotial colour in all isolates except for RS 24 was yellowish brown to dark reddish brown, whereas the rice isolate was dark yellowish brown. Most of the sclerotia were found scattered in the colony and highest sclerotial aggregation was observed in RS11. The sclerotial production was surface and aerial in most of the isolates including rice isolate, while it was aerial in RS2. Together, by using principal component analysis (PCA), the divergence of morphological, cultural and sclerotial characters of isolates was assessed and grouped into nine similar groups. Molecular characterization of genetic diversity (RAPD markers) of the test isolates with UPGMA exhibited two major clusters with 66–92 % genetic similarity. Though most of the isolates showed grouping specific to the host variety, they did not exhibit any relationship between morphological and genetic similarities (Madhavi et al. 2015b).

Grain mould of sorghum is a serious global disease affecting grain quality and its acceptability to the consumer. Grain mould pathogen infects at the time of flowering and causes severe losses when post-flowering wet weather prevails. *Fusarium*, *Curvularia*, *Alternaria* and *Bipolaris* spp. are some of the fungal pathogens that cause floret infection. Contrary to fungi that infect sorghum floret, fungi infecting grains (and cause grain weathering) are saprophytes and occurrence of the later is extremely weather dependant. Studies have revealed that frequency of floret infection was about 25 % with natural inoculums. Floret or milk stage grain infection frequency played significant role in deciding grain mould severity on postmaturity sorghum. Therefore, management of floret infection might cause ultimate grain mould severity. Genotypic variability in floret infection has been studied. The use of host resistance and fungicides has been reported for effective management of floret infection in sorghum. Spraying of fungicides at anthesis significantly reduced seed-borne infection at milk stage and grain mould severity at maturity. Propiconazole was found the most effective for minimization of floret infection. Through this new approach, more than 60 % each of *Fusarium* spp., *Curvularia* spp. and total fungal infection could be reduced on milk stage grain compared to control. Visible mould score was also significantly reduced from 7.0 to 4.0 on matured grain, thus producing clean grain (Das et al. 2015).

7.6 Other Crops

F. oxysporum f. sp. *zingiberi* causing yellows of ginger is one of the most important pathogens of ginger in Himachal Pradesh, India. Studies were carried out to see morphological variations with respect to mycelial colour, conidial size and formation of chlamydospores and pathogenic variation in terms of disease incidence among different isolates. The mycelial colour varied from white to dull white with slightly pinkish tinge. The microconidial size varied from 3.16 × 3.16 μm (I19) to 9.13 × 5.44 μm (I7), whereas macroconidial size varied from 11.77 × 3.16 μm (I19) to 24.60 × 5.91 μm (I7). Chlamydospore dimensions also varied in all the 19 isolates of the test pathogen. All isolates formed chlamydospores on PDA medium. Pathogenic variability was revealed in terms of lesion size that varied from 8.50 to 18.00 mm after 10 days of inoculation and incubation period varied from 11 to 19 days. Genetic variation was also analysed by using 40 ten-mer oligonucleotide RAPD primers, out of which two primers yielded informative, strong and reproducible DNA amplicons of *F. oxysporum* f. sp. *zingiberi* and thus were selected for studying the variation among isolates. The dendrogram produced from computerized cluster analysis of the DNA fingerprints revealed 0–80 % variation among isolates as expected after DNA banding pattern analysis of gel image. All isolates were grouped into two different major groups, each comprising of ten and nine isolates (Gupta 2015).

T. harzianum (K2 isolate) was selected for the evaluation against chilli wilt caused by *Fusarium solani* in two areas of north Kashmir (Pattan and Magam). These areas are found to be hot spots of *Fusarium solani* causing diverse losses to chilli crop. Combinations of different locally available carriers with different proportion were found effective in maintaining colony-forming unit (cfu) load for the period of 120 days. The maximum inoculum load of 6.22×10^8 cfu g^{-1} was observed in cow dung-charcoal-molasses (3:2:1 v/v) which was followed by cow dung-ash-molasses (3:2:1 v/v) with an inoculum load of 6.18×10^8 cfu g^{-1} formulation after 30 days of storage. After 120 days, cfu count dramatically decreased in all the combinations with maximum load of 6.2×10^6 cfu g^{-1} observed in cow dung-charcoal-dalweed-soil (3:2:1:1 v/v) followed by cow dung-ash-dalweed-soil (3:2:1:1 v/v) with an inoculum load of 5.62×10^6 cfu g^{-1} of formulation. In the field evaluation of *Trichoderma* sp. K2, seed treatment was found effective in managing chilli wilt (37 %) followed by seedling dip (29 %). Lowest disease control (12 %) was found with the seed treatment with carbendazim (0.1 %) followed by seedling dip in captan (0.2 %). However, disease management was found highest at Magam area of Kashmir as compared to Pattan (Mohiddin et al. 2015).

Grey mildew of cotton caused by *Ramularia areola* is an economically important disease in Andhra Pradesh, India. The effect of weather factors on the development of grey mildew disease in susceptible Bt cotton hybrid Jadoo was investigated during *kharif* 2013–2014 in Vertisols at Regional Agricultural Research Station, Lam, Guntur and Andhra Pradesh under rain-fed conditions. Scoring of disease was done at weekly intervals by adopting 0–4 scale on randomly labelled plants up to mid-February. The development of disease expressed as % disease intensity (PDI), along with meteorological data (maximum temperature, minimum temperature, morning relative humidity, evening relative humidity, rain fall and sunshine hours), was recorded. Correlation between progress of grey mildew and weather factors was calculated. The study indicated that the disease appeared during 44th meteorological week (29 Oct–4 Nov 2013), with mean maximum temperature 31.3 °C, mean minimum temperature 24.1 °C, mean morning relative humidity 94 %, mean evening relative humidity 62 %, sunshine hours 6.4 h day^{-1} and rainfall 1 mm $week^{-1}$ at flowering stage. The grey mildew increased progressively and reached its peak (43 %) during 52nd meteorological week (24–30 Dec 2013), with mean maximum temperature 28.9 °C, mean minimum temperature 16.4 °C, mean morning relative humidity 93 %, mean evening relative humidity 49 %, sunshine hours 6.4 h day^{-1} and no rain fall at boll maturity stage. The data on % disease index was subjected to multiple linear regression, and the following equation was obtained as $Y = 349.698 – 7.237$ max T – 4.307 min T + 1.0 RH II ($r^2 = 92.26$ %). The partial regression coefficient for evening relative humidity was significant and positively correlated with PDI. This study is useful to plan the management strategies against this important disease and take up preventive and/or protective measures with recommended fungicides like 0.3 % wettable sulphur or 0.1 % carbendazim (Bhattiprolu et al. 2015).

White root rot in apple caused by *Dematophora necatrix* (perfect state *Rosellinia necatrix*) is becoming a major threat in the cultivation of crop. The disease occurs in severe form both under nursery and orchard conditions leading to huge economic losses. Integration of cultural and biological methods and use of novel chemicals have been quite effective in reducing the disease incidence and increasing the production of the crop. These serve as the best way to manage this soil-borne disease of apple in the orchards (Verma and Sharma 2015).

References

Ajithkumar K, Biradar SA, Rajanna B et al (2015) Management of Linseed (*Linum usitatissimum* L.) Powdery mildew by sulphur through soil and foliar application. In: Compendium of abstracts of the 2nd international conference on bio-resource and stress management. ANGRAU & PJTSAU, Hyderabad, 7–10 January 2015

Bhaik A, Ranjith Kumar L, Sanjay Sharma (2015) A new strobilurin fungicide galileo® (picoxystrobin 25SC) for the management of blast disease of rice in India. In: Compendium of abstracts of the 2nd international conference on bio-resource and stress management. ANGRAU & PJTSAU, Hyderabad, 7–10 January 2015

Bhat N, Vennila MS, Sardana HR et al (2015) Disease scenario of tomato in semi – arid region of karnataka as influenced by rainfall and dry spell. In: Compendium of abstracts of the 2nd international conference on bio-resource and stress management. ANGRAU & PJTSAU, Hyderabad, 7–10 January 2015

Bhattiprolu SL, Bharathi S, Ratna Kumar S et al (2015) Grey mildew disease in cotton as influenced by weather parameters and its management. In: Compendium of abstracts of the 2nd international conference on bio-resource and stress management. ANGRAU & PJTSAU, Hyderabad, 7–10 January 2015

Bhuvaneswari V, Krishnam Raju S, Satyanarayana PV et al (2015) Compatibility of fungicides and insecticides targeting leaf, neck blast and major rice pests. In: Compendium of abstracts of the 2nd international conference on bio-resource and stress management. ANGRAU & PJTSAU, Hyderabad, 7–10 January 2015

Dabbas MR, Kumar S, Tiwari P (2015) Variability of different fungicides on *F. oxysporum f.sp. pisi*. In: Compendium of abstracts of the 2nd international conference on bio-resource and stress management. ANGRAU & PJTSAU, Hyderabad, 7–10 January 2015

Das IK, Govardhan C, Patil JV (2015) Management of floret infections of grain molds in sorghum. In: Compendium of abstracts of the 2nd international conference on bio-resource and stress management. ANGRAU & PJTSAU, Hyderabad, 7–10 January 2015

Gopi R, Kalita H, Avasthe RK et al (2015) Organic management of tomato late blight in Sikkim Himalaya. In: Compendium of abstracts of the 2nd international conference on bio-resource and stress management. ANGRAU & PJTSAU, Hyderabad, 7–10 January 2015

Gupta M (2015) Variability among isolates of *F. oxysporum f.sp. zingiberi* causing yellows of ginger. In: Compendium of abstracts of the 2nd international conference on bio-resource and stress management. ANGRAU & PJTSAU, Hyderabad, 7–10 January 2015

Hanuman LN, Madhavi BG, Narayana F (2015) Compatibility of different fungicides with *Pseudomonas fluorescens*. In: Compendium of abstracts of the 2nd international conference on bio-resource and stress management. ANGRAU & PJTSAU, Hyderabad, 7–10 January 2015

Jalender P, Bhat NB, Vijayalakshmi K et al (2015) Studies on host range of cucumber mosaic virus in tomato. In: Compendium of abstracts of the 2nd international conference on bio-resource and stress management. ANGRAU & PJTSAU, Hyderabad, 7–10 January 2015

Jayalakshmi SK, Mahalinga DM, Sumangala E et al (2015) Management of Phytophthora blight (*Phytophthara drechslerif.*sp. cajani) in pigeonpea. In: Compendium of abstracts of the 2nd international conference on bio-resource and stress management. ANGRAU & PJTSAU, Hyderabad, 7–10 January 2015

Jayale ST, Khare UK, Nemade PW et al (2015) Studies on collar rot of chickpea caused by *Sclerotium rolfsii* Sacc. In: Compendium of abstracts of the 2nd international conference on bio-resource and stress management. ANGRAU & PJTSAU, Hyderabad, 7–10 January 2015

Khan H, Viswanatha KP (2015) Inheritance of resistance to bacterial leaf blight disease in cowpea (*Vigna unguiculata* L. Walp.). In: Compendium of abstracts of the 2nd international conference on bio-resource and stress management. ANGRAU & PJTSAU, Hyderabad, 7–10 January 2015

Kuldhar DP, Suryawanshi AP, Nikam PS (2015) Bioefficacy of bioagents against *Sclerotium rolfsii* causing stem or pod rot of groundnut. In: Compendium of abstracts of the 2nd international conference on bio-resource and stress management. ANGRAU & PJTSAU, Hyderabad, 7–10 January 2015

Kulkarni VV, Shankergoud I, Govindappa MR et al (2015) Identification of sunflower powdery mildew resistant sources under artificial screening. In: Compendium of abstracts of the 2nd international conference on bio-resource and stress management. ANGRAU & PJTSAU, Hyderabad, 7–10 January 2015

Kumar MVN, Shankar VG, Ramya V et al (2015) Fusarium wilt resistance in castor: An overview of the recent advances and future strategies for genetic improvement. In: Compendium of abstracts of the 2nd international conference on bio-resource and stress management. ANGRAU & PJTSAU, Hyderabad, 7–10 January 2015

Madhavi GB, Bhattiprolu SL, Hanuman LN et al (2015a) Induction of systemic resistance against turcicum blight of maize using chemical inducers and *Pseudomonas fluorescens*. In: Compendium of abstracts of the 2nd international conference on bio-resource and stress management. ANGRAU & PJTSAU, Hyderabad, 7–10 January 2015

Madhavi M, Narayan Reddy P, Ranga Reddy R (2015b) Morphological and genetic characterization of *Rhizoctonia solani* isolates causing banded leaf and sheath blight in maize. In: Compendium of abstracts of the 2nd international conference on bio-resource and stress management. ANGRAU & PJTSAU, Hyderabad, 7–10 January 2015

Mohiddin FA, Bhat FA, Hamid B et al (2015) Development of *Trichoderma* based bio-formulations for the management of chilli wilt. In: Compendium of abstracts of the 2nd international conference on bio-resource and stress management. ANGRAU & PJTSAU, Hyderabad, 7–10 January 2015

Nikam PS, Suryawanshi AP, Chavan AA (2015a) Pathogenic, cultural, morphological and molecular variability among the isolates of *Alternaria solani*, causing early blight of tomato. In: Compendium of abstracts of the 2nd international conference on bio-resource and stress management. ANGRAU & PJTSAU, Hyderabad, 7–10 January 2015

Nikam PS, Suryawanshi AP, Kuldhar DP (2015b) Physiological and nutritional variability in *Alternaria solani*, causing early blight of tomato. In: Compendium of abstracts of the 2nd international conference on bio-resource and stress management. ANGRAU & PJTSAU, Hyderabad, 7–10 January 2015

Pagoch SS (2015) Unexplored pathways in plants unveil the possibility to microbial pathogen control. In: Compendium of abstracts of the 2nd international conference on bio-resource and stress management. ANGRAU & PJTSAU, Hyderabad, 7–10 January 2015

Patro TSSK, Jyothsna S, Sandhya Rani Y et al (2015) Development of blast resistant lines in finger millet. In: Compendium of abstracts of the 2nd international conference on bio-resource and stress management. ANGRAU & PJTSAU, Hyderabad, 7–10 January 2015

Puri S, Kumar V, Saraf RK (2015a) Screening of lentil lines for resistance to Fusarium wilt under field conditions. In: Compendium of abstracts of the 2nd international conference on bio-resource and stress management. ANGRAU & PJTSAU, Hyderabad, 7–10 January 2015

Puri SV, Agam MN, Pansambal SA et al (2015b) Studies on in vitro evaluation fungicides against of wilt of soybean caused by *F. oxysporum f. sp. glycine*. In: Compendium of abstracts of the 2nd international conference on bio-resource and stress management. ANGRAU & PJTSAU, Hyderabad, 7–10 January 2015

Raghavendra G, Ameer Bashaa S, Nagesh Kumara MV et al (2015) Search for potential *Pseudomonas fluorescens* strains against pigeon pea wilt pathogen. In: Compendium of abstracts of the 2nd international conference on bio-resource and stress management. ANGRAU & PJTSAU, Hyderabad, 7–10 January 2015

Rajeswari B, Devi GU (2015) Effect of disease incidence on yield attributes of oyster mushrooms. In: Compendium of abstracts of the 2nd international conference on bio-resource and stress management. ANGRAU & PJTSAU, Hyderabad, 7–10 January 2015

Rakhonde PN, Mane SS, Ukey NS et al (2015) Genetic diversity of *F. oxysporum f. sp. ciceri* isolates in India. In: Compendium of abstracts of the 2nd international conference on bio-resource and stress management. ANGRAU & PJTSAU, Hyderabad, 7–10 January 2015

Ramya V, Nagesh Kumar MV, Gouri Shankar V et al (2015) Identifying sources of resistance for Fusarium wilt in castor (*Ricinus communis* L.). In: Compendium of abstracts of the 2nd international conference on bio-resource and stress management. ANGRAU & PJTSAU, Hyderabad, 7–10 January

Sadarunnisa S, Reddy RVSK, Begum H et al (2015) Identification of resistant sources for bacterial wilt caused by *Ralstonia solanacearum* in brinjal. In: Compendium of abstracts of the 2nd international conference on bio-resource and stress management. ANGRAU & PJTSAU, Hyderabad, 7–10 January

Supriya SM, Kulkarni VV, Shankergoud I et al (2015) Screening of sunflower genotypes for powdery mildew resistance. In: Compendium of abstracts of the 2nd international conference on bio-resource and stress management. ANGRAU & PJTSAU, Hyderabad, 7–10 January 2015

Suvarna, Savitha AS, Lokesha R (2015) Evaluation of germplasm for Alternaria leaf spot disease resistance in niger (*Guizotia abyssinica*). In: Compendium of abstracts of the 2nd international conference on bio-resource and stress management. ANGRAU & PJTSAU, Hyderabad, 7–10 January 2015

Verma S, Sharma IM (2015) Present Status and Strategies in the Management of white root rot of apple in Himachal Pradesh. In: Compendium of abstracts of the 2nd international conference on bio-resource and stress management. ANGRAU & PJTSAU, Hyderabad, 7–10 January 2015

Wagh SS, Kuldhar DP, Falake AR (2015) Integrated management of potato (*Solanum tuberosum* L.) leaf spot, caused by *Alternaria solani*. In: Compendium of abstracts of the 2nd international conference on bio-resource and stress management. ANGRAU & PJTSAU, Hyderabad, 7–10 January 2015

Biotic Stress: Insect Pests

8

Ratikanta Maiti and Aruna Kumari

Abstract

The productivity of crops is influenced by abiotic and biotic factors. Insects are the major pests contributing to severe yield losses. This chapter makes a concise review of the researches undertaken on insect incidence, influence of photoperiod and environment on insect multiplication and the specific control measures in various crops, viz., rice, maize, sorghum, chickpea, green gram, pigeon pea, black gram, safflower, sunflower, castor, groundnut, cotton, cabbage, okra, chilli, tomato, tobacco, banana, mango and sugarcane. It also discusses the various rearing media employed for the study of biology of insects, the integrated pest management, the compatibility of insecticides with the fungicides and various weed control measures.

8.1 Introduction

Agriculture is under pressure to reduce reliance on chemicals and focus on more sustainable methods of production. In recent decades, IPM has shown great potential for reducing the dependence of crop protection on chemical control methods. IPM reflects the idea that pest management requires a coordinated approach, integrating diverse tactics, including cultural, ecological, biological and chemical control. The productivity of crops is influenced by abiotic and biotic factors. Insects are the major pests contributing to severe yield losses.

India is a tropical country with climate conditions being very conductive to insect and pest breeding. Pesticides are an important component in agriculture efficiency and improved food quality by making them pest-free. Use of pesticides introduced and increased with the Green revolution in India. But environmental exposure of pesticides to humans can occur through improper agricultural practices, consumption of contaminated food materials or polluted air. Functional response is an important behavioural response to reveal different aspects of prey-predator interactions.

R.K. Maiti (✉)
Facultad de Ciencias Forestales, Universidad Autónoma de Nuevo León, Carr. No 85, km 145, NL 67700 Linares, Mexico
e-mail: ratikanta.maiti@gmail.com

A. Kumari
Department of Crop Physiology, Agricultural College, Professor Jayashankar Telangana State Agricultural University, Polasa, Jagtial, Karimnagar 505 529, Telangana, India
e-mail: arunasujanagcjgl@gmail.com

R.K. Maiti et al. (eds.), *Bioresource and Stress Management*, DOI 10.1007/978-981-10-0995-2_8

It has been established that indiscriminative use of pesticides is hazardous for human and animal health. Hence, pesticide residue has also become an important parameter of quality assessment of food crops for both domestic and export market. Significant advances have been attained on the physiology of plants and mechanisms of resistances to various abiotic stresses. Several papers are presented on these various aspects in the second International Conference on Bio-resource and Stress management in Hyderabad, January 2015.

8.2 Rearing Media for Biocontrol Agent's Production

A wide variety of rearing medium is employed and is under use for the production of the biocontrol agents. The rearing media is also employed to study the biology of insect pests under laboratory conditions. The development and reproductive performance of *Corcyra* evaluated under 12 different combinations of food containing maize, wheat, Italian millet and scented rice – alone and fortified with 3 % dextrose and yeast as rearing media – indicated that the Italian millet (alone or fortified) increased the net reproductive rate, innate capacity, number of hypothetical females, daily finite rate and weekly multiplication. The potential fecundity was highest and true generation time and doubling time was shortest in the Italian millet rearing medium. The performance was better in wheat alone (second to Italian millet) than its fortification with dextrose and yeast and also superior to maize and scented rice. Hence, Italian millet alone or fortified with 3 % dextrose and yeast was the most suitable rearing media for mass production of *Corcyra* (Chaudhuri et al. 2015).

8.3 Parasitization of Parasitoids

Different parasitoids, viz., *Campoletis chlorideae*, *Goniophthalmus halli*, *Rogas aligarhensis*, *Bracon* sp. and *Agathis* sp., evaluated (Niharika and Chaturvedi 2015) for % parasitization by natural enemies, eco-friendly and synthetic insecticides revealed that % parasitization varied from 8.1 to 38.62 %. The significantly highest % parasitization was observed in untreated control (38.62 %) followed by other eco-friendly treatments, viz., Thiamethoxam 70 WS 0.3 % seed treatment followed by release of *Chrysoperla carnea* at 10,000 first instar larvae ha^{-1} + *Trichogramma chilonis* at 1.5 lakh ha^{-1} + mechanical collection. The parasitization was (32.78 %) with Bt at 1 kg ha^{-1} spraying, (32.52 %) with NSE 5 % spraying and (28.41 %) with endosulfan 35 EC 0.07 % spraying along with the combination of above treatments. The lowest % parasitization was observed in dimethoate 30 EC 0.03 % followed by carbaryl 50 WP 0.2 % and quinalphos 25 EC 0.05 % sprayings (8.1 %). However, the lowest parasitization was observed due to the application of synthetic insecticides.

8.4 Insecticides Against Pod Borers

Pod borers are a serious pest of crops leading to a decreased yield due to the damage of pods and grains. The pod borer's damage is more inflicting the grain yield in unprotected condition. In pigeon pea, the yield losses were estimated 35.96 % due to pod borers. Indoxacarb 14.5 SC at 50 g a.i. ha^{-1} evaluated (Joshi et al. 2015) against pod borer complex, viz., *Maruca vitrata*, *Helicoverpa armigera*, *Exelastis atomosa* and *Melanagromyza obtusa* in pigeon pea variety Asha revealed its effectiveness against these pests. The larval population per plant, % pod damage and grain damage due to pod borer complex, viz., *Maruca vitrata* (4.07, 3.2 and 2.09 %), *Helicoverpa armigera* (2.93, 4.45 and 2.49 %), *Exelastis atomosa* (3.67, 3.07 and 1.94 %) and *Melanagromyza obtuse* (2.53, 3.30 % and 2.35 %) in pigeon pea variety Asha were minimum. Further, in protected condition with Indoxacarb spray, there was reduced pod damage and grain yield loss.

8.5 Insecticide Resistance of Boll Worm

The bollworm, *Helicoverpa armigera* Hübner (Lepidoptera: Noctuidae) is a polyphagous pest of worldwide occurrence inflicting annual crop damage in India worth US $1 billion. In India this insect occurs as a major pest in many economically important crops, including cotton, pigeon pea, chickpea, tomato, okra, black gram, maize, sorghum and many other crops, inflicting substantial crop losses every year. The ability of insect species to thrive on diverse host plants is an adaptive advantage for their better survival in the ecosystem.

H. armigera is also characterized by its high mobility and fecundity and it has shown great capacity to develop resistance to synthetic insecticides used in its management. The versatility of this species may be due to the presence of a strong genetic variability governing the behaviour of *H. armigera* making it a serious pest on several crops.

The occurrence of insecticide-resistant strains can be reduced or delayed by reducing the selection pressure, by using alternate insecticides with novel mode of action. The pyrethroids and organophosphorus combination of insecticides were found to be effective against the resistant insect pest population of *H. armigera*.

The toxicity of all the test insecticides against insect pest is done by bioassay topical application method using Hamilton micro applicator. One microlitre of the respective insecticidal solution is applied on the dorsum of second thoracic segment by micro applicator. The mortality at 72 h after treatment can be considered as end point for the assessment of toxicity of test insecticides. The amount of insecticide present in one microlitre of test concentration is calculated and expressed as (LD_{50}) dose in μg μl^{-1}.

Bioassay of *Helicoverpa armigera* collected from different crops of red gram, cotton and Bengal gram, from different places, viz., Raichur, Mahaboobnagar and Nagpur evaluated (Upendhar et al. 2015) against the degree of resistance acquired *by H. armigera* to insecticides, viz., cypermethrin representing synthetic pyrethroids and methomyl representing the carbamate group of insecticides revealed that the Mahaboobnagar district strain to methomyl recorded a LD_{50} of 3.651 μg $larva^{-1}$ which rose sharply to 10.287 μg $larva^{-1}$ at LD_{90} at F_1. Further these values rose to 3.872 and 5.829 μg $larva^{-1}$ at LD_{50} and LD_{90}, respectively, at F_2. Interestingly at F_3 the resistance recorded was 3.934 and 4.668 at LD_{50} and LD_{90}. The LD_{50} and LD_{90} values of methomyl for Raichur population of *H. armigera* were 3.630 and 10.417 μg $larva^{-1}$, respectively, at F_1 which increased to 3.659 and 5.471 μg $larva^{-1}$ at LD_{50} and LD_{90}, respectively, at F_2 and to 3.851 and 4.708 at LD_{50} and LD_{90}, respectively, at F_3. The resistance of toxicity of methomyl to Nagpur population of *H. armigera* showed that the LD_{50} and LD_{90} values were 2.652 % and 7.214 %, respectively, at F_1 which rose to 2.844 and 3.716 μg $larva^{-1}$ at LD_{50} and LD_{90}, respectively, at F_2 and to 2.944 and 3.279 at LD_{50} and LD_{90}, respectively, at F_3 (Upendhar et al. 2015).

Among the three populations of *H. armigera*, the population of Mahaboobnagar had developed 1.006- and 0.988-fold relative resistance at LD_{50} and LD_{90}, compared to the Raichur population and still higher levels of relative resistance by 1.377- and 1.426-fold compared to the Nagpur population at LD_{50} and LD_{90}, respectively. Raichur population recorded 1.369- and 1.444-fold resistance at LD_{50} and LD_{90}, respectively, in comparison with Nagpur. In comparison with base line data of Nagpur, susceptible three populations of Mahaboobnagar, Raichur and Nagpur acquired 121.700-, 121.000- and 88.400-fold at LD_{50} and 62.345-, 63.133- and 43.721-fold resistance to methomyl at LD_{90} levels, respectively. The cypermethrin resistance ratio appeared to be high because of the comparison with most susceptible strain as base line (LD_{50} = 0.007 μg $larva^{-1}$), but in reality it is indicated that there is decrease in the levels of resistance to cypermethrin in *H. armigera* compared to latest reports, which may be probably due to decreased selection pressure with significant decrease in the use of cypermethrin and increasing the area under Bt cotton. The methomyl investigations have indicated that there was a decrease in the levels of methomyl resistance in *H. armigera* compared to latest reports, which may be probably due to significant decrease in the use of methomyl in managing the pest due to reduced bollworm incidence and due to increased area under Bt cotton (Upendhar et al. 2015).

8.6 Pesticide Residue in Foods

India is a tropical country with climate conditions being very conductive to insect and pest breeding. Pesticides are an important component in agriculture efficiency and improved food quality by making them pest-free. Use of pesticides introduced and increased with the Green revolution in India. But environmental exposure of pesticides to humans can occur through improper agricultural practices, consumption of contaminated food materials or polluted air.

It has been established that indiscriminative use of pesticides is hazardous for human and animal health. Hence, pesticide residue has also become an important parameter of quality assessment of food crops for both domestic and export market. Pesticides are tested and approved by the Environmental Protection Agency (EPA), which establishes "tolerances", that limit the amount of a given pesticide that can safely remain in or on a food.

Residues of pesticides in food are influenced by the storage, handling and processing technologies that occur between harvesting of raw commodities and consumable foods. Processing technologies, washing and peeling for vegetables, milling for cereals and pulses and fat reduction for animal food can reduce the residue content for some extent.

Selected whole or processed cereals and pulses, viz., jowar, maize, red gram and black gram grown in the Southern Telangana zone of Andhra Pradesh analysed for acephate, carbofuran, chlorpyrifos, endosulfan, monocrotophos and quinalphos pesticide standards (Niharika and Chaturvedi 2015), have shown that the pesticide residues in these were below the detectable limits (BDLs) due to strict observance of postharvest measures (safety limiting period) by the farmers. Though the pesticide usage was reportedly high, as cereals and pulses are stored before consumption, the dissipation of pesticides might have resulted in lower levels of pesticides residues. The best way to avoid consuming synthetic pesticides along with our food is to eat organic foods.

8.7 Rice

8.7.1 Brown Plant Hopper

8.7.1.1 Antixenosis Affect

The brown plant hopper (BPH), *Nilaparvata lugens* (Stal.) (Homoptera: Delphacidae), is one of the most destructive monophagous insect pests of rice throughout the rice-growing countries in Asia. Heavy dependence on chemical pesticides for the control of this pest leads to many adverse effects like harmful effects on natural enemies, development of insecticide resistance, environmental pollution and high cost of production.

Cultivation of resistant rice varieties is the most economical and efficient method for the management of BPH. Understanding the mechanisms underlying mechanisms of rice resistance to BPH is very essential for developing varieties with durable resistance. In general, resistant rice plants exhibit two strategies against BPH: antixenosis and antibiosis.

The antixenosis affect studies were conducted (Bhanu et al. 2015a) in 12 resistant rice cultures, viz., NLR 3090, NLR 3093, MTU 1075, WGL 401, WGL II 218-5-1, MTU PLA 99-1-3-1-2, NLR 20131, BPT 2404, RDR 34, RGL 7001, RGL 7002 and MTU IJ 206-7-4-1 along with resistant check (Ptb 33) and susceptible check (TN1) on insect settling, colonization, oviposition and feeding of BPH at Maruteru, West Godavari district. The settlement of adults after their release at 24, 48 and 72 h on rice cultures assessed by conventional seed box method revealed that the mean number of nymphs and adults settled on these resistant rice cultures after 72 h of their release varied significantly with the rice cultures and were lower than susceptible check, TN1 (12.01 nymphs seedling^{-1} and 20.56 adults per ten plants). Among the tested rice cultures, MTU IJ 206-7-4-1 exhibited high level of antixenosis for colonization of BPH (3.22 nymphs seedling^{-1} and 1.12 adults per ten plants) and recorded lowest number of eggs (79.67). The other rice cultures were also least preferred adults of BPH for oviposition and recorded less number of eggs compared to susceptible check, TN1 (608.00).

8.7.1.2 Functional Response of BPH

In tropical rice, it is well established that BPH as a secondary pest problem is caused by ecological disruptions, particularly due to use of insecticides that reduce natural biological control. Among the natural enemies of BPH, green mirid bug, *C. lividipennis* (Hemiptera: Miridae), is one of the important insect predators of rice hoppers, predaceous on nymphs as well as on eggs of BPH. Functional response is an important behavioural response to reveal different aspects of prey-predator interactions. The term "functional response" shows the response of individual natural enemy to varying prey density.

An investigation on the functional response of female green mirid bug on the varying egg densities of BPH (Bhanu et al. 2015b) indicated that the logistic regression showed a type II functional response for female green mirid bug. The number of eggs attacked per day per predator increased from 1.0 at a prey density of 12 to a maximum of 23.0 at a prey density of 71.0, and thereafter the number of eggs attacked per day was decreased gradually with increase in prey density. The parameters (using Rogers' model), the search rate (a) and the handling time (Th), were estimated to be 0.318 and 0.0679 (days), respectively. Based on the estimates of handling time, the maximum number of eggs that one female *C. lividipennis* was able to consume per day was 14.73 (Bhanu et al. 2015b).

8.7.1.3 Granular Insecticides

The efficacy of granular insecticides, viz., fipronil 0.3G, carbofuran 3G, phorate 10G, cartap hydrochloride 4G, fipronil + imidacloprid 80WG, Rynaxypyr 0.4G and chloropyriphos 10G, evaluated (Kharbade et al. 2015a) against brown plant hopper-infesting rice has shown that among these fipronil + imidacloprid 80WG proved to be most effective against *N. Lugens* recording 8.42 hopper per hill followed by Rynaxypyr 0.4G at (9.13), fipronil 0.3G (11.33) and cartap hydrochloride (11.64) hopper population per hill. Chloropyriphos, carbofuran and phorate recorded higher hopper population of 13.23, 14.56 and 15.53 per hill noticed as against 22.61 hoppers per hill in untreated control.

8.7.1.4 Ecological Engineering

Agriculture is under pressure to reduce reliance on chemicals and focus on more sustainable methods of production. In recent decades, IPM has shown great potential for reducing the dependence of crop protection on chemical control methods. IPM reflects the idea that pest management requires a coordinated approach, integrating diverse tactics, including cultural, ecological, biological and chemical control.

Maintenance of pests below economic injury levels while minimizing hazards to humans and environment including natural enemies is a key component of IPM. Even with the implementation of IPM, modern agro systems are often inhospitable to natural enemies because of decreasing landscape heterogeneity, frequent disturbance, agrochemical inputs, decreasing genetic diversity and increasingly homogeneous vegetation. Due to several problems associated with inundative and inoculative biological control, this new approach of ecological engineering came to the forefront. The main concept involved in ecological engineering is enhancing the build-up and activity of natural enemies through changes in cultural practices having low dependence on external and synthetic in puts, reliance on natural processes and habitat management via vegetation structure exerting resource concentration effects (Reddy et al. 2015).

8.7.2 Phototrophic Insect Pests

The major phototropic insect pest species of paddy are rice gundhi bug, green leaf hopper, army worm and rice butterfly. Standard design of Jawahar light trap used to analyse (Muchhala et al. 2015) the effect of various weather parameters on the seasonal activity of these pests indicated that the major activity period of rice gundhi bug, *Leptocorisa acuta* (Thunberg), was observed from August to December with two distinct peaks during the 39th and 42nd SW; green leaf hopper, *Nephotettix virescens* (Distant), was active from August to December with three distinct peaks during the 37th, 39th and 42nd SW, respectively; army worm, *Mythimna separata* (Walker) was active from September to November with two distinct peaks during the 39th and 44th SW; and *Melanitis leda ismene* Cramer was active from August to December with three distinct peaks during the 38th, 42nd and 44th SW, respectively. Among all the parameters, maximum temperature was found to be a significantly favourable weather factor for development of these pest species.

Some of the major predatory species *Prothyma* sp., *Coenagrion* sp. and parasitic species, namely, *Myrmicaria brunnea* Saunders and *Enicospilus purgatus* (Say) are also phototrophic. Standard design of Jawahar light trap used to analyse the effect of seasonal activity (Sharma and Muchhala 2015) of these major predatory and parasitic species and their relationship with various weather factors indicated that the major activity period of *Prothyma* sp. was observed from July to October with three distinct peaks during the 28th, 37th and 41st SW, respectively; *Coenagrion* sp. was active from June to October with two distinct peaks during the 37th and 39th SW; *Sirthenea carinata* (Fabricius) was active from August to November with two distinct peaks during the 36th and 39th SW; *Myrmicaria brunnea* Saunders was observed from July to August; and *Enicospilus purgatus* (Say) was active from June to November with distinct peak in the 30th SW and the 27th and 42nd SW, respectively. Among all the parameters, minimum temperature was found significantly favourable for *Prothyma* sp., while morning relative humidity has significantly negative correlation with the activity of *Coenagrion* sp.

8.7.2.1 Insect Pest Incidence in Different Systems of Rice Cultivation

In India, currently rice is being cultivated in about 44 m ha, through different methods (transplanting, direct seeding, broadcasting, SRI, mechanized transplanting, etc.). An assessment of pest incidence in different systems of rice cultivation (Nadimpally and Rumandla 2015a) revealed that during *kharif* 2008, significantly lower whorl maggot incidence (2.93 %) was noticed in SRI than conventional method (4.58 %). Higher stem borer incidence (both in terms of dead hearts and white ears) was observed in conventional method (16.94 % and 4.84 %, respectively) compared to SRI (11.04 % and 1.65 %, respectively) in 2008. During 2009, dead heart incidence varied from 13.46 % in conventional to 4.97 % in SRI. Conversely, incidence of hispa and thrips was high in SRI method during both the seasons. Significantly higher grain yields were realized in SRI (4.9 t ha^{-1}) than in conventional system (4.4 t ha^{-1}) during *kharif* 2008, but yield differences were non-significant across systems during 2009. In Direct seeding and conventional transplanting (*kharif* 2010 and 2011) during *kharif* 2010, hispa damage was significantly lower (0.19 %) in direct seeding than conventional method (2.56 %). Higher stem borer incidence (white ears) was observed in direct seeding (15.25 %) than conventional method (10.07 %) in 2008. However, during 2011, stem borer and hispa was lower in direct seeding. Higher grain yields (12–13 %) were realized in direct seeding during both the seasons. The incidence of whorl maggot and stem borer (dead hearts) did not differ significantly across the cultivation systems. However, BPH incidence was significantly low in direct seeding with drum seeder and broad casting systems compared to other systems. Mechanized transplanting and direct seeding with drum seeder registered 25 and 23 % increased yields over farmers practice. This information would be highly useful in decision-making on pest management options based on rice cultivation systems.

8.7.2.2 Lesser Grain Borer

The growth and development of lesser grain borer, *R. dominica*, assessed (Muthukumar et al. 2015) in different paddy varieties, viz., MTU-1075, CM-9 (2012), NLR-34449, BPT-5204, MTU-1001, WGL-14, WGL-3943, WGL-44 and WGL-32100 of varied biochemical composition (protein, carbohydrates, amylose and phenol), indicated that the biochemical composition (protein, carbohydrate, amylose and phenol) of paddy varieties was correlated with the % of damage and % weight loss caused by *R. dominica*.

The susceptibility index of the test paddy varieties calculated (Muthukumar et al. 2015) based on the parameters of observation, viz., adult emergence, mean developmental period, % weight loss and the % of damage caused by *R. dominica*, revealed that there was a significant difference among the treatments with regard to adult emergence and susceptibility index. The maximum number of adults of 75.83 emerged from WGL-3943 and minimum was recorded in CM-9 (2012) (17.53). The average susceptibility index was 10.74 ranging from 8.05 to 12.45. The maximum % weight loss and the % of damage were recorded in WGL-44 (7.30 % and 10.67 %, respectively) and minimum was observed in CM-9 (2012) (0.28 % and 0.71 %, respectively). There was significant positive and negative correlation between the protein and phenol contents of the paddy varieties with respect to % weight loss and the % of damage caused by *R. dominica*. The infestation of *R. dominica* on paddy varieties resulted in significant increase in protein and phenol contents after 3 months period of storage.

8.7.2.3 Population Dynamics

Population dynamics of stem borer and leaf folder as well as their predators and parasites investigated (Chatterjee et al. 2015) through sweeping method at weekly intervals revealed that from the seedbed sweeping during boro season, the most abundant rice insect pests were *Cofona* sp., *Recelia* sp., *Hieroglyphus* sp. and *Hydrellia* sp., whereas *Micraspis* sp., *Mesovelia* sp., *Paederus* sp., *Ophionea* sp. and *Agriocnemis* sp. *Tetrastichus* sp., *Telenomus* sp., *Copidosomopsis* sp. and *Opius* sp. were the parasites whose population was found maximum from boro seedbed. The transplanted field sweeping during boro season indicated that the most abundant insect pests were *Leptocorisa* sp., *Baliothrips* sp., *Cofona* sp., *Scirpophaga incertulas*, *Sogatella* sp. and *Nephotettix* sp. In earlier studies, the most abundant arthropods noticed were green leaf hopper, white leaf hopper, spiders, dragonfly and damselfly. The maximum temperature played a positive effect on population build-up in seedbed of yellow stem borer, leaf folder, pest, predator and parasite. The minimum temperature showed a significant positive relation with yellow stem borer, pest and predator and parasite population. The significant negative correlation was found with relative humidity (morning) and relative humidity (evening) to the population build-up of YSB. Rainfall and sunshine hours failed to show any significant effect with the population of insects.

8.7.2.4 Resistance to Leaf Folder

Rice is the staple food for more than one half of the world's population. The major reason for dismal state of rice production and productivity worldwide is due to biotic stresses. Among insect pests, rice leaf folder (*Cnapholocrocis medinalis*), earlier considered as a minor pest, has gained the status of major pest with the widespread cultivation of HYV and the accompanying changes in cultural practices. Due to lack of resistant donors and varieties, farmers are solely dependent on insecticides to combat this pest. So, there is an urgent need to identify resistant donors and develop resistant varieties against leaf folder.

A large amount of variability exists between the genotypes for leaf folder resistance and its contributing traits which can be utilized in breeding programmes to develop rice varieties with strong resistance to leaf folder coupled with good grain quality and higher yield under typical tropical irrigated ecosystem. This was evident from a study conducted (Rao et al. 2015) to study the variability and molecular genetic diversity for leaf folder resistance in

30 genotypes including susceptible check TN1 under natural conditions. Based on damage score (25, 50 and 75 DAT), 21 genotypes were resistant, 8 were moderately resistant and TN1 was highly susceptible. Correlation studies among different traits revealed that the trait damage score recorded significant negative correlation with leaf width ($r = -0.180^{**}$) and SPAD chlorophyll meter reading ($r = -0.136^{**}$) and non-significant negative correlation with leaf length. Molecular diversity studied through 60 microsatellite markers resulted in detection of a total of 127 polymorphic alleles. Polymorphic information content (PIC) values of the markers ranged from 0.061 to 0.62 with an average of 0.30. UPGMA analysis has grouped the 30 rice genotypes into three clusters. Cluster I had only one genotype, i.e. TN1, the susceptible check. Cluster II had 11 genotypes and Cluster III had 18 genotypes.

8.7.2.5 Yellow Stem Borer

Among the major pests attacking rice crop, the yellow stem borer, *Scirpophaga incertulas* (Walker), is the serious pest which attacks the crop both at vegetative and reproductive stages. This pest causes 25–30 % damage to rice crop. Insecticide utilization in insect pest management is very important under farmer's field conditions and is a last resort, especially when insect pests assume economically serious proportions. Further, it is necessary to assess the efficacy of some of the recommended insecticides from time to time in addition to new chemicals available in the market for their efficacy against paddy stem borer before their recommendation. Nine available insecticides, viz., phosphamidon 40EC, acephate 75SP, chlorpyriphos 25EC, cartap hydrochloride 50WP, chlorantraniliprole 0.4G, chlorantraniliprole 18.5SC, cartap hydrochloride 4G and carbofuran 3G, evaluated (Kumar et al. 2015c) against stem borer incidence revealed that chlorantraniliprole 18.5 % SC resulted in lowest of 1.30 % dead hearts (DH) followed by chlorantraniliprole 0.4G with 1.64 % DH and cartap hydrochloride 4G with 2.17 % of DH. These were significantly superior over other treatments in controlling the stem borer damage at vegetative phase of the crop. Carbofuran 3G with 2.39 %DH was next to above treatments. Chlorantraniliprole 18.5SC with 1.71 % white ears (WE) followed by Chlorantraniliprole 0.4G (1.76 %WE) and Cartap hydrochloride 4G (2.19 %WE) were significantly superior over other treatments and untreated control in checking the stem borer attack at reproductive phase of the crop. Chlorantraniliprole 18.5SC recorded significantly higher grain yield of 7947 kg ha^{-1}, followed by chlorantraniliprole 0.4G (7023 kg ha^{-1}) and cartap hydrochloride 4G (6596 kg ha^{-1}) and prevented the grain yield loss due to stem borer. Chlorantraniliprole 18.5SC, chlorantraniliprole 0.4G and cartap hydrochloride 4G were found to be effective in checking the stem borer at vegetative as well as at reproductive phase of the crop.

8.7.2.6 Panicle Mite

Five acaricides (diafenthiuron, fenpyroximate, profenofos, dicofol and spiromesifen) evaluated (Nadimpally and Rumandla 2015a) alone and in combination with propiconazole for their efficacy against rice panicle mite and grain discolouration revealed that a combination of spiromesifen 240SC + propiconazole 25EC (at1 + 1 ml l^{-1} of water) and dicofol 18.5EC + propiconazole 25EC (at 5 + 1 ml l^{-1} of water) were found to be the most effective treatments with mean % healthy grains of 69.92 and 64.12, respectively. Spiromecifen + propiconazole and dicofol + propiconazole registered significantly higher grain yields (5131 and 4686 kg ha^{-1}, respectively) as against control (water spray) treatment (3943 kg ha^{-1}).

8.7.2.7 Compatibility of Insecticides and Fungicides

During the crop growth period, sometimes there is simultaneous occurrence of both diseases and pests. The application cost of insecticides and fungicides increase the cost under such events. To reduce the cost of application, the insecticides and fungicides applied as tank mix studied for their compatibility in rice revealed that all the 25 insecticide and fungicide combinations tested were physically compatible. Among the 25 combinations tested, no phytotoxicity either at single or at double dose was observed with combination of profenophos + kresoximethyl, profenophos + tebuconazole, profenophos + kasugamycin, chloropyriphos + tebuconazole, chloropyriphos + kasugamycin, spinosad + hexaconazole or spinosad + trifloxystrobin + tebuconazole. However, combinations of flubendiamide + mancozeb and flubendiamide + trifloxystrobin + tebuconazole showed high phytotoxicity at both the doses. Of all the combinations of chlorpyriphos tested with different fungicides, chlorpyriphos + kasugamycin (2.5 ml +2.5 ml l^{-1}) was the most effective combination followed by spinosad + trifloxystrobin + tebuconazole (0.4 ml +0.6 g l^{-1}) in managing multiple insect pests and diseases (stem borer, leaf blast, neck blast, sheath rot). Among all the combinations tested chlorpyriphos + kasugamycin (2.5 ml +2.5 ml l^{-1}) followed by chlorpyriphos + tebuconazole (2.5 ml +1 ml l^{-1}) were most cost-effective. No phytotoxicity was observed with acephate at 1.5 g l^{-1} and various aforesaid fungicidal combinations. Acephate was equally effective in all the combinations tested during dead heart stage with incidence ranging from 2.69 to 5.28 % as against control (13.22 %). However, it was ineffective at white ear stage. Acephate at 1.5 g and tricyclazole + mancozeb at 2 g l^{-1} or acephate at 1.5 g + kresoxim-methyl at 1 ml l^{-1} or acephate at 1.5 g and tebuconazole + trifloxystrobin at 0.4 g l^{-1} could effectively control leaf blast (2.42–3.72 %) as against control (39.28 %). They have also realized significantly higher grain yield (5182–5120 kg ha^{-1}) than control (3744.9 kg ha^{-1}). Among these, acephate and tricyclazole + mancozeb combination was the most cost-effective against stem borer and blast. Similarly, acephate at 1.5 g and tricyclazole + mancozeb at 2 g l^{-1}, acephate at 1.5 g and tebuconazole + trifloxystrobin at 0.4 g l^{-1}, acephate at 1.5 g + kresoxim-methyl at 1 ml l^{-1} recorded highest net profit of INR16685, 16,125 and 16,039 ha^{-1}, respectively, over untreated control (Nadimpally and Rumandla 2015b).

Three fungicides and ten insecticides at recommended concentrations were evaluated (Patil and Lande 2015) as tank mix in various fungicide and insecticide combinations for their efficacy against leaf blast, neck blast, brown plant hopper and stem borer and to investigate their compatibility as tank mix application for the purpose of reducing the application cost in the event of simultaneous occurrence of both diseases and pests during crop growth period. Among the different combinations tested, pymetrozine 50 % WG at 0.5 g l^{-1} in combination with Isoprothiolane 40% EC at 1.5 ml l^{-1} recorded less leaf blast incidence (44.87 %), neck blast incidence (79.29 %) and also lesser number of plant hoppers (1.35 $hill^{-1}$) followed by dinotefuran 20% SG at 1.0 g l^{-1} + tricyclazole at 0.6 g l^{-1} (52.48 %, 75.28 %, 0.78 $hill^{-1}$ and white ears 0.20 $hill^{-1}$) and combination product (imidacloprid + ethiprole 80 % WG) at 0.8 g l^{-1} + tricyclazole 75 % WP at 0.6 g l^{-1} (47.52 %, 82.10 % 2.72 $hill^{-1}$) compared to untreated check where the incidence of leaf blast and neck blast was 80.90 % and 100 %, respectively. The number of plant hoppers and white ears in untreated check plot was 224.38 and 0.63 per hill. There was no reduction in the efficacy of these insecticides and fungicides when used as tank mix and phytotoxicity symptoms were observed first in some treatments but recouped after short period. Thus, all the insecticide and fungicide combinations used in the investigation were compatible with each other and can be safely combined as tank mix for the control of rice pests and diseases thus saving labour costs.

8.7.2.8 Stored Grain Pests

Public distribution system (PDS) plays a pivotal role in West Bengal by providing food and nutritional security, maintaining price stability and ensuring micro level self sufficiency to the poor and downtrodden section of the society. According to Annual Administrative Report 2010–2011, Food and Supply Dept., Govt. of W.B. 1, 89, 63,621 are Below Poverty line consumers (BPL); 74, 67,406 Antodaya Anno Yojona (AAY) and 63,855 Annapurna beneficiary (all belong to poorest section of the society) are directly benefited from the public distribution system. According to the ministry of Food and Civil Supplies, every year almost 10 % of total food grain productions are lost in storage, eaten by rodents and insects or spoiled by moisture. Insects are the major cause of loss of stored food grain products. As early as 1967, an FAO study estimated that worldwide annual losses in storage due to insect damage was as high as 10 % of all stored cereals which amounted to 13 million tons grain lost in that year alone (Wolpert 1967) cited in (Patil and Lande 2015). Rice weevil (*Sitophilus oryzae*) and lesser grain borer (*Rhyzopertha dominica*) are the important stored-grain insect pest of procured par-boiled rice stored in government godowns for distributions through PDS.

An experiment conducted (Goswami 2015) at a godown at Purba Midnapore, West Bengal, for management of rice weevil and lesser grain borer with the application of malathion chemical, safe to human beings with different doses in storage bags just before storing of "Swarna" variety rice established that treatment of storage bags with malathion applied as 8.0 ml l^{-1} of water was the best treatment against the target insects. It also suggested that proper malathion dosage in storage bags plays an important role as an effective insecticide against stored-grain insect pests of rice. It also helps in saving public money with high cost-benefit ratio.

8.8 Maize

Popcorn is an important specialty corn type that not only provides nutritious food in the form of popular snacks but also generates income and employment. The distinguishing popping trait is the formation of large flakes after subjecting popcorn kernels to heat treatment.

8.8.1 Storage Pest

Rice weevil (*Sitophilus oryzae* L.) is a serious misery for maize grain storage especially in developing countries leading to postharvest loss in quantity and quality of different specialty corn types including popcorn. A set of 21 popcorn hybrids evaluated (Zunjare et al. 2015) for their responses to stored-grain weevil infestation at three locations, viz., IARI, New Delhi, CSK-HPKV, Bajaura and GBPUAT, Pantnagar, revealed the presence of wide genetic variations for grain weight loss (5.40–48.77 %), kernel popping expansion volume (6.09–85.21 ml) and kernel popping (4.33–66.67 %) after infestation. Mean grain weight loss (19.31 %) was low in the Delhi location, as compared to other locations (Bajaura, 26.77 %; Pantnagar, 24.94 %). Kernel popping expansion and kernel popping were 57.90 % and 30.21 %, at Delhi, while the same were 46.23 % and 19.56 % at Bajaura and 49.42 % and 20.72 % at Pantnagar, respectively. This suggested that environment possibly influences the level of weevil infestation and its effects on kernel popping traits. Grain weight loss was negatively correlated with both kernel popping expansion volume and kernel popping percentage. Popping expansion volume showed positive correlation with popping percentage. Despite decrease in grain weight and popping quality upon weevil infestation, resistant genotypes with least damage to the popping attributes were identified.

8.9 Sorghum

8.9.1 Shoot Fly

8.9.1.1 Physico-chemical Characteristics for Resistance

Sorghum shoot fly, *Atherigona soccata* is an important pest of sorghum, and host plant resistance is one of the important components for minimizing the losses due to this pest. A diverse array of sorghum genotypes evaluated (Raut et al. 2015) to identify the physico-chemical characteristics conferring resistance to *A. soccata*. Susceptibility to shoot fly was associated with high amounts of total sugars, soluble protein, total phenols, crude fat and leaf surface wetness (RSV 1226, RSV 1251, RSV 1357 and Susceptible check DJ 6514), and seedling vigour, leaf glossiness and trichome density showed resistance to shoot fly. High enzyme activity was observed in resistant genotypes (RSV 1098, RSV 1188, RSV 2312, RSE 3 and IS 18551). Correlation studies with oviposition and % dead hearts as dependent character indicated the importance of seedling height at 14 and 28 DAE, leaf length and breadth ratio on 14 and 28 DAE, seedling vigour, glossiness, LSW, % moisture content, soluble protein, total sugars, fats, total phenols and polyphenol oxidase and peroxidase enzyme activity are important traits conferring resistance against shoot fly.

8.10 Chickpea

8.10.1 Pulse Beetle

The pulse beetle *Calasobruchus chinensis* is an economically important pest of stored chickpeas producing losses ranging from 30 to 100 %. Though conventional methods are being used for protection of stored chickpeas against bruchid species, ecologically sound methods including resistant genotypes are needed for an integrated approach to pest management. There is no information on susceptibility of different varieties of chickpea in southern zone to bruchid damage. In this connection pre-release genotypes and popularly grown varieties by farmers were tested for their reaction to pulse beetle (Kamakshi et al. 2015). Among the 18 chickpea entries tested, 6 were kabuli chickpeas and remaining were *desi* chickpeas. Each germplasm was artificially infested with *Calasobruchus chinensis* and was subjected to antixenosis and antibiosis test. Of the 18 genotypes tested, none of the genotypes exhibited resistance reaction against pulse beetle in storage. Vihar was found highly susceptible with the % seed damage of 76.7 % and the remaining 17 genotypes were under the category moderately susceptible (10–60 % of seed damage). The minimum seed damage was recorded in the genotypes NBeG 49 (16.61 %), WR 315 (17.2 %), K- 850 (18.7) and JAKI −9218 (19.9 %). The developmental period exhibited a negative correlation with % seed damage by recording 29 days in Vihar and 31.5 days in K-850.

8.10.2 Alternative Synthetic Insecticide: Pulse Beetle

Callosobruchus chinensis L. is a serious pest of chickpea during storage. Various vegetables oil, viz., mustard, ground nut, castor, coconut and palm oil evaluated (Rathod et al. 2015a) against pulse beetle during storage of chickpea, revealed a maximum % mortality of bruchid in the chickpea treated with 0.75 and 0.5 % groundnut oil, while there was a minimum % of grain damage and grain weight loss in stored chickpea in the grain treated with groundnut oil at 1.0 and 0.75 %. The minimum number of eggs 18.0 eggs per 100 grains and 20.5 eggs per 100 grains were recorded in the grain treated with groundnut oil at 10 ml (1.0 %) kg^{-1} grain and castor oil at 10 ml (1.0 %) kg^{-1} grain. This shows that groundnut oil can be used as an alternative to synthetic insecticide against pulse beetle.

8.10.3 Transgenic Chickpea Lines

Evaluation of transgenic chickpea lines, viz., BS5A.1(T2) 18-1P1, BS5A.1(T2) 18-2P1, BS5A.2(T2) 19-1P2, BS5A.2(T2) 19-2P1, BS5A.2(T2) 19-3P1 and BS5A.2(T2) 19-3P2, for resistance to *Helicoverpa armigera* under glasshouse conditions using cage technique was conducted during 2011–2014 at ICRISAT along with two non-transgenic chickpea lines, ICC506 EB (resistant check) and Semsen (control) revealed that the transgenic line, BS5A.1(T2) 18-1P1, suffered significantly lower leaf damage (DR: 2.4) as compared to non-transgenic line, Semsen (DR: 7.8) and ICC 506 EB (DR: 6.7). The larval survival and the weight gain by the *H. armigera* larvae was significantly greater on ICC 506EB (75.8 % and 14.1 mg $larva^{-1}$) and Semsen (72.9 % and13.1 mg $larva^{-1}$) than on the transgenic chickpea lines tested. The weight gain by *H. armigera* larvae on other transgenic lines ranged from 5.1 to 8.7 mg $larva^{-1}$, with significantly greater weight gain on BS5A.2(T2) 19-3P2 (8.7 mg $larva^{-1}$). Significant differences in grain yield were observed between transgenic and non-transgenic plants when infested with *H. armigera* larvae for 10 days. The dry matter weight was significantly higher in BS5A.2 (T2) 19-3P1 (6.5 g per three plants) than in Semsen (3.4 g per three plants) and ICC 506 EB (3.7 g per three plants). The weight of the pods in transgenic chickpea lines ranged from 3.3 to 1.6 g per three plants as compared to 1.0 g per three plants in Semsen and 1.3 g per three plants ICC 506 EB. Seed weight was significantly greater in BS5A.2 (T2) 19-2P1 (2.7 g per three plants) compared to Semsen (0.7 g per three plants) and ICC 506 EB (1.0 g per three plants). Number of seeds formed was highest in BS5A.1 (T2) 18-1P1 (18) as compared to that on non-transgenic chickpeas. Since leaf feeding damage was less in transgenic chickpea plants, the dry matter weight, pod weight, seed weight and number of seeds formed were significantly more than on non-transgenic chickpea plants (Shaila et al. 2015).

8.11 Green Gram

8.11.1 Efficiency of New Insecticides Against Thrips

The insect pests, viz., thrips, *Maruca vitrata*, *Spodoptera exigua* and galerucid beetles are causing considerable economic loss to green gram crop. The pod borers are the major pests at flowering and pod development stage and are the key impediments for its low production. The existing control measures are not highly effective against the insect pests of green gram. There is a need to evaluate new molecules with novel modes of action and the farmers are using freely without having the knowledge of their efficacy.

New insecticides, viz., spiromesifen 240SC, lufenuron 5EC, novaluron 10EC, chlorantraniliprole 20SC, flubendiamide 480SC, metaflumizone 22SC, emamectin benzoate 5WSG, spinosad 45SC and acephate 75SP, were evaluated (Kumar et al. 2015a) for their efficacy against thrips and *Maruca vitrata* in green gram. Spinosad 45SC at 0.3 ml l^{-1} was superior in controlling the thrips population and recorded a 64.32 % reduction, while novaluron 10EC at 0.75 ml l^{-1} led to a reduction of 52.83 % in thrips population over control. Further, spinosad 45SC at 0.3 ml l^{-1} recorded least mean pod damage, i.e. 6.25 %, and recorded highest yield, i.e. 983 kg ha^{-1}, followed by chlorantraniliprole 20SC at 0.15 ml l^{-1}, with 8.37 % pod damage and yield of 875 kg ha^{-1}. The other treatments, viz., emamectin benzoate 5 WSG, flubendiamide 480SC, novaluron 10EC, lufenuron 5EC and metaflumizone 22SC, recorded yields of 826 kg ha^{-1}, 784 kg ha^{-1}, 751 kg ha^{-1}, 728 kg ha^{-1} and 692 kg ha^{-1}, respectively. The treatment acephate 75SP recorded yield of 657 kg ha^{-1} followed by the control plot, i.e. 546 kg ha^{-1}.

8.12 Pigeon Pea

8.12.1 Surveillance for Pod Borer

Pigeon pea, *Cajanus cajan* (L.) Millsp, is one of the most important pulse crops in India. Insect pests feeding on flowers and pods are the most important biotic constraint affecting yield of pigeon pea. In India it is attacked by nearly 250 species of insect pests. The pod borers are the key impediments for the low productivity. In Maharashtra, pod borer *Helicoverpa armigera* (Hubner), plume moth *Exelastis atomosa* Walsingham, pod fly *Melanagromyza obtusa* (Malloch) and spotted pod borer *Maruca testulalis* Geyer are considered important in causing economic losses to the farmers. These pod borers' damages amount to 57 % pods and 35 % seeds with final yield loss up to 28 %.

A survey carried out under Crop Pest Surveillance and Advisory Project (CROPSAP) in eight districts of Marathwada against pod borers' damage on pigeon pea in four succeeding years from 2010–2011 to 2013–2014 has shown that during 2010–2011, the pod borer damage (%) ranged from 2.90 to 9.44 %. The maximum pod borer damage (%) was noticed in Aurangabad district (9.44) followed by Parbhani (5.04) and Osmanabad (4.71). The pod borer damage was meagre in Latur (2.90). Similarly during 2011–2012, pod borer infestation ranged from 0.59 to 4.36 % with highest in Aurangabad (4.36) followed by Nanded (3.66) and Beed (3.61) and lowest in Osmanabad (0.59). During 2012–2013 the pod borer damage (%) was noticed to the tune of 1.10–3.05 % with a maximum in Jalna (3.05) followed by Aurangabad (2.58) and Beed (2.55). Similarly during 2013–2014 average pod borer infestation (%) ranged from 0.69 to 2.30 % with maximum in Jalna (2.30) followed by Aurangabad (2.13) and Parbhani (2.12) and minimum in Nanded (0.69). On the basis of 4-year surveyed data, the pod damage was severe during 2010–2011 in all districts and maximum pod borer damage (%) was recorded in Aurangabad (4.63) followed by Jalna (3.08), Parbhani (2.98) and Beed (2.97). On the basis of the above study, it could be concluded that Aurangabad, Jalna, Parbhani and Beed are identified as hot spots for pod borer (Badgujar et al. 2015).

8.12.2 Pod Borer Insecticidal Resistance

Helicoverpa armigera (Hubner) collected from major pigeon pea-growing districts of Vidarbha, viz., Akola, Amaravati, Buldhana, Yavatmal and Washim, was evaluated (Salunke et al. 2015) for its resistance against commonly used insecticides. The LDP during 2012–2013 indicated LC_{50} value of cypermethrin in the range of 0.074–0.461 % with maximum in Yavatmal (0.461 %) and LC_{90} value within the range of 0.190–1.840 %. LC_{50} value of quinalphos was in the range of 0.018–0.071 % with maximum in Yavatmal (0.071 %) and LC_{90} value within range of 0.058–0.128 %. LC_{50} value of methomyl was in the range of 0.026–0.083 % with maximum in Amaravati (0.083 %) and LC_{90} value within range of 0.071–0.130 %. LC_{50} value of indoxacarb was in the range of 0.389–4.785 %. The maximum LC_{50} value (4.785 %) was noticed in Yavatmal. LC_{90} value was within the range of 6.631–51.548 %. LC_{50} value of spinosad was in the range of 0.028–0.067 % with maximum in Amaravati (0.067 %) and LC_{90} value within range of 0.058–0.123 %.

Ravages of lepidopteron pod borers during flowering and pod formation stage and pod fly during pod formation stage are the major bottlenecks in attainment of desired productivity levels of pigeon pea. A cost-effective and farmer-friendly module designed (Sreekanth et al. 2015) for the management of pod borer complex in pigeon pea by evaluation of 11 modules along with a control revealed superiority of the module 4 (spinosad 3.0 ml per 10 l at 50 % flowering stage, flubendiamide 480SC 2.0 ml per 10 l at 10 days after first spray, chlorantraniliprole 20SC 3.0 ml per 10 l at 20 days after second spray), followed by module 8 (azadirachtin 10,000 ppm 10 ml per 10 l at 50 % flowering stage-flubendiamide 480SC 2.0 ml per 10 l at 10 days after first spray-chlorantraniliprole 20SC 3.0 ml per 10 l at 20 days after second spray), which registered lowest pod damage due to pod borer complex with highest monetary returns of INR33399 and INR27573 per ha, respectively. However, the highest incremental cost-benefit ratio of 1:5.7 was obtained with module 8, followed by module 4 (1:5.1).

The insecticidal property of sweet flag rhizome (*Acorus calamus* L.) formulations investigated (Latha and Naganagoud 2015) against *Callosobruchus analis* on pigeon pea in storage revealed that the sweet flag rhizome powder with talc as a carrier at 2 % concentration reduced the egg laying with 17.67 eggs per 10 seeds and seed damage with 29.33 % as against 98.33 % in untreated check after 120 days of treatment (DAT). The seed weight loss was also less in the same treatment with 2.42 % seed weight loss after 120 days. The number of adults emerged per 100 g of seeds after 120 days was 55 adults.

8.13 Black Gram

The productivity of black gram is constrained by various abiotic and biotic stresses, resulting in drastic reduction in mean yield. White fly, is the key impediment for its low production by transmitting YMV. There were huge losses due to yellow mosaic virus in pulses for the past few years due to which most of the farmers switched over to maize cultivation. As the whitefly, the vector of this virus, is polyphagous and very few numbers of flies are able to transmit the virus to vast number of plants, control of whitefly is the most important measure to avoid losses due to YMV.

New insecticides, viz., spiromesifen 240SC, diafenthiuron 50WP, acetamiprid 20SP, imidacloprid 70WG, thiamethoxam 25WG, buprofezin 25SC, triazophos 40EC and acephate 75SP, evaluated (Kumar et al. 2015b) against YMV transmitting whiteflies (a vector of yellow mosaic) in black gram revealed that imidacloprid 70WG at 0.1 g l^{-1} was found to be superior over all other treatments in controlling the whitefly population by recording least mean whitefly population, i.e. 4.33 followed by spiromesifen 240SC at 0.4 ml l^{-1} with a mean whitefly population of 6.33 and others, viz., diafenthiuron 50WP at 2.0 g l^{-1}, thiamethoxam 25WG at 0.2 g l^{-1}, acetamiprid 20 SP at 0.2 g l^{-1} and buprofezin 25SC at 1.7 ml l^{-1} recorded a mean whitefly population of 8.66, 10.67, 13.33, 16.00 and 20.00, respectively, while there was highest mean whitefly population of 25.00 in untreated control followed by the treatment acephate with 20.00. Highest grain yield was noticed in imidacloprid 70WG at 0.1 g l^{-1}, i.e. 1172 kg ha^{-1}, while least grain yield was noticed in the control plot, i.e. 707 kg ha^{-1}.

Newer insecticidal molecules with novel modes of action, viz., imidacloprid, thiamethoxam alone as seed treatment and in combination with other recently introduced low dose molecules, viz., spirotetramat, buprofezin, spiromesifen, triazophos as foliar spray against the field population of thrips and another set of insecticides, viz., acetamiprid and triazophos imposed as exclusively foliar sprays evaluated (Jyothi and Kumar 2015) against thrips in black gram, revealed that imidacloprid + spiromesifen and imidacloprid + spirotetramat were proved to be highly effective in reducing the thrips population and similarly, imidacloprid + spiromesifen was found to be the safest on predators among foliar sprays. The mean population of predators was the lowest in all insecticide treated plots over untreated control plots. The seed treated plots with imidacloprid and thiamethoxam recorded maximum population of predators, i.e. spiders and coccinellids, whereas thiamethoxam + buprofezin and thiamethoxam + spirotetramat were relatively more toxic to spiders followed by thiamethoxam + triazophos and triazophos + acetamiprid (as scheduled spray) on the seventh day only after the third round of insecticidal spray. The seed treatment with thiamethoxam in combination with buprofezin, spirotetramat and triazophos was found to be more toxic to coccinellids at the third and seventh day after third insecticidal spray than other treatments. However, among all the treatments tested, seed treatment with imidacloprid and thiamethoxam alone and imidacloprid + spiromesifen were the safest treatments to the natural enemies.

8.14 Safflower

8.14.1 Compatibility of Insecticides and Fungicides

Safflower is attacked by aphids and major fungal diseases like wilt, root rot and *Alternaria* blight. The yield loss due to these biotic constraints can be up to 50–100 % depending on the susceptibility of the cultivar used. Both pests and diseases may occur simultaneously in the field. To bring down the cost of pesticide application, farmers tend to mix both the insecticides and fungicides. However, the compatibility of the newly recommended insecticides like thiamethoxam, clothianidin and acetamiprid with the recommended combination fungicide, (carbendazim+mancozeb) and *Trichoderma asperellum* as seed treatment against diseases like wilt, root rot and *Alternaria* blight is not known.

The compatibility of insecticides, viz., clothianidin, thiamethoxam, acetamiprid and dimethoate with a combination of fungicide (carbendazim+mancozeb) studied (Srinivas and Prasad 2015) for their efficacy against aphids and diseases, has shown that the physico-chemical changes like breaking of fluid or sedimentation or change in colour or temperature of the fluid were not noticed when clothianidin or thiamethoxam or acetamiprid or dimethoate mixed with (carbendazim+mancozeb). In SSF-658 safflower cultivar, in a field trial, aphids were released on the plants and the compatibility of these studied through their application as tank mix sprayed only once during the crop growth period has shown that the number of aphids in top 5 cm twig before treatment were around 20. There were no symptoms of phytotoxicity on safflower due to combined application of clothianidin or thiamethoxam or acetamiprid or dimethoate with (carbendazim+mancozeb).

No disease incidence was observed during the season. There were no significant differences in insecticide efficacy whether they applied solely or tank mixed with carbendazim+mancozeb with a reduction of aphids by 92.7–99.2 % over control. The study suggested that tank mixing with (carbendazim+mancozeb) did not affect the efficacy of tested insecticides. There was no difference in yield among the treatments.

8.15 Sunflower

8.15.1 Insect Management Tactics

Insect pests vary in their ecology. In sunflower disease-tolerant hybrid, an attempt was made to find suitable management tactics (Mane et al. 2015). The management of defoliators and head borer indicated that seed treatment with imidacloprid 70WS at 5 g kg^{-1} of seed checked the sucking insect pests. Profenophos 50EC at 2 ml l^{-1} of water, noticed minimum, i.e. 0.85 larvae of semi looper $plant^{-1}$, resulted to less defoliation, i.e. 6.17 and 6.00 % at 45 and 65 days after emergence. For the management of head borer, spinosad 45SC at 0.1 ml l^{-1} of water recorded minimum 0.22 larvae of head borer $plant^{-1}$. Profenophos 50EC at 2 ml l^{-1} of water and spinosad 45SC at 0.1 ml l^{-1} of water translated maximum yield 1433 kg ha^{-1} and 1389 kg ha^{-1}, respectively. Spraying of HaNPV, also checked the larval population of head borer and recorded 0.62 larvae of head borer per plant which was found at par with module comprised of spraying NSKE 5 % (0.73 larvae of larvae of head borer $plant^{-1}$). Spraying of HaNPV and NSKE 5 % proved safer to natural enemies and pollinating bees. These resulted in good yields of 1207 and 1130 kg ha^{-1}, respectively.

8.16 Castor

8.16.1 Integrated Pest Management

New insecticides and fungicides are being introduced into the market for the management of insect pests and diseases. In some situations both insect pests and diseases are occurring at the same time. Farmers are mixing the insecticides and fungicides to manage pests and diseases without knowledge if their combinations are compatible or not. There are reports of crop damage due to phytotoxic effect after spraying with combinations of chemicals. Hence it is imperative to generate information on compatibility of new insecticides and fungicides commonly used by farmers, which will in turn help the farmers to cut down their spraying costs and at the same time aid in use of effective combinations for management of both insect pests and diseases. The phytotoxic effect of four insecticides, viz., chlorantraniliprole, spinosad, profenophos and thiamethoxam when mixed with fungicides, viz., saaf and nativo, both at recommended and double the recommended dose along with control on castor were studied (Jyothsna et al. 2015a). Treatments were imposed as foliar sprays thrice at an interval of 15 days with hand compression knapsack sprayer, and the amount of spray fluid used was 500 l ha^{-1}. All the combinations of insecticides and fungicides were compatible with each other at both doses with the phytotoxicity scale ranging from 0 to 1.

Integrated pest management (IPM) module evaluated (Jyothsna et al. 2015a) in three popular cultivars of castor, viz., (DCH-519, DCH-177, DCS-107), along with farmer's practice. IPM module comprised of mechanical control of gregarious stages of tobacco caterpillar and older larvae of semilooper, economic threshold level-based application of thiodicarb 0.075 % against tobacco caterpillar (at 25 % foliage damage), indoxacarb 0.015 % against capsule borer (at 10 % capsule damage) and dimethoate 0.05 % against leafhopper (at 10 % hopper burn damage), superior in reducing the insect pest population over farmer's practice. In the IPM module, the mean population of semilooper, tobacco caterpillar, leafhopper and capsule borer damage was 0.3 larvae $plant^{-1}$, 0.8 larvae $plant^{-1}$, 16.2 leafhoppers $three\text{-}leaves^{-1}$ $plant^{-1}$ and 10.3 % capsule damage as against 0.6 larvae $plant^{-1}$, 2.3 larvae $plant^{-1}$, 20.7 leafhoppers $three\text{-}leaves^{-1}$ $plant^{-1}$ and 16.5 % capsule damage in farmer's practice, respectively. IPM module was found safer to its larval parasitoid, *Snellenius* (*Microplitis*) *maculipennis* as compared to farmer's practice. The higher mean castor seed yield (1348 kg ha^{-1}) and avoidable yield loss (67.6 %) was obtained from IPM module over farmer's practice with seed yield of 1008 kg ha^{-1} and avoidable yield loss of 55.5 %. Further, higher net profit of INR 26,889 ha^{-1} and incremental cost-benefit ratio (ICBR) of 1:5.95 was recorded in IPM module.

8.17 Groundnut

8.17.1 Nano-silica Against Bruchids

Nanotechnology has become one of the most promising new technologies in the recent decade. The use of nanoparticles as pesticide is an alternative strategy to combat pests which have become resistant to conventional pesticides. Therefore nanoparticles especially silica nanoparticles have an excellent potential against stored-grain as well as seed protecting agent if applied with proper safety measures. The efficacy of nano-silica evaluated (Jyothsna et al. 2015b) against peanut bruchid, *C. Serratus*, tested in groundnut cultivar K6, revealed that nano-silica applied at 1000 and 500 ppm kg^{-1} of peanut pods provided complete protection up to 9 months of storage and did not record any fecundity, whereas the lowest concentration of nano-silica (250 ppm kg^{-1}) protected the pods for only 6 months. The scanning electron microscopy (SEM) images also clearly revealed that the amorphous uniform-sized flat nano-silica particles resulted in uniform distribution of the nanoparticles over the body, resulting in abrasion of the insect cuticle thereby resulting in loss of water from the body and also caused defacing of the mandibles which in turn resulted in the death of the insect. This hypothesis for the physical mode of action makes the case for the use of nanocides stronger. The nanocides can be removed by conventional milling process without leaving residues on the stored grain unlike sprayable formulations of conventional pesticides.

8.18 Cotton

The pink bollworm, *Pectinophora gossypiella*, is a serious pest on cotton infesting the green bolls. The infestation of pink bollworm (PBW) in green bolls and larval population was effectively checked due to spray of deltamethrin+triazophos at 450 g a.i. ha^{-1} and was found promising in minimizing the loculi damage in green bolls, recorded minimum open boll and loculi damage due to PBW at harvest and resulted in the highest seed cotton yield (914 kg ha^{-1}) than thiodicarb at 750 g a.i. ha^{-1} and lambda-cyhalothrin (25 g a.i. ha^{-1}). Spinetoram at 36 g a.i. ha^{-1} and flubendiamide at 60 g a.i. ha^{-1} were least effective in reducing pink bollworm infestation. Further, highest ICBR was recorded in deltamethrin+triazophos (450 g a.i. ha^{-1}). Thus, use of deltamethrin+triazophos at 450 g a.i. ha^{-1} can form cost-effective pink bollworm management option in cotton (Choke et al. 2015).

8.18.1 Transgenic Cotton

Transgenic crops hold great promise for improving agricultural output, but the potential unwanted effects of transgenic crops on soil microbial diversity is still not fully understood. The majority of studies addressing the risks of transgenic crop cultivation have addressed only aboveground effects. However, recent methodological advances in soil microbial ecology have allowed researchers to focus on the underground part and gain knowledge of transgenic crop-driven effects on the microbial communities and processes in soil that are essential to key terrestrial ecosystem functions. Genetically modified plants have the potential to significantly change the microbial dynamics, biodiversity and vital ecosystem functions such as nutrient mineralization, disease incidence and carbon turnover and plant growth.

The effect of transgenic cotton evaluated by (Mandal et al. 2015) on activities and diversities of microbes of the rhizosphere soils of transgenic cotton-based cropping system, viz., in soybean, red gram, wheat and vegetables under vertisols, revealed that soil enzyme activities and microbial population counts of cellulose decomposers, phosphate solubilizers and nitrogen fixers were found higher under Bt cotton than non-Bt cotton system. The study clearly indicated that the higher activities of beneficial microbes and diversity under Bt cotton than non-Bt cotton cropping system might be due to greater rhizodeposition, leaf fall and root biomass accumulation serving as source of bioenergy for native microbes. It was further concluded that the Bt cotton-based cropping system is having no adverse effect on soil beneficial microbes.

8.19 Cabbage

8.19.1 Diamondback Moth

The relative toxicity of chemical insecticides, viz., flubendiamide, chlorantraniliprole, emamectin benzoate, spinosad, lufenuron and novaluron evaluated (Nikam and Chandele 2015) against diamondback moth (DBM), *Plutella xylostella* L., a pest on cabbage, has shown that among the chemical insecticides that were tested for their larvicidal action against DBM, spinosad was the most effective, followed by emamectin benzoate, chlorantraniliprole, flubendiamide, lufenuron and novaluron. The LC_{50} values recorded were 0.00009 % and 0.00011 % for spinosad, 0.00016 % and 0.00022 % for emamectin benzoate, 0.00076 % and 0.00099 % for chlorantraniliprole, 0.00119 % and 0.00148 % for flubenadimide, 0.00280 % and 0.00284 % for lufenuron and 0.01557 % and 0.01888 % for navaluron, respectively, against third and fourth instar larvae of *P. xylostella*. Under field conditions, spinosad has emerged as "most effective one in reducing the larval count and increasing the marketable cabbage head. It is followed by emamectin benzoate, chlorantraniliprole, flubendiamide, lufenuron and novaluron. The larval count and yield of marketable cabbage heads in the different treatment ranged from 0.27–0.89 to 250.88–189.60 q ha^{-1}, respectively, as against 9.16 and 103.56 q ha^{-1} in untreated control. Spinosad has recorded less number of larvae (0.27 larvae plant^{-1}) with the highest yield (250.88 q ha^{-1}).

8.20 Okra

8.20.1 Shoot and Fruit Borer

Okra [*Abelmoschus esculentus* (L.) Moench], a member of family Malvaceae, is an important vegetable crop grown for its tender green fruits during the summer and rainy season in India. The shoot and fruit borer, *Earias vittella* (Fabricius), is a highly destructive pest causing serious damage. The studies on consumption, digestion and utilization of fruit blocks of selected varieties on fresh weight basis by the fourth instar larvae (5–7 days) of spotted bollworm conducted (Sultani et al. 2015) at a constant temperature of 28 ± 1 °C in BOD have shown that the survival of larvae was less on fruit blocks (67.00–82.66 %) and on whole fruits (33.01–51.66 %) of HBT-1-19-1-1-2 and HB-03-29-7B in comparison to that of HBT-1-18-1-C-1 (82.66 % and 51.66 %) genotypes, respectively. Longer larval period (15.33–14.66 days), pupal period (9.66 and 10.33 days), lowest pupal weight (0.07 and 0.08 mg) and lowest % adult emergence (93.00–93.33 %) were observed on fruit blocks of HBT-1-19-1-1-2 and HB-03-29-7B genotypes than on HBT-1-18-1-C-1 (10.33 days, 8.66 days, 0.11 mg and 96.33 %), respectively. The adverse effects were found on various food consumption and utilization indices of *E. vittella* larvae on HBT-1-19-1-1-2 and HB-03-29-7B genotypes.

8.20.2 Seasonal Influence on Sucking Pests

Seasonal incidence of leafhopper and aphids studied (Dudhale et al. 2015) revealed that population dynamics of sucking pests and weather parameters are correlated. Leafhopper population ranged from 0.6 to 18.3 per three leaves with its peak in 40th SMW. Aphid's population varied from 0.2 to 15.7 per three leaves with its peak in 40th SMW. Population of ladybird beetle was in the range of 0.2–15.10 per three leaves with its peak in 40th SMW. Shoot and fruit borer (number and weight basis) was with its peak in 40th SMW. The correlation studies revealed that the population of leaf hopper aphids, fruit and shoot borer and ladybird beetle was negatively correlated with maximum temperature. Minimum temperature with population of leafhopper, aphids, ladybird beetle and shoot and fruit borer was negatively correlated and highly significant. Morning humidity and rainfall showed the negative and non-significant correlation with pests of okra. Evening humidity had negative significant correlation with aphids and leafhopper activity.

Okra (var. Akola Bahar) crop was highly susceptible to sucking pests, i.e. aphids, jassids and whiteflies. The foliar application of different doses of newer insecticides evaluated for their effectiveness and economic management of sucking pests revealed that the 1.25 % and 1.50 % doses over recommended dose of treatment with foliar spray applications (at three DAT and seven DAT) were found to be most effective against aphids, jassids and whitefly. The recommended dose with foliar spray applications of thiamethoxam 25 WG at 0.006 % of treatment was found most effective against average par leaf population of aphids, jassids and whitefly. Lambda-cyhalothrin 5EC at 0.004 %, followed by thiamethoxam 25WG at 0.008 % was proved most effective against aphids, jassids and whiteflies. The recommended dose with foliar spray applications of thiamethoxam 25WG at 0.006 % of treatment has emerged as the most economical recording highest yield, net profit and highest ICBR. The highest cost-benefit ratio has been recorded in treatment thiamethoxam 25WG at 0.006 % followed by triazophos 40EC at 0.025 % and lambda-cyhalothrin 5EC at 0.004 %. The recommended dose with foliar spray applications of thiamethoxam 25WG at 0.006 % of treatment followed by lambda-cyhalothrin 5EC at 0.004 % and triazophos 40EC at 0.025 % was proved to be safer and conservative to predators.

8.21 Brinjal

8.21.1 Shoot and Fruit Borer

Fifteen brinjal genoypes screened (Patil et al. 2015a) for shoot and fruit borer incidence has shown that minimum shoot infestation as well as fruit infestation on number and weight basis was recorded on genotype AB 7–2 with 5.97 %, 9.87 % and 9.53 %, respectively, followed by genotype GBL 1 which showed 6.17 % shoot infestation as well as 10.28 % and 9.96 % fruit infestation on number and weight basis, respectively. The highest infestation of shoot (16.34 %) and fruit infestation on number (48.75 %) and weight basis (48.35 %) noted on genotype JBGR 1. The remaining genotypes, viz., JBL 8–8, GJB 2, AB 8–5, GOB 1, JBL 10–20, AB 9–1, JDNB 19, Surati Ravaiya, JBGR 6–7, JBL 10–203, JBGR 8–6 and JBL 10–197 registered shoot infestation ranging from 7.65 to 15.10 % while 16.74–44.77 % fruit infestation on number basis and 16.29–44.47 % on weight basis.

8.21.2 Mass Trapping Technology

Field experiments conducted to study the impact of mass trapping technology on the infestation of brinjal shoot and fruit borer in comparison with farmers' practices in the fields around the town of Bagalkot, Karnataka, India (Mansa et al. 2015), indicated that the total number of moths caught during the 10 weeks of study period from ten traps was 861. The highest mean number of male moths trapped per week was 29.6 during the tenth week and the lowest was 1.7 during the third week. Highest percentage of fruit damage (6.6 %) was reported in the ninth week, whereas lowest percentage was 0.2 in mass trapping during the seventh week, whereas in farmers practices highest % fruit damage (27.3 %) was observed in the ninth week, lowest (0.9 %) was found in the fifth week. In mass trapping block, highest shoot damage (6 %) was recorded during first week of trap installation which was on par with second week, i.e. 5.7 %. The lowest % shoot damage was observed during tenth week of traps installation, i.e. 0.4 %. Highest cost-benefit ratio (8.6) was recorded in the mass trapping technology when compared to farmers' practices.

8.21.3 Predatory Potentiality of Mites

The predatory mites of the family Phytoseiidae plays an important role in regulating and controlling the population of various species of phytophagous mites throughout the world.

The predatory potentiality of two predatory mites *Amblyseius indicus* and *Amblyseius tetranychivorous* on the egg, larva and adult stages of phytophagous mite *Tetranychus neocaledonicus* on brinjal host plant studied (Singh and Singh 2015) revealed that *Amblyseius indicus* proved to be a better predator with mean consumption of 8.23 eggs, 3.95 larva and 2.01 adult as compared to *Amblyseius tetranychivorous* with mean consumption of 6.85 eggs, 3.63 larva and 1.90 adult during 72 h. The mean number of *Tetranychus neocaledonicus* consumed increased as the population of the vegetable mite increased. These predatory mites can be effectively used in biological control of phytophagous mites in greenhouse and field conditions.

8.22 Chilli

8.22.1 Pest Management

In integrated pest management, different insecticides evaluated (Jadhav et al. 2015a) against major pests of chilli and the safety of these insecticides to natural enemy population in chilli ecosystem revealed that acetamiprid 20SP at 10 g ha^{-1} and spinosad 45SC at 135 g ha^{-1} were effective in reducing thrips and whitefly population. However, the highest yield (51.30 g ha^{-1}) was also obtained from treatment of acetamiprid 20SP at 10 g ha^{-1} then Spinosad 45SC at 135 g ha^{-1} (48.40 q ha^{-1}).

8.22.2 Acaro-insecticides

The bioefficacy of acaro-insecticides, viz., bifenazate 240SC, fenazaquin 10EC, thiamethoxam 25WG, chlorfenapyr 10SC, hexythiazox 10EC, clothianidin 50WG, dicofol 18.5EC (standard check) and abamectin 1.9EC (treated check) along with botanicals, viz., neem oil 5000 ppm and horti-impact evaluated (Kharbade et al. 2015b) against *Polyphagotarsonemus latus* (Bank) on capsicum in *rabi* season revealed that these acaro-insecticides were promising in suppressing the mite's population on capsicum. Abamectin 1.9EC was observed to be most promising by recording lowest cumulative mean mite population of 3.21 mites per three leaves and was on par with chlorfenapyr 10SC (3.58 mites per three leaves), fenazaquin 10EC (3.91 mites per three leaves) and bifenazate 240SC (4.13 mites per three leaves) followed by dicofol 18.5EC (4.56 mites per three leaves). Next effective treatments were thiamethoxam 25WG, clothianidin 50WG and hexythiazox 10EC which recorded 8.00, 8.77 and 9.35 mites per three leaves, respectively, and were noticed at par with each other followed by horti-impact (15.38 mites per three leaves) and neem oil 5000 ppm (16.59 mites per three leaves).

8.22.3 Compatibility of Insecticides and Fungicides

The compatibility of four insecticides, spinosad 45SC at 0.25 ml l^{-1}, fipronil 5 SC at 2 ml l^{-1}, diafenthiuron 50 WP at 1.5 g l^{-1}, chlorfenapyr 10EC at 2 ml l^{-1} and one fungicide, copper oxychloride (COC) at 3 g l^{-1} and their efficacy against chilli thrips, *Scirtothrips dorsalis* (Hood) and leaf spot, *Cercospora capsici* studied (Pathipati et al. 2015) revealed that the four insecticides and fungicide were physically (without forming any precipitation, curding, heating, bulbing) and chemically compatible (without any blasting, foaming). There was no phytotoxicity effect on any part of the plant from 1 to 14 days after spray. Lowest number of thrips $leaf^{-1}$ was recorded with spinosad+ COC (4.77) followed by fipronil + COC (5.91), diafenthiuron (6.03) and chlorfenapyr + COC (7.42). Lowest disease scale was recorded with spinosad+ COC (1.1) followed by diafenthiuron (1.5). This study helps to manage the thrips and leaf spot of chilli at same time and reduce the cost of plant protection.

8.23 Tomato

8.23.1 Jassid

Incidence of jassid (*Empoasca binotata* Walsh.) and its botanical management in tomato assessed for two consecutive seasons (2011–2013) has shown that a lower jassid population level was recorded during 42nd standard week to 2nd standard week and 17th–22nd standard (0.06–0.50 leaf^{-1}). The higher population level was maintained during 3rd standard week to 16th standard week and peak population (2.23 leaf^{-1}) was recorded on 10th standard week. In management trial acetamiprid was found to be most effective against jassid providing more than 80 % suppression. However, neem+spilanthes extract gave a satisfactory jassid control, recording more than 70 % suppression. Highest yield, 30.15 t ha^{-1} were recorded from acetamiprid-treated plots followed by neem+spilanthes (27.55 t ha^{-1}), spilanthes extract (26.67 t ha^{-1}) and polygonum extracts (26.32 t ha^{-1}). Other than control plots, the lowest yield was recorded from garlic-treated (23.11 t ha^{-1}) plots closely followed by tobacco extract (24.02 t ha^{-1})-treated plots. Though acetamiprid-treated plots gave satisfactory yield, it is not feasible for use as it is harmful when consumed by human beings. Therefore it can be suggested to use neem+spilanthes to control the jassid population effectively (Subba et al. 2015).

8.23.2 Morphological Characters on Fruit Borer

The influence of morphological characters in relation to larval population and fruit infestation of *H. armigera* in tomato studied (Archana et al. 2015) has shown that the correlation of plant height ($r = 0.7775$), branches per plant ($r = 0.7874$) and fruits per plant ($r = 0.7388$) with infestation of *H. armigera* was significantly positive, while the effect of fruit pericarp thickness ($r = -0.9576$) exhibited significantly negative association with the damage of tomato fruit borer. However the stem diameter, fruit diameter, locules fruit^{-1} and calyx fruit^{-1} showed positive but non-significant association with fruit infestation recorded 0.4855, 0.3048, 0.3413 and 0.3943 correlation values, respectively.

8.23.3 Biopesticides

The efficacy of biopesticides, viz., *Beauveria bassiana* (2.5 kg ha^{-1}) and *Verticillium lecanii* (2.5 kg ha^{-1}), and novel insecticides, viz., imidacloprid 30.5SC (100 ml ha^{-1}) and thiamethoxam 25 WG (75 g a.i. ha^{-1}), evaluated (Bharani et al. 2015) against the thrips, *Thrips tabaci* Lind and safety of biopesticides and novel insecticides against predatory coccinellids in tomato ecosystem revealed that lowest thrips population was recorded in insecticidal treatments imidacloprid 30.5SC and thiamethoxam 25WG (1.37 and 1.35 nymphs plant^{-1}), respectively. Biopesticides *Beauveria bassiana*, *Verticillium lecanii* (3.26 and 3.12 coccinellids per five plants) were found safer to predatory coccinellids than insecticides. Imidacloprid 30.5SC and thiamethoxam 25WG recorded (2.64 and 2.79 coccinellids per five plants) in tomato ecosystem.

8.24 Chrysanthemum

8.24.1 Incidence of Aphids and Thrips

Fifteen germplasm accessions of chrysanthemum (*Dendranthema grandiflora* Borkh.) consisting of yellow, white and red flowered cultivars assessed for relative incidence of aphid and thrips (Charan et al. 2015) indicated that the genotype Red gold was least affected by aphids and recorded the lowest population of 19.10 aphids per apical shoot. The highest mean aphid population was recorded on PAU-B-107 (31.35 aphids per apical shoot) followed by Poonam (29.89 aphids per apical shoot), closely followed by Raichur (29.06 aphids per apical shoot) and Akitha (28.27 aphids per apical shoot). Among the five cultivars which recorded highest mean aphid population PAU-B-107, Ratlam selection, Akitha were white flowered; Poonam and Raichur were yellow coloured cultivars. Similarly of the five cultivars which recorded lowest aphid population count, Red gold and Priya were red flower cultivars, IIHR-6 and Kadapa local were white coloured and Aparjitha was the yellow coloured cultivar indicating no clear influence of colour on aphid incidence. With respect to thrips incidence, the genotype Redstone recorded lowest thrips count (14.06 per ten plants) closely followed by Kadapa local (14.80 thrips per ten plants). The maximum thrips population was found in Raichur (53.28) closely followed by Poonam (48.06 thrips per ten plants), CO-3 (46.71 thrips per ten plants) and Geethanjali (41.61 thrips per ten plants). The top five cultivars which recorded highest thrips population were yellow flowered, and among the five cultivars which received least thrips count, four were red coloured and only one cultivar was white flowered showing clearly that the colour of the flower influenced the thrips incidence.

8.25 Tobacco

8.25.1 Chitin Synthetic Inhibitors

The efficacy of chitin synthesis inhibitors evaluated (Gurve and Chormule 2015) against *S. litura* on soybean revealed that lufenuron emerged as the most effective treatment on the basis of larval count, % leaf damage and grain yield, followed by flufenoxuron, novaluron and diflubenzuron. The average larval count, % pod damage and grain yield in the different treatments ranged from 0.32 to 3.00 and 1076.83–2064.35 kg ha^{-1}, respectively, as against 9.005 and 826.88 kg ha^{-1}, respectively, in untreated control. Lufenuron 0.006 % dominated over all other treatments, recorded highest grain yield (2064.35 kg ha^{-1}): and registered 0.32 average larval plant^{-1} and 2064.35 kg grain yield ha^{-1}. All the treatments except chlorpyriphos were effective in controlling the *S. litura* population. Chitin synthesis inhibitors due to their specificity and safety towards non-target organisms are ideal products and could be used in alternations with conventional chemical insecticides. The findings are useful in determining the most effective concentration as well as interval between treatments with chitin inhibitors for the management of *S. litura*.

8.25.2 Anthranilic Diamide Insecticides

DuPont™ Cyazypyr™ (DPX-HGW86, cyantraniliprole) is the second anthranilic diamide insecticide (IRAC Group 28) registered in India. The compound is the first in its class to control a cross-spectrum of important species of insect pests on agronomic crops. The pest groups include piercing-sucking pests such as aphids, psyllids, leafhoppers, plant hoppers and whiteflies, as well as biting-chewing pests such as Lepidopteran caterpillars, leaf miners, fruit flies, sawflies, weevils and beetles and rasping pests such as thrips. The mode of action of Cyazypyr™ is as an agonist of the insect ryanodine receptor agonist. Cyazypyr™ resulted in rapid disruption of pest muscle function, immediate cessation of pest feeding, impaired locomotion and other activities that are controlled by the insect muscle action. Cyazypyr™ acts mainly through ingestion and less so through contact action. Although it exhibits the unique selectivity to arthropods and has excellent pest control features, Cyazypyr™ still has remarkably low impact on pest natural enemies since their exposure to the product is very limited. DuPont™ Cyazypyr™ has been designated the coveted reduced-risk product status by the US-EPA. Therefore, products powered by DuPont™ Cyazypyr™ are great technologies in integrated pest management (IPM) and insecticide resistance management (IRM) programmes in agriculture and valuable tools for agronomic crop producers (Rajiv Rathod et al. 2015b).

8.26 Banana

8.26.1 Postharvest Shelf Life

Banana is considered as staple fruit worldwide because it is easily grown, nutritious, a cheap source of energy and vitamins, easy to peel and the most edible everyday fruit relished from cradle to grave. In developing countries like India, there is a need for an alternative non-sophisticated technology for extension of postharvest shelf life and maintaining the quality at ambient temperature. Such non-expensive technology is postharvest treatment of fruits with 1-MCP, a non-toxic gaseous product that has been used as a tool to extend postharvest life, delay softening and to improve postharvest shelf life and storage quality of different climacteric fruits.

Banana cv. Grand Naine evaluated (Jyothi et al. 2015) for the effect of 1-Methylcyclopropene (1-MCP) on shelf life and postharvest quality parameters during storage has shown that 1-Methylcyclopropene was found to be effective in maintaining the peel colour, firmness, TSS, ascorbic acid content and titrable acidity under ambient conditions. The effectiveness increased with increase in the concentration and duration of exposure. 1-MCP at 750 ppb for 24 h was found to be effective in maintaining the quality of banana cv. Grand Naine under ambient conditions.

8.27 Mango

8.27.1 Population Dynamics of Mango Hopper

The mango, *Mangifera indica* L. is grown in Telangana in an area of 53,000 ha producing 3.22 tonnes and is considered as king of all fruits. Among the mango pests, mango hoppers are the most serious and widespread pests throughout the country. *Idioscopus clypealis* and *Idioscopus nagpurensis* are the most commonly observed destructive species of hoppers in Telangana which cause heavy damage to mango crop. The hopper damage manifests in the form of poor or no fruit set and premature dropping of fruits resulting in total loss of crop in severe outbreaks. Proper timing of pesticide application is very important in managing this pest which is dependent on the population build-up of the pest.

Field studies conducted (Anitha et al. 2015) to study the seasonal incidence and influence of abiotic factors on population dynamics of hoppers on mango var. Banganpalli at Fruit Research Station, Sangareddy, revealed that the incidence of mango hopper was observed on the mango panicles during 47th to 12th standard week from 2001 to 2011. A distinct difference in hopper population peaks were observed in different years corresponding to the rise in temperature and decreasing relative humidity in different months coinciding with the flowering season. An attempt was made to develop weather-based pest forewarning models using linear regression models and the correlation between minimum and maximum temperature, forenoon relative humidity and hopper incidence was worked out.

8.27.2 Gall Midge

Nine mango cultivars screened (Jadhav et al. 2015b) for their field reaction against mango gall midge *P. Matteiana* (Kieffer and Cecconi) (Cecidomyiidae: Diptera) in Gujarat has shown that infestation of mango gall fly was highest (46.75 %) in Kesar while it was lowest (18.43 %) in Totapuri. The % infestation of mango gall midge in Rajapuri (44.00 %), Begumpali (36.16 %), Sonpari (34.28 %), Local (32.02 %), Amrapali (32.45 %), Mallika (31.07 %) and Dasheri (28.39 %), respectively. Totapuri and Kesar were categorized as least and most susceptible cultivars for mango gall midge.

8.28 Sapota

8.28.1 Seasonal Incidence of Fruit Borer

The seasonal incidence studies (Vijayaraghavendra et al. 2015) of sapota fruit borer, *Phycita erythrolophia* Hampson, (Pyralidae: Lepidoptera) infesting four genotypes of sapota, viz., cricket ball, Kalipatti, DSH-1 and DSH-2, has revealed that the fruit borer was active throughout the year with a varying degree of infestation. Among the four different genotypes of sapota, the incidence of fruit borer damage was more in cricket ball (10.24 %) followed by Kalipatti (9.64 %) varieties, whereas on hybrids it ranged from 7.33 to 8.07 %. The mean percentage of fruit damage across the genotypes was highest (16.42 %) in March and lowest (3.79 %) in August. Pest incidence was more during dry period, i.e. from November to April and less during rainy season (July to October). There was a significant and positive correlation between fruit borer damage and maximum temperature in all genotypes. Rest of the weather factors, viz., minimum temperature, relative humidity and rain fall had no influence on pest population during the period of study.

8.29 Sugarcane

8.29.1 Shoot and Internode Borer

Among the economically important borer species infesting sugarcane, only two species, viz., shoot borer, *Chilo infuscatellus* Snellen, and internode borer, *Chilo sacchariphagus indicus* Kapur, are important in Andhra Pradesh under stress conditions. While the early shoot borer (ESB) damages the crop at early stage, the internode borer (INB) infests the crop from the cane formation and its activity continued unabated till harvest. Infestation by early shoot borer was found high under moisture stress conditions. The infestation was found to range from 15.5 to 60.5 % of the shoots produced and a maximum of 90 % clumps have been affected in North Coastal Zone of Andhra Pradesh. It causes losses of between 22 and 33 % in cane yield, 12 % in sugar recovery and 27 % in jaggery. All commercial varieties were more or less equally susceptible to the early shoot borer, when the dead hearts were 15 %; the reduction in yield was only half a tonne per hectare. The damage due to internode borer infestation results in both yield and quality loss.

The efficacy of soil application of granular insecticides over foliar sprays against the ESB as well as the INB studied (Bhavani et al. 2015) revealed that all the insecticidal treatments significantly reduced sugarcane borers and were superior over untreated control. Among the test insecticides, soil application of chlorantraniliprole 0.4G at 22.5 kg ha^{-1} at the time of planting and 60 days after planting significantly reduced the incidence of early shoot borer (14.86 %) and internode borer intensity (6.12 %) and registered highest cane yield (99.10 t ha^{-1}) as compared to untreated control (43.15 %; 20.01 %; 64.35 t ha^{-1}). However, spraying of chlorantraniliprole 18.5SC at 375 ml ha^{-1} at 30 and 60 days after planting was also found effective against internode borer (6.20 %) with superior cane yield (97.78 t ha^{-1}) and it was on par with chlorantraniliprole 0.4G at 22.5 kg ha^{-1} at the time of planting and 60 days after planting. It was concluded that scheduled soil application of 0.4G at 22.5 kg ha^{-1} at planting and 60 days after planting or spraying chlorantraniliprole 18.5SC at 375 ml ha^{-1} at 30 and 60 days after planting are effective against early shoot borer and internode borer and to reap sustainable cane yield in sugarcane.

8.30 Weed Control

8.30.1 Biopesticidal Impact of *Lantana camara*

Lantana camara is an aggressive invader of natural ecosystem due to the presence of harmful allelo chemicals in its leaf, stem and root which has been found to exhibit an adverse impact on other plants growing in its vicinity and on agronomic crops. It has been found to inhibit dicot weeds in the field when the dry matter of this plant was buried in soil before first rain.

The biopesticidal impact of callus of *Lantana camara* investigated (Singh et al. 2015) on the growth of an aquatic fern *Salvinia molesta* by use of D.S. Mitchell bioassay has shown that callus extract (1 % conc.) showed toxicity to the growth of *S. molesta* after 7 days. The leaves and flowers of this species serve as a source of allelopathic compounds, both exhibited high potential to act as strong biopesticide and suppressed a number of organisms including bacteria, fungi, aquatic and terrestrial weeds in its aqueous leachate or extract in earlier studies. Further, this plant contains 14 phenolic acids including salicylic acid which makes it a safe and potent pesticide. This study highlights the ability of callus of *L. camara* to produce allelochemicals under in vitro conditions that also exhibited phytotoxicity.

8.30.2 Selective Post-emergence Weedicide in Sugarcane

The efficacy of metribuzin, a selective post-emergence weedicide studied (Devi et al. 2015) in control of weed flora in sugarcane, revealed that post-emergence spraying of metribuzin at 1.0 kg ha^{-1} + 2,4-D at 2.0 kg ha^{-1} twice at 20 and 90 days after planting controlled the weeds effectively and registered higher mean cane yield (90.9 t ha^{-1}), was found on par with metribuzin at 1.5 kg ha^{-1} + 2,4-D at 2.5 kg ha^{-1} sprayed at 20 and 90 DAP (89.3 t ha^{-1}) or recommended practice of 2,4-D at 4.5 kg ha^{-1} + Gramoxone at 2.5 l ha^{-1} (88.0 t ha^{-1}) and found superior to all other weedicide combinations. Metribuzin alone at 1.5 kg ha^{-1} as pre-emergence application on third day after planting registered lowest cane yield of 77.6 t ha^{-1} (during 2013–2014 season only). Hence, metribuzin, a selective herbicide, proved to be applied safely at 1.0 kg or 1.5 kg ha^{-1} in combination with 2, 4-D at 2.0 kg or 2.5 kg ha^{-1} in suppressing the weed growth and found alternative to 2, 4-D + Gramoxone, a recommended practice which is to be sprayed with care because of its non selectivity to sugarcane.

8.30.3 Post-emergence Herbicides in Groundnut

Studies carried out on post-emergence herbicide application at 20 days after emergence in groundnut for finding out an appropriate post-emergence herbicide for maximum weeds control and yield enhancement under *kharif* groundnut (Mahatale et al. 2015) have shown that the lowest weed index (3.26 %) and minimum weed dry weight was observed under post-emergence herbicides propaquizafop 10EC at 100 g a.i. ha^{-1} followed by quizalofop ethyl 5 EC at 100 g a.i. ha^{-1} (9.74). Maximum weed control efficiency (96.4 %) was recorded in post-emergence herbicides propaquizafop 10EC at 100 g a.i. ha^{-1} and minimum weed control efficiency recorded under pre-emergence application of pendimethalin 30 EC at 1000 a.i. ha^{-1} (81.64 %). This clearly indicated that weeds were controlled effectively under post-emergence herbicide. The highest dry pods yield (2927 kg ha^{-1}) was recorded with hand weeding (20 and 40 days after emergence) and the lowest (1520 kg ha^{-1}) was under weeded check. Among the herbicidal treatments, in post-emergence herbicides, propaquizafop 10EC at 100 g a.i. ha^{-1} recorded maximum gross monetary return and net monetary return. Herbicidal treatments resulted in considerably lower cost of cultivation. The B:C ratio was found maximum with post-emergence herbicides propaquizafop 10EC at 100 g a.i. ha^{-1} followed by quizalofop ethyl 5EC at 100 g a.i. ha^{-1}.

8.30.4 Pre- and Post-emergence Herbicides

8.30.4.1 Tomato

On farm evaluation carried out to know the efficacy of different pre- and post-emergence herbicides neem cake 200 kg ha^{-1} at sowing by oxyfluorfen 0.1 kg ha^{-1} as pre-emergence, 3 DAP; neem cake 200 kg ha^{-1} at sowing of pendimethalin 1.0 kg ha^{-1} as pre-emergence, 3 DAP; neem cake 200 kg ha^{-1} at sowing of metribuzin 0.5 kg ha^{-1} as pre-emergence, 3 DAP; ethoxysulfuron at 50 g ha-1 as pre- and post-emergence at 45 DAT; sulfosulfuron at 25 g ha^{-1} at 15 and 45 DAT) in the farmer field of Chennvelli village in Chevella Mandal of Ranga Reddy district during *rabi* 2013 in tomato (Rani et al. 2015) has shown that *Orobanche* infestation was not observed up to 60 DAP in any of the treatments except in control plot. Percent infestation in the field varied from 0.5 to 4.0 in different treatments. Application of either metribuzin 0.5 kg ha^{-1} or oxyfluorfen 0.1 kg ha^{-1} or pendimethalin 1.0 kg ha^{-1} as pre-emergence along with 200 kg neem cake ha^{-1} or sulfosulfuron at 25 g ha^{-1} at 15 and 45 DAT was found to be effective in delaying the *Orobanche* infestation and in getting higher fruit yield in tomato during *rabi* season.

8.30.4.2 Chickpea

A field experiment conducted to find out the effective and economical weed control methods in chickpea on vertisols, with eleven treatments four pre-emergence herbicide, three post-emergence herbicides, three cultural treatments and weedy check, has shown that weed-free treatment recorded significantly lowest dry weed weight and highest weed control efficiency, grain, straw and biological yield. It was at par with of one Hoeing + two hand weedings, mechanical weedings and pendimethalin (PE) at 0.75 a.i. kg ha^{-1} and significantly superior over rest of the treatments. Treatment weedy check recorded the lowest grain and straw yield. Among the chemical weed control treatment application of pendimethalin (PE) at 0.75 a.i. kg ha^{-1} was found beneficial higher grain, straw yield, weed control efficiency and highest net monetary returns and B:C ratio and found most economical and effective in controlling weeds and increasing the yield of chickpea (Patil et al. 2015b).

8.30.4.3 Rice

Rice (*Oryza sativa* L.) is the most important major cereal food crop in agriculture and economy of India. One of major problems in rice cultivation for low productivity is weed infestation. Infestation of weeds in transplanted rice not only results in yield reduction but quality of produce is also impaired. About 100 weed species are known to be associated with this crop in India. Among them, *Echinochloa colona* is the most troublesome and widely distributed. Weeds not only cause quantitative but also hamper the quality of produce due to competition for nutrient, moisture, light and, to some extent, for space. They harbour many insect pests and pathogens resulting in poor crop growth. In early stage of the crop, grasses are predominant as compared to others, but at later stage, sedges and broadleaf weeds create interference in crop growth. No doubt that hand weeding is the established effective method of weed control. But nowadays, high cost involved and unavailability of labours makes weed management more difficult. Chemical weed control is regarded to be better than hand weeding due to the drudgery of weeding and meagre availability of labour at peak period of weed infestation. In this respect, application of new and wide spectrum herbicides alone or in combination may give satisfactory weed control. At 15 DAT, application of AE 1887196+AE F 095404 at 45 + 22.5 g ha^{-1} proved significantly superior in reducing total weed density. At 30, 45, 60 and 75 DAT, significantly lower weed density was recorded under two hand weedings (20 and 40 DAT). At 45, 60 and 75 DAT, two hand weedings were found effective to minimize all the weeds species. The weed density was significantly highest in unweeded check. The two hand weedings (20 and 40 DAT) resulted in the lowest weed dry matter production at all the time intervals of observations, except 15 DAT. However, at 15 DAT, AE 1887196 + AE F 095404 at 45 + 22.5 g ha $^{-1}$ resulted in the lowest weed dry matter production. The highest weed control efficiency at 75 DAT was witnessed under two hand weedings (20 and 40 DAT) (87.72 %) followed by bispyribac sodium at 20 g ha^{-1} (67.30 %) and AE 1887196+AEF 095404 at 45 + 22.5 g ha^{-1} (65.40 %). Two hand weedings (20 and 40 DAT) registered significantly highest grain yield (4.63 t ha^{-1}) over the rest of the treatments except treatments AE 1887196 + AEF 095404 at 35 + 17.5 g ha $^{-1}$, AE 1887196 + AEF 095404 at 40 + 20 g ha $^{-1}$, AE 1887196+AE F 095404 at 45 + 22.5 g ha^{-1}, fenoxaprop-p-ethyl + (chlorimuron-ethyl+ metsulfuron-methyl) at

$60+4$ g ha^{-1} and bispyribac sodium at 20 g ha^{-1} which were at par with two hand weedings (20 and 40 DAT). The highest straw yield was produced under two hand weedings (20 and 40 DAT), and it was at par with all the treatments except butachlor at 1,250 g ha $^{-1}$ and pretilachlor at 625 g ha^{-1} (Khwaja et al. 2015).

References

Anitha V, Anitha kumari D, Devaki et al (2015) Effect of abiotic factors on population build up of hopper of mango in Sangareddy, Telengana. In: Compendium of abstracts of the 2nd international conference on bio-resource and stress management, ANGRAU & PJTSAU, Hyderabad, India, 7–10 January 2015

Archana T, Ambule, Radadia GG, Patil DL et al (2015) Morphological characters of tomato in relation to resistance against tomato fruit borer, Helicoverpa armigera (Hubner). In: Compendium of abstracts of the 2nd international conference on bio-resource and stress management, ANGRAU & PJTSAU, Hyderabad, India, 7–10 January 2015

Badgujar AG, Bhede BV, Shinde ST et al (2015) Status of pod borer damage on pigeon pea In Marathwada region. In: Compendium of Abstracts of the 2nd international conference on bio-resource and stress management. ANGRAU & PJTSAU, Hyderabad, 7–10 Jan 2015

Bhanu KV, Satyanarayana Reddy P, Satyanarayana J (2015a) Antixenosis effects of certain rice cultures on brown planthopper, *Nilaparvata lugens* (STAL.). In: Compendium of abstracts of the 2nd international conference on bio-resource and stress management. ANGRAU & PJTSAU, Hyderabad, 7–10 Jan 2015

Bhanu KV, Satyanarayana Reddy P, Satyanarayana J (2015b) Functional response of cyrtorhinus (Reuter) on eggs of brown planthopper, *Nilaparvata lugens* (STAL.). In: Compendium of abstracts of the 2nd international conference on bio-resource and stress management. ANGRAU & PJTSAU, Hyderabad, 7–10 Jan 2015

Bharani GN, Sivasubramanian P, Kohilamba H (2015) Comparative efficacy of bio pesticides and insecticides against tomato thrips, *Thrips tabaci* Lind. and their natural enemies. In: Compendium of abstracts of the 2nd international conference on bio-resource and stress management, ANGRAU & PJTSAU, Hyderabad, 7–10 Jan 2015

Bhavani B, Visalakshi M, Bharathalakshmi M et al (2015) Chemical management of sugarcane early shoot borer, (*Chilo infuscatellus* Snellen) and internode borer (*Chilo sacchariphagus indicus*). In: Compendium of abstracts of the 2nd international conference on bio-resource and stress management. ANGRAU & PJTSAU, Hyderabad, 7–10 Jan 2015

Charan MS, Anitha V, Sridevi D et al (2015) Relative incidence of aphids and thirps in selected cultivars of chrysanthemum (*Dendranthema grandiflora* Borkh.) in Telangana. In: Compendium of abstracts of the 2nd international conference on bio-resource and stress management. ANGRAU & PJTSAU, Hyderabad, 7–10 Jan 2015

Chatterjee S, Dana I, Gangopadhyay C et al (2015) The population dynamics of insect-pests of rice with special reference to *S. incertulas* and *C. medinalis* along with their predators and parasites throughout sweeping method during boro. In: Compendium of abstracts of the 2nd international conference on bio-resource and stress management. ANGRAU & PJTSAU, Hyderabad, 7–10 Jan 2015

Chaudhuri N, Ghosh J, Senapati SK (2015) Performance of different rearing media of corcyra cephalonica stainton (lepidoptera: pyralidae) through life fecundity table study. In: Compendium of abstracts of the 2nd international conference on bio-resource and stress management, ANGRAU & PJTSAU, Hyderabad, 7–10 Jan 2015

Choke NK, Patik L, Chormule AJ et al (2015) Bioefficacy of newer insecticides against Pectinophora gossypela LA (Saunders) on cotton. In: Compendium of Abstracts of the 2nd international conference on bio-resource and stress management, ANGRAU & PJTSAU, Hyderabad, India, 7–10 January 2015

Devi TC, Kumari MBGS, Gouri V et al (2015) Metribuzin-an effective post-emergence weedicide in sugarcane. In: Compendium of abstracts of the 2nd international conference on bio-resource and stress management, ANGRAU & PJTSAU, Hyderabad,, 7–10 Jan 2015

Dudhale DR, Damre AS, Raut AH et al (2015) Seasonal incidence of sucking pests complex of okra. In: Compendium of abstracts of the 2nd international conference on bio-resource and stress management, ANGRAU & PJTSAU, Hyderabad, 7–10 Jan 2015

Goswami R (2015) Management of stored grain insect-pests of rice in government godown through chemical treatment. In: Compendium of abstracts of the 2nd international conference on bio-resource and stress management, ANGRAU & PJTSAU, Hyderabad, 7–10 Jan 2015

Gurve SS, Chormule AJ (2015) Efficacy of chitin synthesis inhibitors against tobacco leaf eating caterpillar, *Spodoptera litura* (Fabricius) (Noctuidae : Lepidoptera). In: Compendium of abstracts of the 2nd international conference on bio-resource and stress management, ANGRAU & PJTSAU, Hyderabad, 7–10 Jan 2015

Jadhav KM, Gaikwad BB, Bhede BV et al (2015a) Integrated pest management of different insecticides against major pests of Chilli. In: Compendium of abstracts of the 2nd international conference on bio-resource and stress management, ANGRAU & PJTSAU, Hyderabad, 7–10 Jan 2015

Jadhav KM, Patel RK, Gaikwad BB et al (2015b) Screening of different mango cultivars against gall midge

Procontarinia matteiana (Kieffer & Cecconi). In: Compendium of abstracts of the 2nd international conference on bio-resource and stress management, ANGRAU & PJTSAU, Hyderabad, 7–10 Jan 2015

Joshi B, Akkabathula N, Rana A (2015) Evaluation of indoxacarb 14.5 SC @ 50 g a.i. ha^{-1} against pod borers in Chhattisgarh plains. In: Compendium of abstracts of the 2nd international conference on bio-resource and stress management, ANGRAU & PJTSAU, Hyderabad, 7–10 Jan 2015

Jyothi C, Kumar DVSR (2015) Efficacy of novel insecticides against Thrips (*Thrips palmi* Karny) of Blackgram (*Vigna mungo* L.) and their influence on natural enemies. In: Compendium of abstracts of the 2nd international conference on bio-resource and stress management, ANGRAU & PJTSAU, Hyderabad, 7–10 Jan 2015

Jyothi J, Sudha Vani V, Jyothi Ket al (2015) Influence of methyl cyclopropene (1-MCP) on shelf life and quality of banana cv. Grand Naine under ambient condition. In: Compendium of abstracts of the 2nd international conference on bio-resource and stress management, ANGRAU & PJTSAU, Hyderabad, 7–10 Jan 2015

Jyothsna M, Rajasri M, Nagesh Kumar MV et al (2015a) Phytotoxic compatibility of pesticides on castor. In: Compendium of abstracts of the 2nd international conference on bio-resource and stress management, ANGRAU & PJTSAU, Hyderabad, 7–10 Jan 2015

Jyothsna M, Vijaya Lakshmi K, Ramesh K et al (2015b) Use of nano silica as grain protectant against peanut bruchid, *Caryedon serratus* Olivier in storage. In: Compendium of abstracts of the 2nd international conference on bio-resource and stress management, ANGRAU & PJTSAU, Hyderabad, 7–10 Jan 2015

Kamakshi N, Jayalakshmi V, Udayasankar A (2015) Evaluation of chickpea germplasm for resistance to the pulse beetle *Callosobruchus chinensis*. In: Compendium of abstracts of the 2nd international conference on bio-resource and stress management, ANGRAU & PJTSAU, Hyderabad, 7–10 Jan 2015

Kharbade SB, Chormule AJ, Patil SC et al (2015a) Field evaluation of granular insecticides against brown plant hopper, *Nilparvata lugens* (Stal.) infesting rice. In: Compendium of abstracts of the 2nd international conference on bio-resource and stress management, ANGRAU & PJTSAU, Hyderabad, 7–10 Jan 2015

Kharbade SB, Tamboli ND, Naik RL et al (2015b) Biorational management of capsicum mite, *Polyphagotarsonemus latus* bank. In: Compendium of abstracts of the 2nd international conference on bio-resource and stress management, ANGRAU & PJTSAU, Hyderabad, 7–10 Jan 2015

Khwaja I, Deva S, Pandey P (2015) Effect of new post-emergence herbicides on weed density, weed dry matter and yield of transplanted rice (*Oryza sativa* L). In: Compendium of abstracts of the 2nd international conference on bio-resource and stress management, ANGRAU & PJTSAU, Hyderabad, 7–10 Jan 2015

Kumar BA, Sheshagiri Rao P, Venkataiah M (2015a) Evaluation of certain insecticides against white flies (a vector of yellow mosaic) in blackgram. In: Compendium of abstracts of the 2nd international conference on bio-resource and stress management, ANGRAU & PJTSAU, Hyderabad, 7–10 Jan 2015

Kumar BA, Sheshagiri Rao P, Venkataiah M (2015b) Studies on the effectiveness with novel modes of action against the insect pests of green gram. In: Compendium of abstracts of the 2nd international conference on bio-resource and stress management, ANGRAU & PJTSAU, Hyderabad, 7–10 Jan 2015

Kumar BA, Swarna Sree B, Sheshagiri Rao P (2015c) Evaluation of insecticides against stem borer in paddy. In: Compendium of abstracts of the 2nd international conference on bio-resource and stress management, ANGRAU & PJTSAU, Hyderabad, 7–10 Jan 2015

Latha HC, Naganagoud A (2015) Effect of sweet flag rhizome, *Acorus calamus* L. formulations against *Callosobruchus analis* Linn. in pigeon pea. In: Compendium of abstracts of the 2nd international conference on bio-resource and stress management, ANGRAU & PJTSAU, Hyderabad, 7–10 Jan 2015

Mahatale PV, Ladole MY, Mahatale YV et al (2015) Effect of post-emergence herbicides in groundnut. In: Compendium of abstracts of the 2nd international conference on bio-resource and stress management, ANGRAU & PJTSAU, Hyderabad, 7–10 Jan 2015

Mandal A, Thakur JK, Sahu A et al (2015) Impact of transgenic cotton on microbial activity and diversity under vertisols of Central India. In: Compendium of abstracts of the 2nd international conference on bio-resource and stress management, ANGRAU & PJTSAU, Hyderabad, 7–10 Jan 2015

Mane PN, Rathod PK, Patel JD et al (2015) Biointensive integrated management of major insect pests of sunflower. In: Compendium of abstracts of the 2nd international conference on bio-resource and stress management, ANGRAU & PJTSAU, Hyderabad, 7–10 Jan 2015

Mansa K, Gangadhar NC, Raja Goud (2015) Impact of mass trapping technology on the infestation of brinjal shoot and fruit borer in comparison with farmer's practices. In: Compendium of abstracts of the 2nd international conference on bio-resource and stress management, ANGRAU & PJTSAU, Hyderabad, 7–10 Jan 2015

Muchhala Y, Sharma AK, Pachori R (2015) Effect of weather parameters on seasonal activity of major phototropic insect pest species of paddy. In: Compendium of abstracts of the 2nd international conference on bio-resource and stress management, ANGRAU & PJTSAU, Hyderabad, 7–10 Jan 2015

Muthukumar M, Ram Kumar DVS, Madhumathi T et al (2015) Effect of biochemical composition of different paddy varieties on the development of lesser grain borer, *Rhyzopertha dominica* (Fab). In: Compendium of abstracts of the 2nd international conference on bio-resource and stress management, ANGRAU & PJTSAU, Hyderabad, 7–10 Jan 2015

Nadimpally RGV, Rumandla J (2015a) Evaluation of certain Acaricides alone and in combination with Propi-

conazole against rice panicle mite, *Steneotarsonemus Spinki* Smiley and associated grain discolouration. In: Compendium of abstracts of the 2nd international conference on bio-resource and stress management, ANGRAU & PJTSAU, Hyderabad, 7–10 Jan 2015
Nadimpally RGV, Rumandla J (2015b) Field evaluation of compatibility of pesticides against rice insect pests and diseases. In: Compendium of abstracts of the 2nd international conference on bio-resource and stress management, ANGRAU & PJTSAU, Hyderabad, 7–10 Jan 2015
Niharika G, Chaturvedi A (2015) Incidence of pesticide residues of selected cereals and pulses in Southern Telangana zone. In: Compendium of abstracts of the 2nd international conference on bio-resource and stress management, ANGRAU & PJTSAU, Hyderabad, 7–10 Jan 2015
Nikam TA, Chandele AG (2015) Efficacy of new chemical insecticides against diamondback moth *Plutella xylostella* (L.): on cabbage. In: Compendium of abstracts of the 2nd international conference on bio-resource and stress management, ANGRAU & PJTSAU, Hyderabad, 7–10 Jan 2015
Pathipati VL, Vijayalakshmi T, Naram Naidu L (2015) Compatibility of certain insecticides with copper oxy chloride (coc) on chilli in Andhra Pradesh. In: Compendium of abstracts of the 2nd international conference on bio-resource and stress management, ANGRAU & PJTSAU, Hyderabad, 7–10 Jan 2015
Patil SR, Lande GK (2015) Evaluation of different doses of some newer insecticides on sucking pests of okra. In: Compendium of abstracts of the 2nd international conference on bio-resource and stress management, ANGRAU & PJTSAU, Hyderabad, 7–10 Jan 2015
Patil DL, Radadia GG, Toke NR et al (2015a) Screening of brinjal genotypes for resistance to brinjal shoot and fruit borer, *Leucinodes orbonalis* Guenee. In: Compendium of abstracts of the 2nd international conference on bio-resource and stress management, ANGRAU & PJTSAU, Hyderabad, 7–10 Jan 2015
Patil PV, Gore AK, Waghmare YM (2015b) Evaluation of pre and post-emergence herbicides in chickpea (*Cicer arietinum* L.). In: Compendium of abstracts of the 2nd international conference on bio-resource and stress management, ANGRAU & PJTSAU, Hyderabad, 7–10 Jan 2015
Rani PL, Yakadri M, Prakash TR (2015) On farm evaluation of different pre and post-emergence herbicides to manage parasitic weed (*Orobanche cernua*) in tomato. In: Compendium of abstracts of the 2nd international conference on bio-resource and stress management, ANGRAU & PJTSAU, Hyderabad, 7–10 Jan 2015
Rao PVR, Divyasri S, Bhanu KV et al (2015) Molecular diversity of lead fodder resistant genotypes in rice (*Oryza sativa* L.) using microsatellite markers. In: Compendium of abstracts of the 2nd international conference on bio-resource and stress management, ANGRAU & PJTSAU, Hyderabad, 7–10 Jan 2015
Rathod PK, Borkar PH, Mane MN et al (2015a) Management of pulse beetle, *Callosobruchus chinensis* L. in stored chickpea with some vegetable oils. In: Compendium of abstracts of the 2nd international conference on bio-resource and stress management, ANGRAU & PJTSAU, Hyderabad, 7–10 Jan 2015
Rathod R, Ranjith Kumar L, Sharma S (2015b) DuPont™ Cyazypyr™ insect control (cyantraniliprole): a reduced-risk, cross-spectrum product for insect pest control in fruit and vegetable crops. In: Compendium of abstracts of the 2nd international conference on bio-resource and stress management, ANGRAU & PJTSAU, Hyderabad, 7–10 Jan 2015
Raut AH, Pawar DB, Kshirsagara SM et al (2015) Physico-chemical resistance mechanism of sorghum genotypes against shoot fly, *Atherigona soccata* (Rondani). In: Compendium of abstracts of the 2nd international conference on bio-resource and stress management, ANGRAU & PJTSAU, Hyderabad, 7–10 Jan 2015
Reddy AV, Malathi S, Varma NRG (2015) Impact of ecological engineering practices against plant brown plant hoppers in irrigated rice. In: Compendium of abstracts of the 2nd international conference on bio-resource and stress management, ANGRAU & PJTSAU, Hyderabad, 7–10 Jan 2015
Salunke PB, Lande SS, Jadhav AJ et al (2015) Evaluation of insecticide resistance to *Helicoverpa armigera* in pigeon pea growing area of Vidarbha). In: Compendium of abstracts of the 2nd international conference on bio-resource and stress management, ANGRAU & PJTSAU, Hyderabad, 7–10 Jan 2015
Shaila O, Ramesh Babu T, Sharma HC (2015) Evaluation of transgenic chickpea lines for resistance to *Helicoverpa armigera* under glasshouse conditions. In: Compendium of abstracts of the 2nd international conference on bio-resource and stress management, ANGRAU & PJTSAU, Hyderabad, 7–10 Jan 2015
Sharma AK, Muchhala Y (2015) Light trap intercepted the effect of weather factors on seasonal activity of major predatory and parasitic fauna of paddy ecosystem. In: Compendium of abstracts of the 2nd international conference on bio-resource and stress management, ANGRAU & PJTSAU, Hyderabad, 7–10 Jan 2015
Singh P, Singh RN (2015) Predatory potential of two predatory mites *Amblyseius indicus* and *Amblyseius tetranychivorous* (Acarina :Phytoseiidae) on different densities of *Tetranychus neocaledonicus* population on Brinjal. In: Compendium of abstracts of the 2nd international conference on bio-resource and stress management, ANGRAU & PJTSAU, Hyderabad, 7–10 Jan 2015
Singh N, Gupta J, Manjula K et al (2015) Allopathic potential of callus extract of *Lantana camara* for controlling weeds. In: Compendium of abstracts of the 2nd international conference on bio-resource and stress management, ANGRAU & PJTSAU, Hyderabad, 7–10 Jan 2015

Sreekanth M, Lakshmi MSM, Koteswara Rao Y (2015) Calendar based application of new insecticides for the management of pod borer complex in pigeon pea. In: Compendium of abstracts of the 2nd international conference on bio-resource and stress management, ANGRAU & PJTSAU, Hyderabad, 7–10 Jan 2015

Srinivas PS, Prasad RD (2015) Stability and efficacy of mixed pesticides against safflower aphid, *Uroleucon compositae* Theobald. In: Compendium of abstracts of the 2nd international conference on bio-resource and stress management, ANGRAU & PJTSAU, Hyderabad, 7–10 Jan 2015

Subba, B, Ghosh SK, Kumar KR, Bhanusree MR, Chetri B (2015) Seasonal incidence of jassid infesting tomato (*Lycopersicon esculentum* (L.) and their sustainable management. In: Compendium of abstracts of the 2nd international conference on bio-resource and stress management, ANGRAU & PJTSAU, Hyderabad, 7–10 Jan 2015

Sultani MS, Singh R, Dhankhar SK (2015) Okra genotypes affecting performance of shoot and fruit borer. In: Compendium of abstracts of the 2nd international conference on bio-resource and stress management, ANGRAU & PJTSAU, Hyderabad, 7–10 Jan 2015

Upendhar S, Satyanarayana J, Singh TVK (2015) Cypermethrin and methomyl resistance in the *Helicoverpa armigera* (Hubner). In: Compendium of abstracts of the 2nd international conference on bio-resource and stress management, ANGRAU & PJTSAU, Hyderabad, 7–10 Jan 2015

Vijayaraghavendra R, Basavanagoud K, Gadad H et al (2015) Seasonal incidence of sapota fruitborer, *Phycitaery throlophia* Hampson. In: Compendium of abstracts of the 2nd international conference on bio-resource and stress management, ANGRAU & PJTSAU, Hyderabad, 7–10 Jan 2015

Zunjare R, Muthusamy V, Choudhary M et al (2015) Effect of rice weevil (*Sitophilus oryzae* L.) infestation on popping traits in maize genotypes. In: Compendium of abstracts of the 2nd international conference on bio-resource and stress management, ANGRAU & PJTSAU, Hyderabad, 7–10 Jan 2015

Research Trends in Abiotic Stress Resistance of Crops

9

Jorge Isaac Sarquís Ramírez and Ratikanta Maiti

Abstract

Several abiotic stresses such as drought, salinity, temperature, heavy metals, herbicides and arsenic affect the productivity of crops. This chapter gives a synopsis of research results on the physiological basis of abiotic stress resistance on drought resistance of various crops such as guar, safflower, soybean, mung bean, sorghum, maize, pearl millet, rice, wheat, oat, chilli, *Cucumis*, linseed, mustard, sesame, ber, sunflower, *Gladiolus*, fodder crops and chilli, and also discusses breeding for drought resistance in crops. In addition, the chapter gives a short resume of research results on salinity tolerance of various crops such as psyllium, finger millet, wheat, maize, rice, jute, beet and fruit crops; in addition it discusses the studies that have been undertaken on the effects of temperature, heavy metals and arsenic on crop productivity, multiple stress resistance and the strategies that have been suggested for abiotic stress management in crops.

9.1 Introduction

With regard to agricultural practice and food production, global warming—once a matter of controversy but today an irrefutable reality—has put the world in a position in which any attempt to dodge directly facing the tremendous challenges this phenomenon poses is—to put it mildly—irresponsible. We can no longer appeal to ignorance, what we did not know 3, 4 or 5 years ago but we know now beyond any doubt: added to the pressure on all natural resources by an overwhelming population of seven billion—one third of which is permanently below the hunger poverty level—a rapidly changing global climate poses a major threat to peace and social stability around the globe, not to speak of the billions of lives at stake from the risk of starvation.

The scientific community cannot afford to underestimate the consequences of any commitment that falls short of totally compromising with an effort to palliate the ever-increasing needs of a very hungry world under conditions in which food production—despite undeniable

J.I. Sarquís Ramírez (✉)
Universidad Veracruzana, Xalapa de Enríquez, Estado de Veracruz-Llave, Xalapa, Mexico
e-mail: jsarquis@uv.mx

R.K. Maiti
Facultad de Ciencias Forestales, Universidad Autónoma de Nuevo León, Carr. No 85, km 145, NL 67700 Linares, Mexico
e-mail: ratikanta.maiti@gmail.com

R.K. Maiti et al. (eds.), *Bioresource and Stress Management*, DOI 10.1007/978-981-10-0995-2_9

progress towards record achievements—must deal with an increasingly extreme and erratic environment.

Up to the seventh decade of the twentieth century, scientific research was conducted in the area of abiotic stress with an emphasis on developing crop resistance to various adverse conditions that would allow food production in areas where population pressure demanded incorporation of new land under cultivation, which would otherwise be left alone or dedicated to other purposes. Advanced technology in soil management, irrigation and crop management helped tame the most hostile environments; together with newly bred varieties of major grain crops for resistance to drought, salinity, and heat and cold stress. This technological progress proved its advantage and allowed the science to merit even a Nobel Prize. However, as much as the Green Revolution may be worthy of celebration and awards, critics are not altogether unjust in condemning it for all the environmental degradation it has been related to over the past four decades on account of the intensive mechanization and intensive use of mineral fertilizers and pesticides to guarantee maximum expression of genotype potential for quantity and quality of produce, as recommended by agronomists—not to mention the alleged unfairness of monopolization of the genetically improved seed industry in the hands of a few huge corporations, which have come to own the world's seed market, ousting traditional agriculture and changing rural landscapes and ways of life altogether for good.

Soil erosion, salinity, desertification and water pollution do not help in averting the deleterious effects of global climate change on plant growth, development and, ultimately, productivity. In order to manage plant stress, significant research input has been directed to counterbalance these stress factors and increase crop productivity in pursuit of more sustainable agriculture. An increasing number of studies in the areas of biotechnology, genetic engineering, green technology, nanotechnology, bio-fortification in bio-resources and stress management, with a special emphasis on pest management, suggest that a lot must still be accomplished to favour the needy and the hungry around our world. However, in order to make progress, everybody will have to come up to the plate—scientists, politicians, policy makers, funding organizations, the United Nations; every one's contribution is required to prevent a catastrophe of terrifying dimensions.

This chapter offers the reader an insight into the recent research advances in abiotic stress resistance presented during the Second International Conference on Bio-resource and Stress Management in Hyderabad, India, in January, 2015.

Global warming and climate change have direct impacts on crop productivity, which have been widely accepted as facts. The changing environments create serious challenges to global agriculture and sustainability of the horticulture industry; various changes of physiological and biochemical processes in plants are used for adverse condition tolerance. Crop development for adverse climatic conditions involves creation of the variation first than selection of the desirable genotype. These is done by various breeding methods such as mutation breeding, marker-assisted selection breeding for gene introgression, mapping QTLs, or identifying new or rare alleles associated with a particular trait. Transgenic plants are developed through genetic engineering, identifying the desirable gene in the plant genome, and used for a resistance line for a breeding programme. Rapid production of doubled haploids (DHs) through androgenesis is an important method for genetic improvement of crop plants. Many stress factors can trigger the reprogramming of microspores, which have been correlated with ultrastructural changes of cells in embryos and finally haploid plants. It has been shown that certain pretreatments such as (i) physical stresses such as cold, heat shock, starvation, drought stress, osmotic pressure, gamma irradiation, oxidative stress and reduced atmospheric pressure, and (ii) chemical treatments such as colchicines, heavy metal, abscisic acid (ABA), CGA, AEC, azetidine and 2-NHA, with either individual or combined effects of more than one stress factor, may positively influence androgenetic efficiency. These techniques have been used for production of crops for adverse climate conditions and sustainable crop production (Panigrahi and Das 2015).

9.2 Drought

In the context of a worsening global climatic disorder, old definitions become inaccurate at best. Such is the case of our conventional concepts of drought, aridity and desertification, all of whose meanings have become rather blurred. While indeed they were never totally exempt from some degree of subjectivity, experts distinguished the more or less permanent climatic conditions described by the term aridity from drought, which was taken for a temporary or transitory condition characterized by a severe reduction of water availability compared to regular standards over a long period of time and over a large region; in turn, desertification used to refer to a long-term, distinctively irreversible process of destruction of soil potential to sustain life (Rossi et al. 1992). But fading annual patterns of thermal and humidity regimes on a local scale are causing food chains and ecosystems to collapse in a kind of chain reaction triggered once a threshold level of damage to the most vulnerable components in the chain is reached, as evidenced by a number of ecological studies (Dobson et al. 2006; MacDougall et al. 2013; Mumby et al. 2011). From an agricultural perspective any environmental imbalance due to abrupt changes in climatic performance is bound to cause changes in a number of factors, including soil microbes, pollinators, disease vectors and evidently the crop itself in terms of plant growth and development.

Recent modelling attempts have shown that by around the 2070s, in critical regions, today's floods and droughts of the century will be an every 5- to 50-year occurrence in northern-northeastern and southern-southeastern Europe, respectively—a projection not to be taken blithely in light of the need for careful planning on ways to alleviate the impact of global climate change on a continental scale (Lehner et al. 2006). Quite the same can be said about practically any other part of the world, although there is no question that there will be very specific large regions where extreme conditions for plant growth will have to be treated as higher priorities in view of the size of the human populations at risk—Central and Southeast Asia is a case in point. Similarly, much of Latin America must be regarded as a geographical region threatened by recurrent dry hot long spells over the usual cropping timeframe.

Inspired by these facts, 2 years ago a group of Asian scientists organized the First International Conference on Bio-resource and Stress Management in Calcutta, India, as a forum to spread the results of ongoing research to tackle the problem in that part of the world. Following up on their steps, a second conference took place in the early days of 2015. In the sections that follow, a view is offered of the works presented on that occasion.

9.2.1 Physiological Basis of Drought Resistance

In reviewing the state of the art in the field, Deva and Pandey (2015a) argue that water scarcity effects such a severe environmental constraint on plant productivity that drought-induced crop yield losses quite likely exceed losses caused by all other biophysical limitations together—both the intensity and the duration of the stress being critical in determining the extent of reduction of leaf size, stem extension, root proliferation and severity of alterations in plant water relations resulting in significant cuts in water-use efficiency.

Plants exhibit a variety of physiological and biochemical responses at the cellular, tissue and whole-plant levels in trying to cope with drought stress. Thus, under such conditions, CO_2 assimilation by leaves is reduced mainly by stomatal closure, membrane damage and disturbed activity of various enzymes, especially those in charge of CO_2 fixation and adenosine triphosphate synthesis. Additionally, increased metabolite flux through photorespiration boosts the oxidative load on the tissues as both processes generate reactive oxygen species; damage caused by reactive oxygen species to biological macromolecules under drought stress is a well documented major deterrent to growth.

The major adaptive mechanisms associated with reduced water loss include increased diffusive resistance, increased water uptake with profuse and deep root systems coupled to more efficient use and smaller and succulent leaves to reduce transpirational water loss. Among nutrients, potassium ions allow for increased osmotic adjustment; silica increases root endodermal silicification and improves cell water balance. Low molecular weight osmolytes such as glycine betaine, proline and other amino acids, organic acids, and polyols have been shown to be either beneficial or even crucial to sustain cellular functions under drought. Plant growth-regulating substances such as salicylic acid, auxins, gibberellins, cytokinin and ABA modulate plant responses to drought. Polyamines, citrulline and several enzymes are known to function as antioxidants and to contribute in reducing the adverse effects of water deficit.

9.2.2 Guar

Guar or cluster bean [*Cyamopsis tetragonoloba* (L.) Taub.] is an annual legume. It is also the source of guar gum. The plant thrives well in arid to semi-arid areas and being a leguminous species, it contributes nitrogen to the soil through symbiotic fixation. Guar is widely used for its high nutritional value; additionally, it is characterized by relatively low calorie content; it is also used as an effective treatment for lowering blood sugar and cholesterol levels. For these reasons guar gum has become a high-economic-value produce and demand for it is rising rapidly due to its industrial use in hydraulic fracturing (oil shale gas). The plant can be cultivated as a sole crop as well as in intercropping or a crop-rotation cycle. Its economic high value has promoted interest in expanding the area of cultivation and, in general, there is increasing interest in conducting research efforts to identify and to develop suitable varieties for different agroclimatic conditions, especially in arid environments. One recent study was undertaken on the growth and yield performance of five cluster bean genotypes in a water-limited environment of a semi-arid region. These bean varieties were evaluated for their yield potential during a summer season under either irrigation or scarce rain conditions. Although moisture stress reduced the total biomass, vegetative and fodder biomass, pod and seed weight, test weight and harvest index in most of the selected genotypes, increased fodder biomass was registered. Genotype RGC-986 maintained the highest total and vegetative biomass, but the smallest impact of stress on harvest index was observed for the RGC-1025 genotype. With a higher biomass and higher seed yield even under water deficit, RGC-1025 was found to be a more suitable variety for regions characterized by a scarcity of rainfall. The second-best performing genotype in these conditions was RGC-986, whose better fodder yield can allow it to be used as a dual-purpose crop (Satyavathi et al. 2015).

9.2.3 Soybean

Patil (2015) undertook evaluation of soybean [*Glycine max* (L.) Merrill] genotypes for drought under induced water stress. They observed that the relative water content (RWC) and total chlorophyll content were least affected by water deficit in soybean genotypes AMS-99-33, TAMS-98-21 and JS-335, resulting in good drought tolerance. Genotypes AMS-99-33 (7.22), TAMS-98-21 (7.85) and JS-335 (7.95) had shown lower stomatal index values in the stress condition at 30 and 50 days after stress initiation, demonstrating higher drought tolerance capacity. Genotype AMS-99-33 in fact recorded the highest RWC (52.44 %) in the stress condition, followed by TAMS-98-21 (51.89 %) and JS-335 (51.10 %). Genotype JS-335 showed the worst symptoms of susceptibility to drought, followed by TAMS-38 and TAMS-98-21. Drought intensity was found to be inversely related to drought tolerance.

9.2.4 Mung Bean

Dutta et al. (2015) studied the water-stress tolerance of mung bean [*Vigna radiata* (L.) Wilczek] and physiological and biochemical changes in leaves during seedling development. The study assessed the effect of PEG-induced moisture stress on physiological and biochemical changes in leaves of cultivars K 851 (drought tolerant) and PDM 139 (drought susceptible) during seedling development. A range of four external water potentials (i.e. −1.0, −2.0, −3.0 and −4.0 bars), besides glass distilled water as a control (0.0 bar), were used. Water stress caused reductions in leaf area, relative leaf water content (RLWC), chlorophyll content, the chlorophyll stability index (CSI), and phenol and ascorbic acid content in both cultivars studied. In contrast, proline content increased with increasing water stress. Drought-tolerant cultivar K 851 registered a better leaf water balance and higher accumulation of proline, phenols and ascorbic acid than PDM 139.

9.2.5 Sorghum

Dhutmal et al. (·2015) studied variability parameters and conducted correlation studies in drought-tolerant post-rainy sorghum (*Sorghum bicolor* L.) genotypes including 48 advanced breeding lines, land races and promising lines. Genetic variability, heritability and genetic advance, correlation and path coefficient analysis for grain yield and its component characters were evaluated. It was observed that grain yield per plant (gm) (35.36):, seed vigour (34.11), panicle dry weight per plant (gm) (24.73), stem dry weight per plant (gm) (23.11), total biomass per plant (gm) (18.45), leaf dry weight per plant (gm) (13.77) and plant height (cm) (12.29) exhibited high genotypic coefficients of variation. The estimates of heritability in a broad sense were high for total biomass per plant (95.1), grain yield per plant (92.5), panicle dry weight per plant (94.7), stem dry weight per plant (92.7), plant height (91.7), SCMR at 50 % flowering (84.0) and RWC (%) (83.4), days to 50 % flowering (72.3), days to physiological maturity (65.2) and leaf dry weight per plant (68.2). High heritability coupled with high genetic advance and high GCV was observed for grain yield per plant, leaf dry weight per plant, stem dry weight per plant and total biomass per plant. A significant and positive association was found between grain yield per plant and panicle dry weight (0.784), days to 50 % flowering (0.770) and total biomass (0.635). Grain yield per plant was strongly correlated with plant height, SCMR, leaf dry weight, 1000 grain weight, RLWC and number of leaves per plant. Total biomass was subject to the highest direct effect (28.33), followed by stem dry weight (17.59), panicle dry weight (15.81), leaf dry weight (2.51) and days to 50 % flowering (0.326). Indirect effects on grain yield per plant were also estimated. It was found that panicle dry weight showed the maximum indirect effect via total biomass, while the indirect effect of total biomass was also positive via stem weight and panicle dry weight.

Nagesh et al. (2015) undertook field screening and studied the biochemical basis of resistance in sorghum [*Sorghum bicolor* (L.) Moench] varieties, for their suitability under scarcity of rainfall in the Nagarjuna Sagar Project and Krishna Western Delta in the Guntur district of Andhra Pradesh. Sorghum is cultivated to an extent of 25,000 ha in rice fallows by the farmers of the Krishna Western Delta tail and in areas in the Guntur district of Andhra Pradesh when the irrigation water from the Nagarjunaa Sagar Project is not available for paddy in the *rabi* season. Sorghum shoot fly, sorghum stem borer and midge are the major insect pests found to be damaging the jowar in this region. Hence, during 2013–2014, 20 ICRISAT varieties or lines of jowar varieties were evaluated against these insect pests at the Regional Agricultural Research Station, Guntur, Andhra Pradesh, for their suitability under rainfed conditions. The germplasm tested showed significant variation in reaction to the natural incidence of shoot fly and stem borer. While ICSV 96011, TAM 2566 and Swarna showed highest susceptibility to shoot fly (14.47–21.6 % dead heart) and stem borer (27.07–31.57 % dead heart), entries ICSB 413, 435, 444, 445, ICSV 700, 705, 197, ICSV 25019, 25022, 25026, IS 2123 and 18551 showed highest resistance to shoot fly (0–6.9 % dead heart) and stem borer (0–9.37 % dead heart). These were statistically on a par with each other. For knowing the biochemical basis of resistance and susceptibility to these pests, four highly resistant lines or varieties (ICSV 25022, 197, 2123 and ICSB 413) and four highly susceptible lines or varieties (TAM 2566, Swarna, ICSV 88032 and 197) were chosen and their chlorophyll a, b and total chlorophyll content, phenol and carbohydrate contents were analysed. Biochemical analysis showed that there were no statistically significant differences with respect to chlorophyll a, phenol and carbohydrate content, whereas with respect to chlorophyll b and total chlorophyll content, these treatments showed statistically significant variation. While the highest yield (39.81 q ha^{-1}) was recorded for ICSB 413, highly susceptible TAM 2566 recorded a lower yield of 18.5 q ha^{-1} only. Hence it can be inferred that higher-resistance lines can be utilized in further resistance variety breeding programmes, while the varieties with higher resistance to shoot fly and stem borer can be used for cultivation by farmers.

Kalpande et al. (2015) studied the relationship between grain yield and some of the index parameters associated with drought tolerance in *rabi* sorghum. The experimental material consisted of 5 lines and 12 testers and 60 hybrids developed by crossing five lines with 12 testers using a line × tester scheme. The experiment was conducted during *rabi* 2006–2007 at the Sorghum Research Unit, Dr. PDKV, Akola (MS). The interrelationships between grain yield per plant (g) and six index parameters associated with drought tolerance, including the CSI (%), harvest index (%), drought index (%), dry matter stress index (%), plant height stress index (%) and stomatal index (%), were studied by calculating the genotypic and phenotypic correlation coefficients. The harvest index designates the partitioning of dry matter into plant parts in relation to the whole aerial plant body. At the genotypic level, it was shown to correlate positively and significantly with the drought index and grain yield per plant. A positive and significant association of the harvest index and grain yield resulted in higher dry matter per plant, which was translocated more efficiently through the stem into developing cobs for proper filling of the grains. The drought index exhibited a positive and significant association with the dry matter stress index and grain yield per plant at both the phenotypic and genotypic levels. The dry matter stress index showed a positive and significant correlation with the plant height stress index and grain yield per plant at both the genotypic and phenotypic levels. The stomatal index exhibited a negative and significant correlation with grain yield per plant at both the genotypic and phenotypic levels. The largest proportion of plant water loss occurs via stomata. Therefore, the frequency of stomata per unit of leaf area and their size are important factors affecting whole-plant water turnover capacity. Data from this study allowed concluding that, in the case of *rabi* sorghum, selection for a lower stomatal index favours chlorophyll stability and a higher harvesting index. Since all index parameters mentioned were shown to play a role in determining the drought tolerance of *rabi* sorghum, it seems that they offer ample opportunity to be taken advantage of for further development of drought-tolerant genotypes in *rabi* sorghum.

Ghorade et al. (2015) used the same experimental material consisting of 5 lines and 12 testers and 60 hybrids developed by crossing five lines with 12 testers by the same line × tester scheme in order to identify promising lines and testers for grain yield and some of the drought tolerance parameters in *rabi* sorghum under drought stress. The estimates of the general combining ability effects indicated that among the five lines used, MS 104A was a good general combiner for grain yield per plant (0.93*) along with five drought tolerance traits, viz leaf area ratio at 75 DAS (0.57*), dry matter at harvest (2.69**), chlorophyll content at 75 DAS (0.21**), specific leaf weight (0.07**) and harvest index (1.17*). Among the testers, M 35–1 showed desirable GCA for grain yield per plant (3.61**) along with seven physiological parameters associated with drought tolerance, viz leaf area ratio at 75 DAS (0.84**), dry matter at harvest (3.62**), chlorophyll content at 75 DAS (0.34**), proline content (2.95**), CSI (−0.07*), specific leaf weight (0.14**) and harvest index (3.74**). Another tester, SPV 504, has also transmitted desirable genes for grain yield per plant (2.07**) along with six physiological parameters like leaf area ratio at 75 DAS (1.05*), dry matter at harvest (3.40*), chlorophyll content at 75 DAS (0.17**), proline content (1.85*), CSI (−0.13**) and stomatal index (−0.58*). Besides these two testers, other promising testers with significant GCA effects for grain yield per plant, along with some of the physiological traits associated with drought tolerance, were CSV-216 R, Ringni, Parbhani Moti and AKSV-13 R. All these six testers and line MS-104A need to be included in future breeding programmes to develop high-yielding drought-tolerant hybrids of *rabi* sorghum.

9.2.6 Maize

Channappagoudar and Sahoo (2015) investigated the physiological basis of drought tolerance in maize inbreds and hybrids in a field experiment during the post-rainy season of 2011–2012 to identify stress-tolerant maize genotypes suitable for rain-fed conditions at the Main Research Station in the University of Agricultural Sciences, Dharwad. The soil in the experiment site was medium black with low available nitrogen, medium phosphorus and high available potassium. In two blocks they evaluated maize hybrids and inbred lines for their tolerance of water stress, which were irrigated up to 25 days after sowing. There was no rain fall during the entire cropping period and the observations on soil moisture status in the stressed block were recorded at weekly intervals after 25 DAS up to maturity. Data were taken on various biometric and physiological parameters. LAI showed significant differences between the genotypes at 60 DAS, with DMIL-181 having higher LAI under both stress and irrigated conditions. It was observed that stress reduced LAI by 12.50 % compared to irrigated conditions. Under the stress treatment, ARBMH-1, followed by DMIL-78, had significantly higher total dry matter (TDM) than all other genotypes. Among hybrids, the maximum reduction in TDM (76.7 %) was observed in DMH-1 and least in ARBMH-1 (22.3 %), while among inbreds, DMIL-43 showed maximum reduction of TDM (70.5 %) in the stress condition. DMH-1 and DMIL-43 did not produce any cobs under the stress condition, indicating their higher susceptibility to it. ARBMH-1 (7.30 %) and DMIL-142 (9.00 %) among hybrids and inbreds, respectively, showed the least reductions in cob length, while ARBMH-9 and CM-501 among hybrids and inbreds, respectively, registered the least reductions in cob girth under the stress condition. Seed rows per cob, number of seeds per cob and test weight declined under stress, and the extent of the decline was less in seed rows per cob, indicating its relatively lower sensitivity to stress in comparison with the other two parameters. There was an overall 54.8 % reduction in yield due to stress with a range from as low as 22.9 % in DMIL-39 to 100 % in DMH-1 and DMIL-43. Soil and plant analysis determination (SPAD) values were lowest (13.4 %) under stress compared to irrigated conditions, followed by RWC at 75 DAS. Drought-resistant genotypes DMIL-39, ARBMH-1 and CM-501 showed the least reductions in physiological parameters, viz the photosynthetic rate, SPAD values and RWC. There was a mean 78.2 % increase in proline content under stress conditions. It was concluded that there existed ample variability with respect to various morphological, physiological and yield parameters among hybrids and inbred lines. Lines DMIL-142, DMIL-39 and DMIL-78, having better performance of these parameters under water-stress conditions, can be incorporated into future breeding programmes for drought tolerance, and the hybrid ARBMH-1 can be adopted and recommended for direct cultivation under rain-fed conditions.

Singh et al. (2015) investigated the physio-morphological basis of selection in maize (*Zea mays* L.) at the Research Farm of Chandra Shekhar Azad University of Agriculture & Technology, Kanpur, India, to search out the best productive inbred lines based on physiological as well as morphological parameters and the correlation. In this respect, 50 inbred lines from the Directorate of Maize Research, New Delhi, and maintained by maize breeder CSA Kanpur, were subjected for study on plant height, chlorophyll intensity, TDM, crop growth rate, leaf area ratio, net assimilation rate, relative growth rate and total grain yield taken at different stages of crop growth, viz at 45 DAS, 60 DAS and maturity. Based on these parameters and their correlated study the best productive inbred lines were selected. Plant height was found to be maximum at maturity in genotype NP-43 (i.e. 143.67 cm) and minimum in NP-3 (89 cm) at the same stage. Dry matter production per plant was observed to be highest in NP-23 (133.83 g) and lowest in NP-7 (55 g) at maturity. Crop growth rate, relative growth rate and net assimilation rate were found to be at a maximum in NP-35 at maturity. Genotype NP-26 gave the maximum grain yield. High heritability estimates in a broad sense were observed for all the characters at all the stages of growth registered. An advancement of 49.36 cm^{-2} g^{-1} per cycle of selection was observed for LAR at 60 DAS, while it was 1.59 g cm^{-2} (leaf area) per day per cycle of selection for NAR at 60 DAS. The highest genetic gain over the mean to the tune of about 108.06 % was expected for TDM at 45 DAS at $K = 2.06$. Grain yield per plant showed a positive and significant correlation coefficient with plant height and TDM production at all the stages. On the basis of these results, it can be concluded that genotypes NP-35, NP-43 and NP-26 can be utilized for development of productive lines or can be utilized in heterotic combinations for production of single cross hybrids to be used for production of a crop.

Sumalini et al. (2015) assessed the general combining ability effects of parents and specific combining ability effects of hybrids for yield and yield-related traits strictly under rain-fed situations which, incidentally, mimicked mid season drought conditions, i.e. moisture stress at the flowering stage for a period of 22 days. Combining ability analysis using a line × tester design was conducted in maize (*Zea mays* L.) inbred lines by growing 135 F_1's generated by crossing 15 lines with nine testers at ARS, Karimnagar, PJTSAU, during the rainy season in 2011. Female lines were derived from the recurrent selection cycle carried out in the drought-tolerant population 'Tuxpeño Sequia' provided by CIMMYT in 1996. Males were developed using pedigree breeding. Analysis of variance showed highly significant genotypic differences for all the traits studied, indicating a wide range of variability among genotypes included in this evaluation. The ratio GCA:SCA was lower than unity, indicating the preponderance of non-additive gene action in the expression of all the characters studied, except flowering traits. Moderate narrow sense heritability was observed for the majority of the characters. Three lines (KMLD-3, KMLD-11 and KMLD-19) and one tester (KML-9) were good general combiners for grain yield and one or more yield-contributing characters. Crosses KMLD-3 × KML-99, KMLD-5 × BML-7, KMLD-5 × KML-801, KMLD-6 × KML-36 and KMLD-11 × KML-29 showed significant positive SCA, standard heterosis and high mean values for grain yield per plant and two or more yield components. These hybrids were found to be early flowering, with significant negative heterosis providing some clue about their usefulness under drought conditions. Therefore, further testing of these new inbred lines for use in a crossing programme is recommended to combine major yield components with high yield to derive climate-resilient hybrids.

9.2.7 Pearl Millet

Patil and Jadeja (2015) evaluated 113 pearl millet *Pennisetum glaucum* (L.) R.Br hybrids under terminal drought stress and optimum (non-stress) conditions to study their responses to drought and to identify the traits that are associated with drought tolerance or escape. All hybrids demonstrated significant differences in yield-contributing traits and yielding ability under stress and non-stress conditions. A wide range in the drought response index (DRI) was observed among the hybrids studied. Yield stability of the genotypes depended upon different characters in the presence and absence of drought stress, with emphasis on the manipulations of different traits for enhancing yield under stress and non-stress conditions. Higher panicle number and greater biomass accumulation were associated with target traits for improving grain yield under stress, but not for enhancing yield under non-stress conditions. Increase in the test weight was related to drought escape, while higher grain numbers per panicle and the harvest index were associated with drought tolerance. Hybrids with a high degree of drought tolerance have been identified during development of drought-tolerant genotypes. Drought escape, stress tolerance and, to a lesser extent, potential yield contributed to determining the grain yield under stress. It is suggested that in selecting a drought-responsive genotype, emphasis should be given to drought stress-responsive traits, viz. root length per plant, proline content and the DRI. This study also indicated that hybrids showing high economic heterotic effects for grain yield also depicted high heterotic responses for at least one or more yield-attributing characters. The presence of both additive and non-additive gene action was observed for major yield-contributing traits in the present investigation, hence a breeding strategy involving biparental matings with reciprocal recurrent selection for simultaneous advancement of yield-attributing characters is important in pearl millet.

9.2.8 Rice

Varthini et al. (2015) undertook a study on the selection of rice genotypes on the basis of the root system. Drought is a most complex and harsh stress, but it can be dealt with by genetic improvement of rice. Plant roots play an important role in water and nutrient acquisition. An efficient root system is known to regulate the amount of water available to the plant, depending on its distribution in the soil. Therefore, breeding rice for drought tolerance can be undertaken by aiming at improving secondary traits, i.e. root traits for identification of the tolerant genotypes. This study was undertaken to evaluate rice root morphology by the modified soil-filled root box experimental approach. Recombinant inbred lines (RILs) developed from Norungan and IR 64 were screened in three runs, each time with 48 RILs and two parental lines from which root sampling data was taken on traits such as root length, root thickness, root volume, root density and the root shoot dry weight ratio. Significant variation was observed between the two parental lines, Norungan and IR 64, as well as RILs. Then RILs exhibiting extreme root phenotypes were screened under natural moisture stress conditions in the field to study the influence of root parameters on field performance.

Traditional rice land races are the reservoir of many tolerant genes for biotic and abiotic stress. Based on this knowledge, Mohapatra et al. (2015) investigated genetic diversity and molecular characterization of rice landraces (*Oryza sativa* L.) using drought stress-linked simple sequence repeat (SSR) and sequence tagged sites (STS) markers. Their study was undertaken on the molecular characterization of specific genomic loci that influence the tolerance of drought stress in rice to facilitate proper selection of genotypes for breeding programmes. Three drought-tolerant controls, viz N22, CR-143-2-2 and Vandana, and drought susceptible controls, viz IR64 and IR 20, were used for genetic diversity and molecular characterization in response to drought stress. The molecular profiling for genetic relationships among the 55 rice accessions was assessed with 23 previously reported drought stress tolerance QTL linked SSR markers and candidate gene-specific markers of STS. Using PCR, amplified product sizes (molecular weights) for each SSR and STS marker were documented and allelic diversity analysed; polymorphism information content (PIC), dendrogram and genetic similarity coefficients were calculated by using NTSYS-pc (version 2.02). Out of 23, 15 SSR markers were revealed to be polymorphic and polymorphism ranged from 0.611 (RM126) to 0.971 (RM314) with an average of 0.879, while in gene-specific STS markers polymorphism ranged from 0.119 (Rab21) to 0.950 (OsCam-F), with an average of 0.66. Highly polymorphic SSR markers could be used for the evaluation of genetic diversity and identification of promising rice genotypes for drought tolerance. The dendrogram was constructed based on Rogers's genetic distance of specific alleles generated in SSR and STS markers from the total rice accessions. The UPGMA cluster tree analysis was used to group the 55 land races of rice accessions into three major groups in SSR and five major groups by STS markers. These findings provide the basis for future efficient use of these molecular markers for the genetic analysis of rice genotypes and development of drought stress-tolerant crops through MAS breeding approaches.

Tripathy and Dash (2015) characterized a set of selected drought-resistant local land races of Odisha, India, and a few drought-tolerant improved cultivars and popular high-yielding upland genotypes using total seed storage protein fingerprinting. Fifteen out of 19 scorable polypeptide bands with molecular weights ranging from 26.0 to 123.0 kD revealed wide polymorphism (78.94 %) based on presence or absence of bands. Additionally, polypeptide markers differed in staining intensity, migration velocity and frequency among the tested genotypes. Genotypes with medium amylose content (21–22 %) are considered to have better cooking quality. Asumakunda, Bastul, Brahmanaki, Kanding, Kalakeri, Padarabank, Salampikit, Somo, CR Dhan 40, Anjali, Vandana and Pandeydhan have been identified as candidate varieties for intermediate amylose content. Cluster analysis of genetic similarity of the protein bands showed a wide genetic variation among the tested materials, with the similarity coefficients varying from 0.47 to as high as 1.00. Six of the tested genotypes together and four paired genotypic comparisons showed complete homozygosity (SI = 100 %) for seed storage protein profile, while Khursudi (Av. SI = 0.58) followed by Khandagiri (Av. SI = 0.62) and Nagina 22 (Av. SI = 0.69) had high genetic dissimilarity compared to most other varieties. Cluster analysis revealed nine distinct groups of rice varieties at the similarity coefficient level of 0.91. Khursudi maintained high genetic distance from Bastul, Brahmanaki, Zhu 11–26 and Vandana, whereas N22 showed appreciable genetic dissimilarity from Khandagiri and Zhu 11–26. Additionally, Khandagiri was shown to be genetically distant from Asumakunda, Brahmanaki, Hiran, Khursudi, N22, Padarabank, Anjali and Vandana. These genotypes may be used in a hybrid formation programme for characteristic upland situations. In addition, a few genotype-specific polypeptide bands identified in the present pursuit should be useful for identification of rice germplasm.

9.2.9 Wheat

Singh and Singh (2015) conducted pot experiments to elucidate the effect of plant growth-promoting rhizobacteria (PGPR) (namely *Bacillus subtilis*, *Pseudomonas* sp. Iso13 and *Microbacterium resistens*) and their EPS on physiological parameters of soil and wheat plants under drought. They found drought stress had significant adverse effects on the growth of wheat seedlings. Seed bacterization of wheat with EPS-producing bacterial strains in combination with their respective EPS improved soil moisture contents. Under drought stress, the inoculated plants showed increased RWC, protein and sugar, although proline content and the activities of antioxidant enzymes, namely superoxide dismutase (SOD), peroxidase (POD) and catalase (CAT), actually decreased. *Pseudomonas* sp. Iso13 strain isolated from a semi-arid region was the most potent PGPR under drought stress. The association of inoculum and their respective EPS showed greater potential in promoting wheat drought tolerance in comparison to plant performance under drought when PGPR inoculum was used alone.

9.2.10 Oats

Singh (2015) evaluated 70 genotypes of oat (*Avena sativa*) germplasm for traits imparting drought tolerance in an intermediate zone and the green fodder potential of its components under rain-fed environments during *rabi* 2010–2011, 2011–2012 and 2012–2013 at RARS, Armoury. The genetic variability, character association and path analysis between green fodder yield and its contributing characters were studied using 70 genotypes of oat. Analysis of variance showed highly significant differences among genotypes for all characters recorded. A high phenotypic and genotypic coefficient of variation coupled with high heritability and genetic advance as a percentage of the mean were reported for green fodder productivity, plant height, tillers per plant and green fodder yield, indicating predominance of additive gene effects in controlling these characters. A highly significant and positive correlation of green fodder yield was reported with green fodder productivity, tillers per plant, plant height and days to flowering. Path coefficient analysis showed positive direct effects of green fodder productivity, plant height, days to flowering and tillers per plant on green fodder yield. Plant height had shown negative indirect effects through most of the characters. Most of the characters showed negative and low-magnitude indirect effects except days to flowering, crop duration and green fodder productivity. Green fodder productivity, plant height, days to flowering and tillers per plant were considered important; therefore, emphasis should be given to these characters in selection and improvement of green fodder yield in oat breeding programmes.

9.2.11 Cotton

Although irrigation is required for profitable yields in cotton and other agricultural crops, it is also a major production cost. Deva and Pandey (2015b) investigated remote sensing of biotic and abiotic stress for irrigation management of cotton. They found that deficit irrigation of short-season upland cotton—primarily produced as an irrigated upland crop—using a LEPA system enhanced lint yield and conserved groundwater. When a plant water deficit is imposed during peak flowering periods, yield is reduced more than when deficit irrigation is imposed either earlier or later in the flowering period. Recent investigations have demonstrated site-specific management (SSM) as an additional means to address water management issues. SSM can adopt satellite-based remote sensing technology and mapping systems to detect specific areas suffering from stress within a field (i.e. water deficit, insect damage or disease incidence). Crop canopy temperature, an effective indicator of plant water stress, could also be useful in developing an efficient irrigation management system. The use of remote sensing to determine canopy temperature has been considered useful in monitoring plant stress by using IRTs mounted on centre-pivot irrigation systems. Canopy temperature between the 100 % (31.8 °C) and 75 % (32.1 °C) ETc irrigation regimes was not significantly different, but the canopy temperature for the 50 % (33.1 °C) ETc regime was significantly higher than for the other two irrigation regimes. Although there were significant differences in canopy temperature between all three irrigation regimes, the differences between 100 % and 75 % ETc were smaller compared to the 50 % ETc regime. The IR camera was able to differentiate biotic (root rot) from abiotic (drought) stress with the assistance of ground truthing. Biotic stress (*Phymatotrichum* root rot) was present within the field and canopy temperatures increased in areas where disease was present.

Rathod et al. (2015) studied the influence of morpho-physiological traits for drought tolerance in different cotton species. Their investigation was conducted during the *kharif* season of 2010 and 2011 in RBD on the experimental field of the Post Graduate Department of Agricultural Botany under non-stress and water-stress conditions in pot culture. Seeds of 31 cotton genotypes (10 *G. arboreum* and 21 *G. hirsutum*) were sown in three replications. The experimental findings are based on 2 years of pooled data under water stress and non-stress conditions. The results indicated significant superiority of *G. arboreum* over *G. hirsutum* in tolerance of water stress on the basis of morpho-physiological characters and seed cotton yield and its attributes. *G. arboreum* genotypes grew consistently taller under stress than under non-stress conditions. At 120 DAS under both stress and non-stress conditions, *G. arboreum* genotypes showed significantly larger specific leaf weight (SLW) than *G. hirsutum*. SLW was found to be the best indicator for drought tolerance. When SLW was compared among *G. arboreum* genotypes, AKA-8 (1.78 g dm^{-2}), Garrohill (1.91 g dm^{-2}), AKDH-1 (1.82 g dm^{-1}), KWA-8 (1.53 g dm^{-2}), HD-110-151 (1.50 g dm^{-2}), and AKA-7 (1.49 g dm^{-2}) were the best performers, while among *G. hirsutum*, PKV-Hy-4 (1.51 g dm^{-2}), PKV-Hy-5 (1.34 g dm^{-2}), Yamuna (1.17 g dm^{-2}) and PKV-Hy-2 (1.10 g dm^{-2}) were found to be the better genotypes with regard to tolerance of water deficit. The correlation coefficient values suggest that SLW is highly and positively correlated with seed cotton yield ($r = 0.741$) and this criterion may be useful for assessing genotypes for drought studies.

9.2.12 Cucumis

Purru and Bhat (2015) investigated assembly, annotation, functional classification of transcriptomes and identification of key genes for moisture stress tolerance in musk melon (*Cucumis melo* L.), an important fruit crop worldwide, known as a very susceptible plant to moisture stress, this being one of the major restraining factors for its production. The main objective of this study was to identify key genes that play a major role in moisture stress tolerance by *de novo* assembly and annotation of the musk melon transcriptome. In this study, a high-throughput, SOLiD sequencing technology was employed to characterize the de novo transcriptome of musk melon. High-quality unique reads for *Cucumis melo* control and stress samples were assembled using CAP3, DNA STAR and CLC de novo assembly programs. Merging of assemblies by CAP3 resulted in larger and robust transcripts instead of using a single program alone. The assembled transcripts were annotated with gene ontology (GO) terms using Blast 2 Go. All the GO terms were classified into functional groups including biological processes, cellular components and molecular function. Comparison of gene expression levels between control and stress transcriptomes by RNA-Seq mapping revealed that 177 genes were differentially expressed, with some genes highly upregulated in stress samples, such as kinases, DREB genes and heat shock proteins; some of these are well known to be involved in critical pathways, such as plant hormone signal transduction, protein processing in the endoplasmic reticulum and inositol phosphate metabolism. Thus, these selected genes will not only facilitate increased understanding of genetic basis of moisture stress response, but also accelerate genetic improvement for drought tolerance in musk melon.

9.2.13 Linseed

Root and other morphological characters are important while breeding for drought tolerance. Rajann et al. (2015) investigated the response of root and other morphological characteristics under water-stress conditions in linseed (*Linum usitatissimum* L.) with an objective to identify moisture stress-tolerant linseed germplasm based on better root and other morphological characteristics among 80 lines during *rabi* 2012–2013 under a randomized block design with two replications using cement root structures which mimic field conditions for root and crop growth. Ten different morpho-physiological characters were evaluated. The germplasm lines 68/56120 (33.8 cm), Bengal-46 (31.5 cm) and ES-13239 (28.5 cm) had showed greater root length and for high root volume, Bengal-46 (4.3 cc), Bengal-70 (3.7 cc), CI-2006 (3.5 cc), ES-13239 (3.2 cc) and CI-1924 (3.1 cc). Based on the mean per se performance, three promising lines, viz Bengal-46, ES-13239 and EC-322659, were identified for significantly higher high root length, root volume, fresh root weight, dry root weight and number of capsules, which may perform more efficiently under moisture stress conditions by being able to explore and successfully utilize moisture from the deeper layers in the soil profile.

9.2.14 Mustard

Puttawar et al. (2015) undertook screening of mustard with the objective to identify superior drought-tolerant crosses for recombination breeding and to study the correlation between the different traits. Thirty crosses obtained by crossing two testers (drought-tolerant donors) with 15 lines along with their parents were evaluated in a factorial randomized complete block design replicated twice with three stress treatments, i.e. control (C), moderate stress (MS) and severe drought stress (SS), in pot experiments during *rabi* 2009–2011. The data recorded included days to first flower, days to maturity, plant height, number of branches per plant, number of siliqua per plant, root length, root volume at maturity, leaf RWC, proline content, yield per plant and drought-tolerant efficiency. The genotypes showed significant differences for all the characters studied except for number of branches per plant under both normal and drought conditions. The genotype × treatment interactions were significant for all the traits except for days to 50 % flowering and number of branches per plant. Severe stress treatment produced flower and siliqua formation only in very few genotypes. The mean performance of all traits were affected either on the higher or lower side due to stress. The number of siliqua per plant and seed yield per plant were the traits most affected due to stress. Moderate stress resulted in 11.7 %, 38.87 %, 34.72 %, 73.69 %, 40.77 %, 7.26 % and 63.67 % reductions in days to maturity, plant height, number of branches per plant, number of siliqua per plant, root volume, LRWC and seed yield per plant, respectively; and 3.11 % and 60.47 % increases in root length and proline accumulation due to stress, respectively. A correlation study under control and stress indicated that seed yield per plant was positively and significantly correlated with plant height, number of siliqua per plant under control conditions and with plant height and proline content under the stress treatment. Parents RH-819, Geeta, Kranti, JD-6 and Pusa Bold and crosses Yaruna × RH-819, Ashirvad × RH-819, JD-6 × Geeta and ACN-9 × Geeta were identified as superior parents and crosses, respectively, for drought tolerance based on high DTE, high mean seed yield under stress, less percentage reduction in yield and number of siliqua per plant and high percentage increase in proline content. These five parents (RH-819, Geeta, Kranti, JD-6 and Pusa Bold) and four crosses (Varuna × RH-819, Ashirvad × RH-819, JD-6 × Geeta, ACN-9 × Geeta) will be exploited in breeding mustard for drought tolerance.

9.2.15 Sesame

In a field experiment, Damdar et al. (2015) studied the effects of irrigation and nitrogen levels on yield and water efficiency (WUE) of summer sesame (*Sesamum indicum*) in Akola, India. The results revealed that yield-contributing characters and moisture were significantly higher with irrigation scheduling at 1.0 IW/CPE (irrigation water amount/cumulative pan evaporation) and nitrogen application at 90 kg N ha^{-1}. Similarly, treatment 1.0 IW/CPE combined with nitrogen application at 90 kg N ha^{-1} recorded higher seed yield than any other combination, while the lowest seed yield was recorded for irrigation scheduling at 0.4 IW/CPE combined with 30 kg N ha^{-1}.

9.2.16 Ber

Ziziphus nummularia, a wild species of ber (Ziziphus), is well distributed in the northwestern parts of India, where it grows into either a tree or a bush in grasslands with an average annual rainfall of even less than 100 mm. Transcriptome profiling was undertaken in *Z. nummularia* genotype Jaisalmer under control and water stress at 0.3 MPa using a standardized in vitro technique to identify transcript expression during drought stress (Sivalingam et al. 2015). Transcripts were sequenced on an Illumina HiSeq 2000 platform and de novo transcriptomes were assembled for both control and stressed plant samples. The trimmed reads were aligned to the assembled transcriptome (length ≥150 bp) using the Bowtie2 program. Differential gene expression analysis was performed using DESeq. Among the unique transcripts identified in stress samples, 283 transcripts were found to be downregulated and 554 upregulated in comparison to controls. The important downregulated transcripts identified are LRR receptor-like serine or threonine protein kinase, oligopeptide transporter OPT family, myrcene synthase, LRR protein kinase family isoform1, ATP binding cassette transporter, phytochrome kinase, pectin esterase inhibitor like-35, auxin efflux facilitator isoform 1, multidrug resistance protein, pectin methyl esterase 3 and ABC transporter B family. Similarly, some of the upregulated transcripts are Raffinose synthase family protein, wall associated receptor kinase like, acyl transferase like protein, cadmium or zinc transporting ATPase 3 like, sucrose synthase 6-like, cytochrome P-450 like, phosphatase 2C family protein and TT12-2 MATE transporter. The results will undoubtedly be very useful in identifying novel genes and their involvement in the drought tolerance response by *Z. nummularia*.

9.2.17 Safflower

Goud et al. (2015) studied the role of the root and its related traits in utilization of available moisture under water-stress conditions in safflower (*Carthamus tinctorius* L.). Fifty-five safflower lines were included in the study. Root and other morphological characters are important for breeding for drought tolerance. The experiment was carried out to identify moisture stress-tolerant safflower germplasm based on better root characters and other morphological characteristics, at the Main Agricultural Research Station, Raichur, during *rabi* 2012–2013. A total of 55 lines were evaluated using a randomized block design with two replications under specially constructed cement root structures which mimicked field conditions for root and crop growth for ten different morpho-physiological characters. The germplasm sample included lines GMU-1811 (48 cm), GMU-2293 (45.8 cm), GMU-3828 (44 cm), GMU-4914 (43.3 cm), GMU-4021 (43 cm), GMU-174 (40.5 cm) and GMU-589 (40.3 cm), all checked for higher root length, and lines GMU-4093 (17.4 cc), GMU-3828 (17 cc), GMU-3148 (15 cc), GMU-4454 (14.7 cc), GMU-2866 (31.2) and GMU-7324 (14.6 cc), checked for higher root volume. Based on the mean per se performance, 12 promising lines, viz GMU-3148, GMU-2576, GMU-4480, GMU-7324, GMU-1811, GMU-4021, GMU-174, GMU-3828, GMU-4454, GMU-609, GMU-4965 and GMU-5197, exhibited high significant per se performance for plant height, root length, root volume, fresh root weight, dry root weight, number of branches per plant and yield per plant. High GCV coupled with high heritability were recorded for dry shoot weight (43.34 and 90, respectively), root volume (31.75 and 85, respectively), dry root weight (30.93 and 79, respectively) and fresh root weight (30.04 and 91, respectively). These data indicated the presence of variability, suggesting that these traits were under genetic control.

9.2.18 Sunflower

An investigation was carried out to understand the effect of drought on physiological characteristics in 12 sunflower genotypes during *rabi* 2009–2010 at College Farm, College of Agriculture, PJTSAU, Rajendranagar, Hyderabad (Geetha et al. 2015). Moisture stress treatments were imposed at the flower bud initiation stage (irrigation withheld from 40 DAS to 60 DAS), while control plots were irrigated at 10-day intervals throughout the cropping experimental period. The results revealed that water stress showed negative effects on RWC, photosynthetic rate, leaf fluorescence, membrane leakage, chlorophyll content and specific leaf area across the genotypes examined. A decline in specific leaf area under water stress is considered an adaptation to water stress; however, genotypic variation was significant for the characters studied. Genotypes SH-177, SH-491 and DSF-111 were considered promising lines and managed to maintain higher RWC, photosynthetic rate, leaf fluorescence and SPAD chlorophyll meter readings with lower membrane leakage and specific leaf area. Thus the above genotypes could be exploited for drought-resistant breeding.

9.2.19 Gladiolus

Ranjan and Hazarika (2015) investigated mechanisms of adaptation to drought by some *Gladiolus* cultivars under rain-fed conditions. The study aimed to investigate the inherent adaptation mechanism of some plant cultivars to cope or be sustained in a given set of environmental stresses and was conducted in the northeastern part of India, at Assam Agricultural University, Jorhat, under rain-fed conditions. Gladiolus shows wide variation and is popular among cut flowers. Any cell modifications are easily detectable. Among the different cultivars of *Gladiolus*, seven popular cultivars, viz Suchitra, Jacksonville Gold, American Beauty, White Prosperity, Blue Isle, Aarti and Apollo, were grown and their adaptation mechanisms were evaluated through the physiological and anatomical modifications recorded in order to understand crop–environment interactions with respect to water availability. There were significant morphological, physiological and anatomical modifications in each cultivar under the given condition. Additionally, cultivars showed significant differences in leaf anatomical parameters among themselves in the rain-fed condition. Among the seven cultivars under study, Aarti stood out significantly as the highest on these parameters, viz maximum number of stomatal cell, maximum pore size, highest thickness of the leaf layer, and highest numbers of palisade cells and spongy cells. Further, the highest leaf xylem and phloem size were recorded in Aarti, followed by the cultivar Suchitra. From the study, it was clear that some cultivars like Aarti and Suchitra undergo significant modifications of various anatomical parameters in response to water stress.

9.2.20 Fodder Crops

Meena et al. (2015) studied the productivity of fodder as influenced by soil moisture conservation techniques and cropping systems under semi-arid environments. A field experiment was conducted from 2011 to 2014 to study the impact of soil moisture conservation techniques on plant growth and yield of fodder crops and to identify the best suitable soil moisture conservation approach for the region. The treatments comprised seven cropping systems, viz sole *Cenchrus ciliaris*, sole pearl millet, sole cluster bean, sole *Clitoria ternatea*, *Cenchrus ciliaris* + cluster bean (1:1), pearl millet + cluster bean (1:1) and *Cenchrus ciliaris* + *Clitoria ternatea* (1:1), and four moisture conservation techniques, viz control, dust mulch, straw mulch (5 t ha^{-1}) and farmyard manure (5 t ha^{-1}). The experiment was laid out in a randomized block design with three replications. Soil moisture samples were taken 48 h after effective rainfall during the cropping season. Farmyard manure was mixed 1 month before sowing and straw mulch was spread out after sowing. Plant growth parameters and forage yield were recorded at the 50 % flowering stage. The results showed that the plant height, number of tillers branches per plant and dry fodder yield of all fodder crops studied was greater under straw mulching, followed by the application of FYM, dust mulching and control. The straw mulching treatment was significantly superior over dust mulching and the control treatment in terms of growth and dry fodder yield. Dry fodder yield of *Cenchrus ciliaris*, cluster bean, pearl millet and *Clitoria ternatea* was significantly greater under straw mulching (3607, 2243, 7041 and 1746 kg ha^{-1}, respectively) than under any other soil moisture conservation treatment. Fodder yield was also found to be significantly higher under the application of FYM (3338, 2046, 6339 and 1604 kg ha^{-1}, respectively) than under dust mulching (2191, 1864, 6094 and 1395 kg ha^{-1}, respectively) or control treatment (2012, 1741, 5718 and 1182 kg ha^{-1}, respectively). The yield difference was non-significant between dust mulching and control but was higher in the first case. This trend was found in all four crops taken under investigation. Maximum soil moisture content (%) was retained under the cropping system of *Cenchrus ciliaris* (4.83) alone, followed by *Cenchrus ciliaris* + *Clitoria ternatea* (4.09), *Cenchrus ciliaris* + cluster bean (3.99), *Clitoria ternatea* (3.58), pearl millet (3.55), pearl millet + cluster bean (3.44) and cluster bean (3.42). Soil moisture content was retained significantly more under the application of straw mulching (4.27) over dust mulching (3.71) and control (3.41) but was found to be on a par with FYM (3.99). The soil moisture percentage was found to be higher in the deeper layers (30–60 and 60–90 cm) within the soil profile. Straw mulching under *Cenchrus ciliaris* cropped with *Clitoria ternatea* was found to be effective for conserving soil moisture and improving forage productivity in the semi-arid environments.

9.2.21 Chilli

Mohan et al. (2015) studied the effect of moisture stress and NAA spray on growth, flower bud drop and yield on chilli (*Capsicum annuum* L.). Among the problems faced by the farmers in chilli cultivation, flower drop is a serious one, which causes a great reduction in yield. Flower bud drop in chilli is a complex phenomenon influenced by abiotic stresses like soil moisture, sunlight, high temperature, salinity and growth regulators. Biotic stresses like attacks by pests and disease incidence also cause flower bud drop. Moisture stress causes abscission of flowers through its influence on the plant hormone balance, rate of photosynthesis and availability of carbohydrates for growth. This may lead to a heavy drop of reproductive structures. A field study was conducted during *kharif* 2011 to understand the effects of moisture stress and NAA application on plant growth, flower drop and yield in chilli at Loyola Academy farm, Alwal, Secunderabad. A factorial randomized block design was used to test on chilli cv. LCA-235. Treatments comprised three irrigation levels: single irrigation at flowering, biweekly irrigation and monthly irrigation; and three hormonal sprays: control (normal water), NAA-10 ppm and NAA-20 ppm. The results of the experiment revealed that moisture stress had significantly reduced the plant morphological and physiological parameters, viz plant height, number of branches, leaves and LAI. Among treatments, the biweekly irrigation level resulted in increased plant growth and LAI compared to single irrigation and monthly irrigation. Moisture stress significantly increased the drop of flower buds and flowers more under single irrigation (20.6 %) compared to biweekly irrigation (12.5 %). The percentage nutrient content (N, P and K) was decreased in the dropped flower buds compared to retained ones on the plants. Enzyme indole acetic acid (IAA)-oxidase content (μg IAA degraded g^{-1} dry weight $hour^{-1}$) in the dropped and retained flower buds was increased with increased moisture stress, which may reflect degradation of IAA. Moisture stress significantly decreased fruit yield. Among irrigation levels, biweekly irrigation recorded the maximum yield of matured fruits (2.75 kg per plot) compared to single and monthly irrigation. Among hormonal sprays, NAA at 20 ppm sprayed at the flowering stage significantly reduced flower bud drop and increased fruit yield by 15 %, compared to NAA at 10 ppm and controls.

9.3 Breeding Approaches for Drought Resistance

In discussing breeding approaches for drought resistance in crops, Namrata et al. (2015) remind us that by the year 2025, 1.8 billion people will live in countries or regions where severe water scarcity will be the rule. The states most affected in India include Rajasthan, parts of Gujarat, Haryana and Andhra Pradesh. Now, all the studies presented at the 2nd International Conference on Bio-resource and Stress Management have evidenced the strong impact of drought stress on the normal physiology and growth of plants, resulting in an overall reduction of agricultural productivity by preventing crop plants from expressing their full genetic potential.

Drought escape, avoidance and tolerance are the three mechanisms involved in drought resistance. Various morphological, physiological and biochemical characters that confer drought resistance show different types of inheritance patterns (monogenic and polygenic) and gene action (additive and non-additive). Breeding for drought should be based on the level and timing of stress in the targeted area. If stress is severe, breeding under stress-free conditions may be unsuccessful and traits that confer survival may become a priority. Traditionally, three breeding approaches for drought resistance have been put to exercise. The first approach consists of breeding for high yield under optimum (water unlimited) conditions. This may be useful provided there is enough genetic variation within the germplasm handled. The second approach deals with the problem that the intensity of drought is highly variable from year to year; as a result, environmental selection pressure on breeding materials may vary quite largely from generation to generation. The third approach is to incorporate the morphological and physiological mechanisms of drought resistance.

At the molecular level, several drought-responsive genes and transcription factors have been identified, such as the dehydration-responsive element-binding gene, aquaporins, late embryogenesis abundant proteins and dehydrins. Plant drought tolerance can be managed by adopting strategies such as mass screening and breeding and marker-assisted selection. Due to low heritability of yield under stress and to the spatial as well as temporal variations in the field environment, marker-assisted selection may prove more beneficial.

9.4 Salinity

World food production relies heavily on irrigated agriculture, which accounts for up 40 % of the yearly global agricultural outcome using only 17 % of available arable land—a fact which, in itself, describes what modern technology can accomplish. Nevertheless, vast areas turned into productive arable land over the latter two thirds of the twentieth century around the world face an increasing risk of becoming unusable land because of salinity and water logging. Over half a billion hectares are at stake in just Australia, India, Pakistan, and Central and Southeast Asia (Asif and Ahmad 2002)—a region that is home to over 3.4 billion people, nearly half the world's entire human population. It goes without saying that what we are dealing with here is a major issue.

Irrigation is by far the number-one user of fresh water, a fact that underlines the urgency to develop management practices that ensure the most efficient use of the resource as it becomes increasingly scarce around the globe, while population and consequently food demand keep growing at a fast rate in relation to the stability and functionality of global environmental mechanisms guarding the homeostasis of our ecosystems.

In a recent review of the literature on plant growth and development under salinity stress, Läuchli and Grattan (2007) looked at our current understanding of the physiological mechanisms by which growth and development of crop plants are affected by salinity. Research has established that early growth reduction occurs by salinity as much as by water stress. In both cases an osmotic effect causes cell elongation to stop as higher water potential in the external solution does not support water absorption by the cells; on the contrary, water tends to go out of the cell following the adverse water potential gradient. There seems to be little genotypic variation in plant response at this early stage. In the stages that follow, however, salinity-tolerant species or genotypes differ from sensitive ones in their ability to prevent salt accumulation to toxic levels in leaf tissues. Ca ions have long been observed to counteract salinity and prevent growth inhibition by keeping membrane-selective permeability to K over Na, thus helping maintain membrane functional stability. Most crop plants have been shown to be decreasingly salt tolerant from germination through emergence and increasingly salt tolerant from the period of vegetative growth through the period of reproductive growth.

Sreenivas et al. (2015) developed drainage technology as a viable tool for reclamation of saline-sodic and waterlogged soils of Godavari, the Western Delta and India. A sub-surface pipe was installed in an 18 ha area in a farmers' fields at the Kalipatnam pilot area in the Godavari Western Delta canal command area in India in the year 2005 to combat the problems of water logging (depth of water table 0–0.85 m), salinity and sodicity (ECe 5.9–44.8 dS m^{-2} and ESP 19.4–56.6). A nylon mesh envelope material was used and 50 m lateral spacing was adopted for the pipe drainage system. The effectiveness of this drainage system in the control of water logging at the pilot area was monitored through a network of 24 observation wells (150 × 150 m grid) and 48 surface grid points (100 × 100 m) for soil samples and crop cut data. SSD was operated for 4003 h for the year under review (summer of 2005 to summer of 2009). A total of 66.8 t ha^{-1} of salts was disposed of through the pipe drainage system during the 4 year period. Summer soil salinity (0–15 cm) decreased 38 %; after *kharif* soil salinity decreased 18 % and ground water quality (salinity) improved 40 % from 2005 to 2009. Four years after installation of the subsurface drainage system the results evidenced a 1.01 t ha^{-1} increase in wet season rice (*kharif*) and a 2.31 t ha^{-1} increase in the dry season (*rabi*) rice yield in the area, in comparison to yields obtained prior to installation of the system. A techno feasibility study indicated that the payback period was 2.18 years.

9.4.1 Sugarcane

Gadakh et al. (2015) developed a rapid screening technique for salt tolerance in sugarcane (*Saccharum* spp.). The investigation was carried out on sugarcane var. Co- 99004 at the Sugarcane Tissue Culture Laboratory and farm of the Main Sugarcane Research Station, Navsari Agricultural University, Navsari, during the 2012–2013 and 2013–2014 cropping cycles. Somaclonal variation in combination with in vitro mutagenesis proved beneficial for the isolation of salinity and drought-tolerant lines in little time by employing in vitro selection. Hence the present study was carried out on in vitro selection of sugarcane mutants for salinity tolerance and isolation of tolerant mutant lines. A tissue culture technique was used to obtain salt-tolerant mutants from EMS (0.5 % for 2 h) treated embryogenic calli cultured on a selective medium containing different levels of NaCl (0, 50, 100, 150 and 200 mM). The 100 mM concentration of NaCl was decided to be the LD_{50} of NaCl on the basis of the percentage survival of calli. Calli treated with EMS (LD_{50}) was placed on shoot regeneration media containing the LD_{50} concentration of NaCl. The plants that regenerated from the tolerant calli were grown in a pot culture system under salinity stress (with the previous levels) as compared to normal plants (the source variety). These results demonstrated that an in vitro technique can be exploited to improve agronomic traits and to increase resistance to abiotic stress in crops like sugarcane, for which conventional breeding programmes take much longer periods of time and never stop being decidedly labour intensive.

9.4.2 Blond Psyllium

Shivran and Meena (2015) studied the impact of soil type and saline water irrigation on soil salinity and productivity of blond psyllium (*Plantago ovata* Forsk).

A 2-year pot culture experiment was conducted at the SKN College of Agriculture, Jobner (Rajasthan), using two soil types (sandy loam and loamy sand) and five salinity levels of irrigation water (0–4 dS m^{-1}) to evaluate the effects on blond psyllium production and soil salinity. These treatments were combined factorially for a total of ten treatment combinations, which were replicated three times and arranged in a completely randomized design. The maximum plant stand at harvest, spikes per plant, seeds per spike, test weight and seed yield per plant varied significantly between the two soil types used, regardless of salinity, as shown when all the data were pooled together. However, the data also showed that electrical conductivity of a saturation extract of soil after harvest of the crop was significantly higher in the sandy loam soil than in the loamy sand soil. An increased plant stand at harvest, spikes per plant, seeds per spike, test weight and seed yield per plant were recorded with irrigation water of 2 dS m^{-1}; this treatment resulted in significantly higher values for spikes per plant and test weight than those recorded for plants grown in fresh water, while they were higher for seeds per spike and seed yield per plant than those found in fresh water and irrigation water of 1 dS m^{-1}. Furthermore, it was observed that application of irrigation water of 4 dS m^{-1} significantly reduced plant stand at harvest, spikes per plant, seeds per spike, test weight and seed yield per plant in comparison to lower water salinity levels. The electrical conductivity of a saturation soil extract after the crop harvest increased significantly with increasing levels of irrigation water EC up to the highest level, i.e. 4 dS m^{-1}, used in these experiments. It may be inferred that blond psyllium can be grown well in loamy sands when irrigated with water up to 2 dS m^{-1} EC.

9.4.3 Finger Millet

Syvarna et al. (2015) made an assessment of salt stress impact on the germination of finger millet (*Eleusine coracana*), an important nutritious minor millet, quite tolerant of drought and salt stress. In a laboratory experiment, salt stress was induced by addition of different salt concentrations to the germination media. The experiment included three varieties of finger millet, viz GPU 28, GPU 45 and GPU 67, and five concentrations of salt, viz 0.2 %, 0.4 %, 0.6 %, 0.8 % and 1.0 % NaCl, besides one absolute control, viz 0 % salt. The experiment was replicated twice. Twenty-five seeds were kept in a petri dish containing filter paper dipped in the listed salt concentrations. Germination was observed from the next day onwards up to 16 days after sowing. On the 16th day, shoot length and root length were measured and shoot and root dry weights were taken. The germination percentage, germination rate, seedling length vigour index and seedling weight vigour index were calculated and recorded. Data were statistically analysed according to the two-factor complete block design. Significant differences for salt concentration and for variety were observed for all characters recorded. Among the three varieties, GPU 67 showed the best salt tolerance, while GPU 28 was most sensitive. In general, the germination percentage and other characters were not affected by up to 0.4 % NaCl in the germination media regardless of variety, thus indicating that this much salt in the media is the threshold for finger millet.

Xue et al. (2004) reported enhanced salt tolerance of finger millet (*Eleuesine coracana*) by over-expression of a vacuolar Na^{+}/H^{+} antiporter resulting in improved yields in saline soils. One of the possible mechanisms by which plants could survive salt stress is by keeping sodium ions away from the cytosol. The accumulation of sodium ions inside the vacuoles provides a twofold advantage: (1) by reducing toxic levels of sodium in the cytosol; and (2) by increasing vacuolar osmotic potential. Transgenic finger millet plants (var. GPU 28) were developed for salt tolerance by over-expression of a vacuolar Na^{+}/H^{+} antiporter gene from *Pennisetum glaucum* (Pg). The PgNHX1 gene was introduced into a pCAMBIA1301 vector with hygromycin as the selection marker and GUS as the reporter gene, mobilized into *Agrobacterium tumefaciens* EHA 105, and used for finger millet transformation. The putative transgenic plants that were generated were confirmed for TDNA integration and transgene expression at the transcription level. Further salt tolerance evaluation of putative transgenic plants was tested in the presence of 250 mM NaCl (including controls). Physiological traits, viz leaf senescence, bioassay, cell membrane stability (CMS), ion content and osmotic adjustment (OA), using both a solution culture system (hydroponics) and a soil system developed at the Central Soil Salinity Research Institute (CSSRI), were monitored. Over-expression of PgNHX1 in finger millet resulted in enhanced salt tolerance as evidenced by higher CMS, lower senescence and lower photosynthetic rate. More detailed analysis revealed less visual leaf injury, higher K^{+}/Na^{+} ratio, higher CMS and improved OA in NHX1 transgenics, as compared to wild types subjected to the same level of salt stress. When performance was evaluated in soil systems, based on yield parameters, transgenics showed a 47 % reduction in yield while the wild type plants exhibited extensive chlorosis and failed to survive. These results demonstrate the potential value of these transgenic plants for agricultural use under salinity stress in the field.

9.4.4 Wheat

Saroha et al. (2015) studied the effect of salinity at four salinity levels (6, 9, 12 and 15 EC) on photosynthetic characteristics of six wheat (*Triticum aestivum* L.) varieties, viz KH-65, KRL-99, PBW-343, PBW-621, HD-2851 and HD-2009, differing in salt tolerance. The cellular level of sodium (Na) and potassium (K) was measured in all plant parts. Leaf temperature, air temperature, photosynthetic rate (Pn), internal

CO_2 concentration (Ci), stomatal resistance (Rs) and stomatal conductance (gs) were also recorded in the flag leaf at 45 DAS in controls as well as salinized plants. Internal CO_2 concentration was found to be negatively correlated with stomatal resistance. Correlation was stronger with increased salinity level, which shows that increased Rs during salt stress limits Ci, which in turn results in reduced Pn. Total biomass and seed yield were significantly compromised by the saline environment. Overall, variety KRL-99 performed better with respect to all the photosynthetic parameters and depicted a better Na/K balance. All photosynthetic functions were significantly correlated with total biomass and seed yield. Thus, it may be concluded that altered photosynthetic response to salinity stress affects the total biomass production in wheat.

9.4.5 Maize

Nandhini and Sundaram (2015) report here on the effect of lipo-chitooligosaccharide on salt tolerance and phytotoxicity to roots and shoots of maize. Research on the use of signalling molecules to enhance crop performance is still incipient in India. LCO (lipo-chitooligosaccharide) is a unique signalling molecule, which, when present at the time of planting, enhances the plant's nutritional capabilities and drives the natural growth processes such as root and shoot development, immediately and irrespective of genotype, soil and other environmental conditions. Germination and early seedling growth are phases during which maize is particularly sensitive to salt stress, which can thus effectively hinder subsequent growth and productivity of this crop. The effect of priming with LCO signalling compound on growth and establishment of maize seedlings grown under induced salinity levels (0, 2, 4, 6, 8 or 10 ds m^{-1}) was assessed, and the analysis of variance showed that salinity levels and LCO priming affected seed establishment parameters of maize significantly; phytotoxicity of salt to unprimed maize seedling was increased while salt tolerance decreased with increasing levels of salinity.

9.4.6 Rice

Somasundaram et al. (2015) report mitigation of salinity stress on rice seed germination and establishment by seed priming. Priming of seeds with salts and growth regulators can lessen the negative effect of salinity on plant growth and yield. Seeds of elite rice variety Anna 4 were primed with ascorbic acid, salicylic acid, α-tocopherol, GA_3, KCl, PEG, NaCl and KNO_3 for 12 h and shade dried to bring moisture back to the initial level. Treated seeds were evaluated for various physiological quality parameters, viz germination, seedling length, vigour index and days to initial 50 % flowering and complete emergence under salinity levels of 0.25 %, 0.50 %, 0.75 % and 1 % NaCl, along with untreated controls. A significant reduction in all seed quality parameters was observed with increased salinity level. All treatments turned out to be effective in enhancing seed quality, when compared with control. However, seeds primed with GA_3 at 50 ppm for 12 h performed well under all salinity levels and it promoted 100 % germination up to 0.5 % NaCl. In addition, KNO_3, ascorbic acid and α-tocopherol were found to be effective in improving seed germination and vigour under salinity conditions.

Pharate et al. (2015) made an assessment of salt tolerance at seed germination and seedling development in rice genotypes. Twelve genotypes of rice were used for experimentation, which typically show differential responses to salt stress. The test material was genotyped with 12 stress-specific SSR primers. These included two categories of primers specific to chromosomal regions associated with seed germination and seedling development. The marker data obtained allowed grouping of the 12 genotypes into two broad categories. Seeds of different lines of rice were placed in petri plates lined with germination paper. Seeds were placed on germination paper, with different concentrations of NaCl (100–250 mM) for seed germination and for seedling development (125–500 mM) being used for experimentation. About 10-day-old seedlings were used for recording the data. At the higher concentrations of NaCl used, the plumule remained stunted and root elongation was also very poor. Hardly any shoot elongation was recorded at 150 mM NaCl. Out of the 12 genotypes tested, only genotype 4 showed some extent of plumule elongation. Root elongation was also inhibited at concentrations of 200 mM and higher. Based on the seed germination data at 100 mM, test genotypes were grouped into two categories. Seedling development was also affected by salt concentrations. Based on seedling growth parameters, at concentrations above 250 mM NaCl, genotypes can be grouped into two classes, depending upon the degree of yellowing of leaves as well as elongation of shoots. Molecular data correlated with phenotypic data. Hence this information may prove helpful in assisting selection of rice genotypes to breed for salinity tolerance.

Tomar et al. (2015) studied alleviation of salt sensitivity in rice (*Oryza sativa* L.) by foliar spray of homobrassinolide. It is believed that the repressive effect of salinity on seed germination and plant growth could relate to a decline in endogenous levels of plant growth regulators. Salinity was imposed on rice-seeded pots by addition of sufficient quantities of the major salts NaCl, $CaCl_2$ and $MgSO_4$ (in the ratio of 7:2:1 Na^+, Ca^{+2} and Mg^{+2} on a meq basis) to get the desired salinity level of 1.5 dS m^{-1} (1:5 dilution method). Foliar application of HBR solution at 1 mg l^{-1} was made at the early vegetative growth stage (20 days after sowing). HBR foliar application at 1 mg l^{-1} improved rice performance and increased seed yield under salinity stress by mitigating effects of salinity stress by as much as over 50 %. The foliar spray of HBR overrode the inhibitory effects of salinity on plant growth.

9.4.7 Jute

Naik et al. (2015) studied in vitro screening of nine *C. capsularis* jute varieties (JRC-698, JRC-321, JRC-517, JRC-7447, JBC-5, JRC-212, JRC-80, JRC-532 and UPC-94) and nine *C. olitorius* varieties (JRO-632, JRO-204, S-19, JRO-524, IRA, JRO-8432, JRO-128JRO-2407 and CP-58) for salt tolerance at sodium chloride (NaCl) concentrations of 0, 100, 160, 240 and 300 mM on seed germination and seedling growth. Increasing NaCl concentration caused reduced seed germination and seedling growth in both species. Higher germination percentage, root and shoot length, fresh and dry weight were observed for the *C. capsularis* jute varieties JRC-698 and JRC-517 and for the *C. olitorius* jute varieties JRO-632 and JRO-128 under 160 mM NaCl, as compared to 240 and 300 mM. A general decline in all growth parameters was observed at 240 mM as compared to other salt concentrations. Seed germination was completely inhibited at 300 mM NaCl. However, none of the interactions was non-significant. *C. capsularis* variety UPC-94 and *C. olitorius* variety JRO-204 were found to be most susceptible to salinity stress. *C. capsularis* varieties resulted in better germination and seedling growth in comparison to *C. olitorius* varieties.

9.4.8 Beet

Rameshchand et al. (2015) studied the effect of salinity (2, 4, 6, 8, 10 EC) on two different varieties of fodder beet, Calixta and Kuber, with special reference to the antioxidant system up to the ninth leaf stage (BBCH 9). At sampling of BBCH 9, maximum increases in catalase (CAT), peroxidase (POD) and SOD activities were recorded at 6 EC in var. Calixta, whereas var. Kuber showed maximum increases in CAT, POD and SOD at 4 EC. The highest total antioxidant activity was also shown at 6 EC for var. Calixta and at 4 EC for var. Kuber. Salinity concentrations of 6 and 4 EC caused increased chlorophyll content and photosynthetic activity in Calixta and Kuber, respectively. Maximum lipid peroxidation in both varieties was observed at 10 EC. In both varieties, Ca^{++}, Mg^{++}, Na^+ and K^+ content reached their highest levels at 10 EC. It can be inferred that most of the antioxidant enzymes and antioxidant systems were enhanced under irrigation with 6 EC and 4 EC; thus, both fodder beet varieties can be grown effectively in moderately saline soils, but variety Calixta seems to be more salinity tolerant than variety Kuber.

9.4.9 Fruit Crops

Bhoyar and Kumar (2015) discussed modern techniques for salt stress management in fruit. Salinity refers to the saltiness or dissolved salt content (such as sodium chloride, magnesium and calcium sulfates and bicarbonates) of a body of water or in the soil solution. Salt stress reflects osmotic forces exerted on plants when they are growing in a salt marsh or under other saline conditions. Salinity remains one of man's oldest environments and horticultural constraints on productivity, challenging scientists. Salinity is caused by various factors such as mineral weathering, use of saline irrigation water, poor rainfall and high evaporation rates.

9.5 Temperature

Like no other abiotic stress, high-temperature stress has become commonplace in the scientific literature in the context of global climate change, which by definition is the disruption of global climatic patterns due to the greenhouse effect we experience as a worsening condition on account of the increment in—mainly—atmospheric CO_2 presence. No longer are we worried about the original cause of it. It is an undeniable reality, and whether it is of anthropogenic origin or not, the fact remains that current rates of fossil fuel use to meet our ever-increasing demand for energy are sending just too much CO_2 into the atmosphere, thus playing a pivotal role in the global environmental temperature rising, something for which fast melting of the ice at the North and South Poles stands as irrefutable evidence.

The transitory experience of supraoptimal environmental temperature is known to every living organism. Supraoptimal temperature experiences lasting long enough constitute a heat stress (HS), as opposed to heat shock—a rather short-term exposure to high temperature—and normally elicit a defined pattern of response in biological systems. Plants and animals alike are capable of what is known as the heat shock response (HSR), which consists of a distinct series of molecular events that trigger the onset of transcription, leading to the biosynthesis of molecular chaperones, a family of low molecular weight proteins whose function is to ensure molecular stability of the nuclear envelope and contents, since HS can break cell homeostasis and cause metabolic dysfunction, growth retardation and even death (Kotak et al. 2007).

Plants experience high temperatures in many different ways, and adaptation or acclimation to high temperature occurs over different levels of plant organization. The adverse effect of temperature stress is experienced by the plant during both the vegetative and reproductive stages of development. The intensity of damage depending on two aspects: (1) the rate of temperature increase, which will determine the capacity of the cell to change the insaturation:saturation ratio in fatty acids in membrane phospholipids rapidly enough to maintain membrane fluidity within functional limits; and (2) the duration of high temperature experienced by the tissue. Once the tissue adaptive response reaches maximum expression, continued HS will go on to cause metabolic disruption.

In plants, HSR can protect tissues provided the high temperature is transient and does not exceed a threshold, which may vary greatly among species. In general, HSR is expressed constitutively in organisms that have evolved in tropical or semitropical climates, having integrated to a greater extent than those species that are typical of more temperate or cool thermal stimuli in their metabolic make-up. Temperate climate species, on the other hand, have been shown to express HSR after an induction process has occurred, and the extent of the response goes a much shorter distance in protecting the tissues for very long. Once the capacity for thermal acclimation is surpassed, plant tissues will suffer both physiological and biochemical damage by prolonged exposure to supraoptimal thermal regimes (Rivero et al. 2001).

9.5.1 Wheat

Satbhai et al. (2015) reported on the protective role of osmolytes and antioxidants during high-temperature stress in wheat. Heat stress is a major threat for wheat production in our changing environmental scenario and its negative effect is already apparent. Plants react by triggering an adaptive response to increase thermo-stability by accumulation of osmolytes and antioxidants. A gradual increase of heat stress from control (25 °C) to 30 °C 1 h, 35 °C 1 h, 40 °C 2 h and 46 °C 3 h was imposed on ten cultivars to investigate the wheat response in terms of its effect on RWC, chlorophyll content, osmolyte (GB and soluble sugar) accumulation, lipid peroxidation and activity of the antioxidative enzymes SOD, CAT, GPX, APX and GR. Continuous increases in glycine betaine, total sugars and NRS were registered during heat stress in thermo-tolerant wheat cultivars, whereas the levels decreased at 40 °C 2 h and 46 °C 3 h stress treatment in thermo-sensitive wheat genotypes. RWC, chlorophyll 'b' content and MSI significantly declined with a steady increase in malondialdehyde content from 30 °C 1 h to 46 °C 3 h heat stress. Heat stress induced the activity of SOD, GPX, CAT, APX and GR from 30 °C 1 h to 40 °C 2 h. However, at 46 °C 3 h the activities of all five enzymes declined in all wheat cultivars. Thermo-tolerant wheat cultivars recorded significantly higher activities of all five antioxidative enzymes. Heat stress-mediated increases in GB and NRS and higher activity of SOD, CAT, GPX, APX and GR enzymes was recorded in NIAW-34 and AKAW-4627 wheat cultivars. GB accumulation declined at 40 °C 2 h and 46 °C 3 h in all thermo-susceptible wheat and to some extent in thermo-tolerant cultivars with a steady increase of oxidative stress markers (MDA content). Heat inactivation of all five ROS-scavenging enzymes was observed at 46 °C 3 h stress in all wheat cultivars except NIAW-34, AKAW-4627 and NIAW-917, in which increments in osmolyte levels probably mitigated the adverse effect of heat.

Dhyani et al. (2015) undertook correlation studies to identify traits suitable as selection criteria for wheat genotypes under heat stress during post-anthesis due to late sowing. Heat is one of the major constraints on productivity of wheat in India as well as in other countries. Stomatal conductance, canopy temperature depression (CTD) and photosynthetic rates have been found to be highly correlated with grain yield apart from biological yield. They wanted to test whether, besides these traits, there may be any other selection criteria for mass scale selection of wheat genotypes for heat stress. A field experiment was thus conducted in a silty clay loam soil, with organic carbon content of 0.78 at the Norman E. Borlaug Crop Research Centre of Pantnagar University. Forty-four wheat genotypes were collected from various sources (mainly from the Uttarakhand state) and were sown in a completely randomized design with two replications on 2 January 2013. The ratio of variable to maximum florescence (Fv/Fm) and the SPAD value, morphological parameters (plant height, tiller number, leaf area index) and various biochemical parameters (proline content, H_2O_2 content, malondialdehyde content) were tested in a late planting. Correlation studies revealed that morphological and biochemical parameters were less correlated with yield than physiological parameters. The Fv/Fm ratio and SPAD value at post-anthesis were highly correlated with yield. Thus these parameters at post-anthesis can be used as characters to screen wheat germplasm for heat resistance.

9.5.2 Chickpea

Kuldeep et al. (2015) undertook identification and evaluation of chickpea germplasm for high-temperature tolerance. Gram or chickpea (*Cicer arietinum* L.), a member of the Fabaceae, is an ancient self-pollinated leguminous diploid annual ($2n = 2\times = 16$), with a genome size of approximately 750 Mbp. Chickpea is not only an important source of protein in the human diet but also plays a significant role in maintaining soil fertility through biological nitrogen fixation. India is the largest producer, accounting for 67 % of world production. The experimental material consisted of 100 genotypes (received from ICRISAT) grown in a randomized complete block design in the All India Co-ordinated Research Project on Chickpea at the Seed Breeding

Farm, Department of Plant Breeding and Genetics, JNKVV, Jabalpur, during *rabi* 2012–2013. Chickpea productivity is constrained by several abiotic stresses, and temperature is one of the most important determinants of crop growth over a range of environments and may limit chickpea yield. In the present investigation, chickpea was grown under a late sowing condition in which plant emergence occurs rapidly. A variety should be developed that is suitable for sowing in late January. Looking to overall performance on the basis of earliness in phenological traits and yield-contributing traits (number of effective pods per plant, seeds per pod, 100-seed weight, harvest index and seed yield per plant), genotypes ICCV 93014, ICCV 96317, ICCV 01302 and ICCV 063301 were found to be most promising for the late sowing condition. Selection of the genotypes based on different experimentation under late sowing conditions should be considered in further breeding programmes.

Singh et al. (2015) studied the effect of temperature stress on physiological and productivity processes of chickpea genotypes. A field experiment was therefore conducted to determine the effect of rising temperature on the crop growth rate and membrane injury index of 20 chickpea genotypes and hence to screen tolerant genotypes grown under different planting dates. The results indicated that Pusa-1103 from the north, KWR-108 from the east, RSG-963 from the west and BDG-72 from the south zones showed higher crop growth rates (CGR) and experienced less membrane injury index (MII) than other genotypes. Variations in MII were noticed among different planting dates, i.e. 61.8 %, 64.6 % and 69.4 % for plantings I, II and III, respectively. This clearly indicated that there exists a negative relationship between these two important parameters. It is therefore emphasized that the productivity of chickpea could be enhanced by selecting genotypes from different zones on the basis of MII, as this is a very simple parameter and less expensive, and a large number of breeding populations could be screened for stress tolerance in a short time.

9.5.3 Groundnut

Vaidya et al. (2015) investigated the high-temperature impact on biomass production, partitioning and yield of groundnut (*Arachis hypogaea* L.) genotypes. Groundnut is an important oil seed cash crop of the semi-arid tropics, where it is exposed to moisture stress and high temperatures. To study the impact of high temperature on the performance of different groundnut genotypes, a field trial was conducted with six groundnut genotypes (JL-24, ICGV91114, Narayani, Abhaya, Dharani and Greeshma). The crop was sown during summer in order to expose it to high temperatures of over 40 °C at the flowering and pegging stages and for comparison with winter sowing. The total biomass of the groundnut genotypes improved at high temperatures, except JL-24 and Narayani, which recorded 11 % and 36 % reductions, respectively. Among the six genotypes, Narayani recorded the highest total vegetative as well as reproductive biomass values during winter, whereas Abhaya did so during summer. However, it was interesting to observe that high temperature improved the vegetative biomass and reduced the reproductive biomass across genotypes. The impacts of high temperature on pod number, pod weight, seed number, seed yield and test weight of the selected genotypes were of different magnitudes. With high temperature, genotype Greeshma recorded a higher number of pods and the lowest reductions in pod weight (7.17 %) and seed number (17.60 %). Under both temperature regimes the per se seed yields of Dharani (29.27 and 18.6 g pl^{-1}) and Abhaya (26.57 and 18.5 g pl^{-1}) were highest, whereas ICGV 91114 (14.6 and 11.2 g pl^{-1}) and Greeshma (19.75 and 16.1 g pl^{-1}) showed lower per se seed yield values as well as lesser reductions under heat stress. Though there were reductions in seed yield of Narayani and Dharani due to high temperature, they maintained their 100-seed weights and Dharani even maintained the highest shelling percentage and HI under both thermal conditions. High temperatures elicited a different pattern of response in the selected groundnut genotypes, and by quantifying these responses, the suitability of genotype for specific environment may be identified.

9.5.4 Fruit Crops

Patil et al. (2015) discussed management of temperature extremities in fruit crops. The decade has witnessed the development of Indian horticulture, which has resulted in appreciable growth and laudable achievements, owing to technological advancement and the remarkable share of the world horticulture scenario attained. Present challenges, such as global climate change, water and soil pollution, water growing scarcity, loss of arable land to urbanization, weed infestation, and attacks of different pests and diseases, are current important issues to be tackled. Serious growing interest among scientists on abiotic and biotic stress research is becoming widespread. All crops grown under natural environments are subject to one form of stress or another. Horticultural crops are omnipresent and require specialized care for high production and quality. In order to sustain our horticultural production in the conditions prevailing due to present-day climate challenges, adequate management is required.

Temperature is one of the most important forms of abiotic stress, with direct or indirect influences on other abiotic stresses like salinity and drought and, of course, on biotic factors, viz insects and pathogens. Because of this behaviour of temperature, the productivity of fruits in Maharashtra has decreased considerably, from 6.7 t ha^{-1} in 2009–2010 to 6.2 t ha^{-1} in 2010–2011. The extremities in temperature can be controlled or managed by adopting various physical (mulching, wind breaks, shelter belts) and chemical (antitranspirants, growth regulators, silicon) methods. The use of drought- and salinity-tolerant rootstocks, as well as varieties in different fruit crops, should be adopted by farmers. The ICAR institutes (NIASM and CAZRI), as well as SAU's, are working on abiotic stress management in fruit crops and have recommended suitable rootstocks and varieties for the dry areas. By using these recommendations, farmers can improve the production and quality of fruits.

9.6 Heavy Metal Toxicity

Sharma and Datta (2015) present a review of recent research on heavy metal pollution. Metals like Cd, Hg, Pb and Ni are significant ecological pollutants, as they have a relatively high density and are toxic or poisonous even at very low concentrations. Their existence in the soil and water can be tracked down to the operation of the many enterprises in the metal, mining, pesticide and fertilizer industries, all of which cause pollution and threaten the most diverse ecosystems. It is well known that excess presence of heavy metal causes severe environmental damage and that the methods to remove such pollutants from the environment are hardly economically viable. Food and fodder plants growing in such contaminated soils can play havoc with human and animal health. Most animal feed is from plants, but some is of animal origin. It includes hay, straw, silage, compressed and pelleted feeds, oil and mixed rations, sprouted grains and legumes. In Rajasthan, mostly crops like alfalfa, guar, bajra, jowar and oats are used as green fodder. Generally, in these plants, oxidative stress is caused by accumulation of excess metals present in the soil and water.

9.7 Multiple Stresses

Environmental factors affecting crop productivity have all been generally investigated separately, despite the obvious fact that they usually present themselves in combination in nature. Although a dry spell may well occur at non-limiting air and soil temperatures, just as a hot day may be a common occurrence in temperate regions under unlimited water availability, the most common scenarios faced by crops involve heat, dryness and—at least in arid and semi-arid regions—salinity, all intertwined together. Researchers believe that any combinations of varying degrees of water deficit, heat and salt will naturally impact normal plant performance in different ways, compared to the effects caused by each factor when acting by itself. Putting this proposition to the test is not as simple as it is to conjecture it. Although it is relatively straightforward to impose varying degrees of any independent environmental factor and even to impose a combined treatment consisting of varying degrees of each factor involved in our experimentation, it would be wrong to expect identical response mechanisms to be at work in either case, or a similar extent of any single response. Perhaps the most successful attempts, however, are being made to infer the nature of plant response to multiple stresses by using molecular approaches, as the research described below may indicate.

9.7.1 Sorghum

Varalaxmi et al. (2015) undertook expression analysis of the SnRK2 family of genes in response to multiple abiotic stresses in sorghum. According to the authors, plants respond to extracellularly perceived abiotic stresses such as water deficit, salinity and high temperature by activation of complex intracellular signalling cascades that regulate acclamatory biochemical and physiological changes. SnRK2 is a plant-specific protein kinase family involved in ABA and abiotic stress signalling. In order to investigate the functional role of the SnRK2 gene family in sorghum (SbSnRK2.1–2.10) this study aimed to detect transcriptional profiles under three abiotic stress conditions, namely water deficit, salinity and high temperature, as well as following exogenous ABA treatment at different time intervals (0, 3, 6, 24 h). In addition, analysis for the presence of *cis* elements in the promoter regions of all the genes was undertaken. All the SnRK2 family genes were differentially upregulated under water deficit stress at different time intervals. The maximum levels of expression were observed at 3 and 24 h after stress induction. However, the levels of expression of these genes were low under salinity and high-temperature stresses. All the SnRK2 family genes were induced by exposure to ABA, except SbSnRK2.9, which was weakly induced, while SbSnRK2.2 was unresponsive. In silico analysis of the upstream regions of all the genes revealed the presence of single or multiple copies of ABA responsive element (ABRE), dehydration-responsive element (DRE/CRT) and low-temperature responsive element (LTRE). The presence of these elements correlates with the expression and role of these genes under different abiotic stresses. These results suggest that the SbSnRK genes might be involved in multiple stress responses of sorghum.

Shireesha et al. (2015) report on grafting used to improve tolerance of biotic and abiotic stress in vegetable crops. Plants exposed to stress exhibit various physiological and pathological disorders leading to stunted growth and severe loss in fruit quality and yield. One way to avoid or reduce losses in production caused by adverse soil chemical and physical conditions and environmental stresses in vegetables would be to graft them onto rootstocks capable of reducing the effect of external stresses on the shoot. Grafting is nowadays regarded as a useful tool to speed up the relatively slow breeding methodology aimed at increasing environmental-stress tolerance of fruits and vegetables. Grafting is not associated with the input of agrochemicals into the crops and is therefore considered to be an environmentally friendly operation of substantial and sustainable relevance to integrated and organic crop management systems. Nowadays, grafting is used to reduce infections by soil-borne pathogens and to enhance plant tolerance of abiotic stresses.

9.8 Herbicide Resistance

Prabhu et al. (2015) used a seed bioassay to assess forage sorghum tolerance of different herbicides. Weed competition in fodder crops at an early stage is a serious problem for farmers. Many works have been carried out at the Indian Grassland Fodder Research Institute, Jhansi. Herbicides are one of the options to control weeds in forage crops. Field experiments are accurate and effective but involve investing large sums of money, require time and resources, and impose demands on space and labour. The efficacy of herbicides is influenced by weather conditions like heavy rain. Hence a seed bioassay experiment was performed after Tal et al. (2000) in petri dishes to determine the selectivity of different herbicides for fodder sorghum in vitro. The method is rapid and as accurate as field experiments. Atrazine (0.50, 0.75 and 1 kg a.i. ha^{-1}), oxadiargyl (0.07, 0.08 and 0.09 kg a.i. ha^{-1}), pendimethalin (0.5, 0.75 and 1 kg a.i. ha^{-1}) and the combination of atrazine + pendimethalin (0.50 + 0.75, 0.75 + 0.50 and 1+1 kg a.i. ha^{-1}) were tested on fodder sorghum seeds. Oxadiargyl at 0.09 kg a.i. ha^{-1} recorded the highest RRL (99.9 %), followed by atrazine at 0.75 and 1.0 kg a.i. ha^{-1} (95.6 % and 94.8 %, respectively) and oxadiargyl at 0.08 kg a.i. ha^{-1} (92 %); both these treatments were significantly superior to the rest. Better root lengths were observed in atrazine at 0.75 (2.52 cm) and 1.0 kg a.i. ha^{-1} (2.40 cm) and in oxadiargyl at 0.09 kg a.i. ha^{-1} (2.10 cm). Hence, it can be concluded that the seed bioassay technique offers a better option to test the selectivity of new herbicides in a rapid and accurate way at a minimum requirement for resources and time.

9.9 Arsenic Resistance

The emerging concern about arsenic (As) in the groundwater of the Gangetic alluvium in the state of West Bengal has been principally focused on the contamination of this toxicant metalloid in the endemic areas. The vulnerability of the agricultural system, precisely the rice–rice system, raised with arsenic-contaminated irrigation water during lean periods has been largely ignored. Rice is a potentially important route of human exposure to arsenic, especially in populations with rice-based diets. However, arsenic toxicity varies greatly with species. The initial purpose of the present study (Sinha et al. 2015) was to take an account of arsenic contamination in summer rice through arsenic-contaminated irrigation water, as well as to evaluate arsenic toxicity in summer rice with special reference to speciation in Indian grain and its implications for human health. The arsenic status of the soil and irrigation water of the experimental site was to the tune of 10.24–19.17 kg ha^{-1} (total soil arsenic), 3.75–6.5 kg ha^{-1} ($NaHCO_3$-extractable soil arsenic), 0.34 mg l^{-1} (arsenic in water discharged from shallow tube well) and 0.03 mg l^{-1} (arsenic in pond water), which were above the WHO permissible limit of 0.01 mg l^{-1} for drinking water and also above the FAO permissible limit of 0.10 mg l^{-1} for irrigation water. It appeared very clear from the study that inorganic arsenic shared the maximum arsenic load in rice straw, while in grains it was considerably low. Species recovered from rice grain and straw are principally As-III and As-V with a little share of DMA and almost non-detectable MMA and As-B. Discussion of the health risk of As in rice has largely been based on its inorganic arsenic content because these species have generally been considered to be more toxic than MMA and DMA and can be directly compared to As in drinking water, assuming equal bioavailability of inorganic As in the rice matrix and in water. The maximum dietary risk of exposure to inorganic arsenic through transplanted boro paddy in the present experiment was calculated to be almost 1706 % of PTWI (provisional tolerable weekly intake) for an adult of 60 kg bodyweight. Reductions in the total As load through organic amendments in boro rice grain and straw samples were manifested predominantly through reduced accumulations of inorganic As species (As-III and As-V), among which As-V accounted for the largest share.

Considering the enormity of human health hazards associated with contamination of the food chain with arsenic, Patra et al. (2015) conducted pot culture experiments in the green house at Bidhan Chandra Krishi Viswavidyalaya, Kalyani, to explore the possibility of its mitigation in rice grains through water and nutrient management. Water management plays an important role by altering the redox conditions of the soil in transforming different forms of inorganic As, viz arsenate (As-V) and arsenite (As-III) under oxidized and reduced conditions, respectively; application of P and Si restricts the uptake of these species of As, respectively, into the plant body. Thus the treatments consisted of four methods of water management, viz maintaining water at saturation throughout the growth periods (ST), saturation during the vegetative growth stages (SV), saturation during the reproductive growth stages (SR) and flooded during the rest of the periods, against continuous ponding of water (CP) as a control; and two levels each of P, viz 25 mg and 50 mg P_2O_5, and 0 and 20 mg silicon kg^{-1} soil. The extent of the decrease in the As content of rice grains with application of both P and Si was 11.5 %; with decreasing water availability from ponding to saturation it was about 38 %, and with combination of water management and nutrient management together it was 65 %. Application of 50 mg P_2O_5 and 20 mg Si kg^{-1} soil and maintaining the soil under saturation during reproductive stages was found to be the best option to mitigate As contamination of rice grains.

9.10 Abiotic Stress Management

Sivaji et al. (2015) undertook abiotic stress management by rhizosphere bacteria. It is known that PGPR colonize the rhizosphere of many plant species, which are thus benefited by increased nutrient availability and protection against diseases caused by plant pathogenic fungi, bacteria, viruses and nematodes. Some PGPR also cause physical or chemical changes related to plant defence mechanisms, a process referred to as 'induced systemic resistance' (ISR). ISR elicited by PGPR has suppressed plant diseases caused by a range of pathogens in both the greenhouse and in the field. However, some reports on PGPR claim they act as elicitors of tolerance of abiotic stresses such as drought, salt and nutrient deficiency or excess. Early studies reported that inoculation with the PGPR *P. Aenibacillus polymyxa* increased drought tolerance of *Arabidopsis thaliana*. Drought stress affects plant hormone balance by inducing increased ABA content in the leaves, which, coupled to reduced endogenous cytokinin levels, leads to stomata closure. Cytokinin produced by *P. polymyxa* affects ABA signalling of plants or rhizobia-elicited nodulation. Ethylene content in tomato seedlings exposed to high salt was reduced by application of *Achromobacter piechaudii*, suggesting that bacterial ACC deaminase was functional. *A. piechaudii*, which produces ACC, increased the growth of tomato seedlings by as much as 66 % in the presence of high salt content. Plant growth promotion by some PGPR is associated with the solubilization and increased uptake of phosphate. PGPR have also been reported to affect nitrate uptake by plants. In addition to eliciting increases in general plant growth, some PGPR promote root development and alter root architecture by the production of phytohormones such as IAA, thereby leading to increased root surface area and an increased number of root tips. Such stimulation of roots can aid plant defence against pathogens and can also relate to IST. PGPR-elicited IST can aid the growth of crops in environmentally unfavourable conditions. Research on the mechanisms by which PGPR elicit tolerance of specific stress factors should improve the use of IST in agriculture by enabling the optimization of microbial mixtures for the production of specific bacterial determinants (e.g. cytokinin, antioxidants, ACC deaminase, VOCs and IAA).

Abiotic stresses are a serious problem for crop or fruit production under dry land conditions in arid and semi-arid regions of the world; these abiotic stresses include high temperature, low temperature, water deficit, salinity, sodicity, alkalinity, acidity, ion toxicity and deficiency. Abiotic stresses arise from extreme climatic phenomena such as drought, flood and cold; from soil toxicity of such elements as Na, Al and Fe; and from soil deficiency of elements like P and Zn. Climate change is likely to increase the pressure on Indian agriculture, in addition to our current problems of decreasing land resources, water, labour and other resources; and globalization (Godawari et al. 2015).

The main objective of dry land horticulture is to minimize the risk of rain-fed farming. Development of orchards in such areas will provide some security for food supply and protect farmers against total or partial crop failure. Once established, orchards will yield higher production and better returns per unit area. Besides being highly nutritive, dry land fruit crops help in checking soil erosion and maintaining ecological balance. Some important considerations to be made while assessing the potential of a certain fruit crop species intended for introduction into any dry zones include the following: (1) it must be able to grow on a wide variety of soils, ranging from gravelly to deep alluvial; (2) it should be drought resistant and tolerant of temperatures up to 47 °C; (3) it should be a species that can use a lot of light and grow best in the open; (4) it should be deep rooted and wind firm; and (5) it should possess good tolerance of slightly alkaline or saline soils.

Pawar et al. (2015) hold the view that the key to success in fruit production lies in the correct choice of the fruit species to be cultivated. The characteristic features of some of the fruit crops suited for dry land conditions are those of ber (drought hardy and withstands salinity well), pomegranate (can grow on shallow soils and tolerates salinity), custard apple (tolerates drought and salinity), guava (does well in a great variety of soils and climates), anola (great tolerance of salinity, alkalinity and sodicity), acid lime (relatively drought tolerant and can grow on shallow soils), bael (drought and salinity tolerant), wood apple (great tolerance of drought and salinity), *jamun* (withstands drought), karonda (drought tolerant), date palm (tolerant of drought and salinity), fig (tolerant of drought and salinity), tamarind (can thrive well in dry climate, poor shallow soils), etc. Due to its geographic diversity, India is rich in agroclimatic conditions. This is the reason why it is a centre of diversity of many fruit species. The wealth in germplasm for these species ensures the presence of ample genetic variability to search for desirable genes, yet we have not utilized this diversity for full exploitation of its potential benefit in breeding suitable varieties.

Keeping in view the scope of broccoli in climate-smart diversification, earliness and multiple harvest, studies were conducted on varietal evaluation to find out the best types and varieties most suitable for the Ladakh region (Kanwar et al. 2015). Performance of seven broccoli genotypes, including Green Head as a check, was evaluated in a randomized block design with three replications. All genotypes of broccoli exhibited significant differences for all quantitative characters under study, except for plant height, leaf length and leaf width. The highest primary head weight, yield per plant and yield per hectare were produced by hybrid Green Pia, followed by Fiesta. Palam Vachitra exhibited a purple head colour, while Palam Kanchan produced yellow heads and the rest produced green heads. The hybrids Green Pia and Fiesta, with green-coloured uniform heads, may be recommended for cultivation in the cold deserts of Ladakh.

9.11 Conclusions

Sixty-five works in the general field of abiotic stress were presented during the 2nd International Conference on Bio-resource and Stress Management held in Hyderabad, India, early in 2015. The spectrum of interests within this most important field of research in plant biology included, on this occasion, water deficit, high temperature and salinity as the major current concerns, though concerns about heavy metal toxicity, arsenic toxicity, herbicide toxicity and the occurrence of multiple stresses were also presented. The plant systems investigated include foremost—naturally—those grains and legumes that constitute the basic diet around the world: wheat, maize, sorghum, oats, chickpea, groundnut and soybean; a number of fruit species, fibres, industrial seeds and stalks also received attention. These studies cover a wide scope of disciplinary perspectives, from agronomic to physiological and biochemical, right down to the molecular level. Each one, equipped with its particular set of experimental tools, has successfully approached its object of study to provide us with fresh information that will prove useful in at least two ways. For one thing, those whose interest lies in designing their own research will find in this chapter inspiring ideas on questions of the utmost urgency. On the other hand, those who are interested in the general area of stress in plant systems from an ecological point of view—as it may concern global climate change in the particular ways in which it threatens ecosystem stability, for example, will find the material most valuable to build on current working models explaining interactions between living organisms and a rapidly changing physical environment, and the prospects of evolution of the ecosystems thereupon.

Thirty years after the initial, almost overwhelming euphoria caused by the advent of modern molecular technologies to study and manipulate plant gene expression, their widespread use has been duly tamed by the complexity of the plant stress response phenomenon. Having assumed its possibilities with more realism, we are ready to make the most of them towards advancement of our understanding of how plants work and just what efforts on our part can help crop plants cope better with adverse environments and actually nudge crop performance, in such situations, to the benefit of mankind. An absolute requirement for that to happen is the adoption of a truly interdisciplinary approach to research.

References

Asif S, Ahmad MD (2002) Using state of the art RS for monitoring waterlogging and salinity. Pakistan. Development of a research programme in irrigation and drainage. International Programme for Technology and Research in Irrigation and Drainage. Development of Natural Resources Management and Environment Department. FAO. Programme formulation report 9, April

Bhoyar MG, Kumar K (2015) Modern techniques for salt stress management in fruit crops. In: Compendium of abstracts of the 2nd international conference on Bio-resource and Stress Management, ANGRAU & PJTSAU, Hyderabad, India, 7–10 January 2015

Channappagoudar BB, Sahoo TR (2015) Physiological basis of drought tolerance in maize inbreds and hybrids in a field experiment during post rainy season. In: Compendium of abstracts of the 2nd international conference on Bio-resource and Stress Management, ANGRAU & PJTSAU, Hyderabad, India, 7–10 January 2015

Damdar RR, Bhale VM, Deshmukh KM et al (2015) Effect of irrigation and nitrogen levels on yield and water use efficiency of summer sesame. In: Compendium of abstracts of the 2nd international conference on Bio-resource and Stress Management, ANGRAU & PJTSAU, Hyderabad, India, 7–10 January 2015

Deva S, Pandey P (2015a) Plant drought stress: effects and management. In: Compendium of abstracts of the 2nd international conference on Bio-resource and Stress Management, ANGRAU & PJTSAU, Hyderabad, India, 7–10 January 2015

Deva S, Pandey P (2015b) Remote sensing of biotic and abiotic stress for irrigation management of cotton. In: Compendium of abstracts of the 2nd international conference on Bio-resource and Stress Management, ANGRAU & PJTSAU, Hyderabad, India, 7–10 January 2015

Dhutmal RR, More AW, Mehetre SP et al (2015) Variability parameters and correlation studies in drought tolerent post rainy sorghum (*Sorghum Bicolor* L.) genotypes. In: Compendium of abstracts of the 2nd international conference on Bio-resource and Stress Management, ANGRAU & PJTSAU, Hyderabad, India, 7–10 January 2015

Dhyani VC, Chaturvedi S, Punetha H et al (2015) Correlation studies to identify suitable traits as selection criteria for wheat genotypes under terminal heat stress. In: Compendium of abstracts of the 2nd international conference on Bio-resource and Stress Management, ANGRAU & PJTSAU, Hyderabad, India, 7–10 January 2015

Dobson A, Lodge D, Alder J et al (2006) Habitat loss, trophic collapse, and the decline of ecosystem services. Ecology 87:1915–1927

Dutta P, Bandopadhyay P, Bera AK (2015) Water stress tolerance of mungbean (*Vigna radiata* L. Wilczek): physiological and biochemical changes in leaves during sdeedling development. In: Compendium of abstracts of the 2nd international conference on Bio-resource and Stress Management, ANGRAU & PJTSAU, Hyderabad, India, 7–10 January 2015

Gadakh SS, Patel DU, Singh C et al (2015) A rapid screening technique for salt tolerance in sugarcane (*Saccharum* spp. Complex). In: Compendium of abstracts of the 2nd international conference on Bio-resource and Stress Management, ANGRAU & PJTSAU, Hyderabad, India, 7–10 January 2015

Geetha A, Sivasankar A, Saidaiah P et al (2015) Effect of moisture stress on key physiological parameters in sunflower genotypes. In: Compendium of abstracts of the 2nd international conference on Bio-resource and Stress Management, ANGRAU & PJTSAU, Hyderabad, India, 7–10 January 2015

Ghorade RB, Kalpande VV, Band P et al (2015) Potential donors for grain yield and the drought tolerance parameters in post rainy sorghum. In: Compendium of abstracts of the 2nd international conference on Bio-resource and Stress Management, ANGRAU & PJTSAU, Hyderabad, India, 7–10 January 2015

Godawari S, Supriya P, Aglawe B et al (2015) Abiotic stress management potential in dryland horticulture. In: Compendium of abstracts of the 2nd international conference on Bio-resource and Stress Management, ANGRAU & PJTSAU, Hyderabad, India, 7–10 January 2015

Goud IS, Rajanna B, Ajithkumar K et al (2015) The role of root and its related traits in utilization of available moisture under water deficit in safflower (*Carthamus tinctorius* L.). In: Compendium of abstracts of the 2nd international conference on Bio-resource and Stress Management, ANGRAU & PJTSAU, Hyderabad, India, 7–10 January 2015

Kalpande VV, Ghorade RB, Band P et al (2015) Relationship among grain yield and some of the index parameters associated with drought tolerance in *rabi* sorghum. In: Compendium of abstracts of the 2nd international conference on Bio-resource and Stress Management, ANGRAU & PJTSAU, Hyderabad, India, 7–10 January 2015

Kanwar MS, Akbar PI, Kunjang Lamo (2015) Evaluating broccoli (*Brassica oleracea* L., Plenck, var. Italica) genotypes for climate smart diversification in cold desert Ladakh region. In: Compendium of abstracts of the 2nd international conference on Bio-resource and Stress Management, ANGRAU & PJTSAU, Hyderabad, India, 7–10 January 2015

Kotak S, Larkindale J, Lee U et al (2007) Complexity of the heat stress response in plants. Curr Opin Plant Biol 10:310–316

Kuldeep RK, Pandey S, Verma HP et al (2015) Identification and evaluation of chickpea germplasm for high temperature tolerance. In: Compendium of abstracts of the 2nd international conference on Bio-resource and Stress Management, ANGRAU & PJTSAU, Hyderabad, India, 7–10 January 2015

Läuchli A, Grattan SR (2007) Plant growth and development under salinity stress. In: Jenks MA, Hasegawa PM, Jain SM (eds) Advances in molecular breeding toward drought and salt tolerant crops. Springer, Dordrecht, pp 1–32

Lehner B, Döll P, Alcamo J et al (2006) Estimating the impact of global change on flood and drought risks in Europe: a continental integrated analysis. Clim Chang 75:273–299

MacDougall AS, McCann KS, Turkington KS (2013) Diversity loss with persistent human disturbance increases vulnerability to ecosystem collapse. Nature 494:86–89

Meena SL, Nagar RP, Meena SS et al (2015) Productivity of fodder as influenced by soil moisture conservation techniques and cropping systems under semi-arid environments. In: Compendium of abstracts of the 2nd international conference on Bio-resource and Stress Management, ANGRAU & PJTSAU, Hyderabad, India, 7–10 January 2015

Mohan KK, Rao VVSC, Saritha D et al (2015) Effect of moisture stress and NAA spray on growth, flower bud drop and yield in chilli (*Capsicum annuum*). In: Compendium of abstracts of the 2nd international conference on Bio-resource and Stress Management, ANGRAU & PJTSAU, Hyderabad, India, 7–10 January 2015

Mohapatra S, Mahender A, Rao GJN et al (2015) Genetic diversity and molecular characterization of rice land races (*Oryza sativa*. L) using drought stress linked SSR and STS markers. In: Compendium of abstracts of the 2nd international conference on Bio-resource and Stress Management, ANGRAU & PJTSAU, Hyderabad, India, 7–10 January 2015

Mumby P, Iglesias-Prieto R, Hooten AJ et al (2011) Revisiting climate thresholds and ecosystem collapse. doi:10.1890/11.WB.002

Nagesh M, Bhattiprolu GR, Venugopal G (2015) Field screening and biochemical basis of resistance in sorghum varieties for their suitability under deficit rainfall in Nagarjuna Sagar Project and Krishna Western Delta in Guntur District of Andhra Pradesh. In: Compendium of abstracts of the 2nd international conference on Bio-resource and Stress Management, ANGRAU & PJTSAU, Hyderabad, India, 7–10 January 2015

Naik MR, Barman D, Kumar M et al (2015) In-vitro screening of jute (*Corchorus capsularis* L and *Corchorus olitorius* L.) against salt stress. In: Compendium of abstracts of the 2nd international conference on Bio-resource and Stress Management, ANGRAU & PJTSAU, Hyderabad, India, 7–10 January 2015

Namrata, Sharma H, Bishen P (2015) Breeding approaches for drought stress resistance. In: Compendium of abstracts of the 2nd international conference on Bio-resource and Stress Management, ANGRAU & PJTSAU, Hyderabad, India, 7–10 January 2015

Nandhini DU, Sundaram ES (2015) Effect of Lipo-Chitooligosaccharide on salt tolerance and phytotoxicity on roots and shoots of maize. In: Compendium of abstracts of the 2nd international conference on Bio-resource and Stress Management, ANGRAU & PJTSAU, Hyderabad, India, 7–10 January 2015

Panigrahi TK, Das PS (2015) Genetic improvement in horticultural crops for abiotic stress. In: Compendium of abstracts of the 2nd international conference on Bio-resource and Stress Management, ANGRAU & PJTSAU, Hyderabad, India, 7–10 January 2015

Patil HE (2015) Evaluation of genotypes for drought stress in soybean (*Glycine max* (L.) Merrill) under induced water stress conditions. In: Compendium of abstracts of the 2nd international conference on Bio-resource and Stress Management, ANGRAU & PJTSAU, Hyderabad, India, 7–10 January 2015

Patil HE, Jadeja GC (2015) Assessment of genetic variation for drought response in pearl millet (*Pennisetum glaucum* (L.) R. Br.) in relation to yield parameters under stress conditions. In: Compendium of abstracts of the 2nd international conference on Bio-resource and Stress Management, ANGRAU & PJTSAU, Hyderabad, India, 7–10 January 2015

Patil RA, Bharad SG, Nagre PK (2015) Management of temperature extremities in fruit crops. In: Compendium of abstracts of the 2nd international conference on Bio-resource and Stress Management, ANGRAU & PJTSAU, Hyderabad, India, 7–10 January 2015

Patra PK, Hembram S, Bhattacharyya K et al (2015) Mitigating arsenic accumulation in rice through nutrient and water management. In: Compendium of abstracts of the 2nd international conference on Bio-resource and Stress Management, ANGRAU & PJTSAU, Hyderabad, India, 7–10 January 2015

Pawar GS, Aglawe SB, Kadam RP et al (2015) Abiotic stress management potential in dryland horticulture. In: Compendium of abstracts of the 2nd international conference on Bio-resource and Stress Management, ANGRAU & PJTSAU, Hyderabad, India, 7–10 January 2015

Pharate AT, Gahukar SJ, Jain N et al (2015) Assessment of salt tolerance at seed germination and seedling development in rice genotypes. In: Compendium of abstracts of the 2nd international conference on Bio-resource and Stress Management, ANGRAU & PJTSAU, Hyderabad, India, 7–10 January 2015

Prabhu G, Srinivasan R, Chaudhary M et al (2015) Seed bioassay to assess forage sorghum tolerance to different herbicides. In: Compendium of abstracts of the 2nd international conference on Bio-resource and Stress Management, ANGRAU & PJTSAU, Hyderabad, India, 7–10 January 2015

Purru S, Bhat KV (2015) Assembly, annotation, functional classification of transcriptomes and identification of key genes for moisture stress tolerance in *Cucumis melo*. In: Compendium of abstracts of the 2nd international conference on Bio-resource and Stress Management, ANGRAU & PJTSAU, Hyderabad, India, 7–10 January 2015

Puttawar MR, Patil AE, Gawande NG et al (2015) Screening of mustard parents and crosses for drought tolerance. In: Compendium of abstracts of the 2nd international conference on Bio-resource and Stress Management, ANGRAU & PJTSAU, Hyderabad, India, 7–10 January 2015

Rajann B, Goud S, Biradar SA et al (2015) Studies on the response of roots and other morphological characteristics in linseed (*Linum usitatissimum* L.) under water stress conditions. In: Compendium of abstracts of the 2nd international conference on Bio-resource and Stress Management, ANGRAU & PJTSAU, Hyderabad, India, 7–10 January 2015

Rameshchand AA, Hetal V, Karmakar Nilma (2015) Effect of salinity stress on two different varieties of fodder beet with special reference to antioxidant system. In: Compendium of abstracts of the 2nd international conference on Bio-resource and Stress Management, ANGRAU & PJTSAU, Hyderabad, India, 7–10 January 2015

Ranjan D, Hazarika GN (2015) Study of the adaptation mechanism of some Gladiolus cultivars under rainfed conditions. In: Compendium of abstracts of the 2nd international conference on Bio-resource and Stress Management, ANGRAU & PJTSAU, Hyderabad, India, 7–10 January 2015

Rathod TH, Tumdam JS, Shamkuwar GR (2015) Influence of morpho-physiological traits on drought tolerance of different genotypes in cotton species. In: Compendium of abstracts of the 2nd international conference on Bio-resource and Stress Management, ANGRAU & PJTSAU, Hyderabad, India, 7–10 January 2015

Rivero RM, Ruiz JM, García PC et al (2001) Resistance to cold and heat stress: accumulation of phenolic compounds in tomato and water melon plants. Plant Sci 160:315–321

Rossi G, Benedini N, Tsakiris G et al (1992) On regional drought estimation and analysis. Water Resour Manag 6:249–277

Saroha L, Sharma PC, Rana JS et al (2015) Effect of salinity on photosynthetic characteristics of six wheat (*Triticum aestivum* L.) varieties differing in salt tolerance. In: Compendium of abstracts of the 2nd international conference on Bio-resource and Stress Management, ANGRAU & PJTSAU, Hyderabad, India, 7–10 January 2015

Satbhai R, Kale A, Rajeev Naik (2015) Protective role of osmolytes and antioxidants during high temperature stress in wheat. In: Compendium of abstracts of the 2nd international conference on Bio-resource and Stress Management, ANGRAU & PJTSAU, Hyderabad, India, 7–10 January 2015

Satyavathi P, Vanaja M, Vagheera P et al (2015) Growth and yield performance of cluster bean genotypes in moisture stress environment of semi-arid region. In: Compendium of abstracts of the 2nd international conference on Bio-resource and Stress Management, ANGRAU & PJTSAU, Hyderabad, India, 7–10 January 2015

Sharma CP, Datta S (2015) Plant growth-studies on the impact of heavy metal pollution affecting fodder plants in Rajasthan. In: Compendium of abstracts of the 2nd international conference on Bio-resource and Stress Management, ANGRAU & PJTSAU, Hyderabad, India, 7–10 January 2015

Shireesha P, Reddy EEP, Madhavi K et al (2015) Grafting-A tool to improve tolerance to biotic and abiotic stress in vegetable crops. In: Compendium of abstracts of the 2nd international conference on Bio-resource and Stress Management, ANGRAU & PJTSAU, Hyderabad, India, 7–10 January 2015

Shivran AC, Meena OP (2015) Impact of soil type and saline water irrigation on soil salinity and productivity of blond psyllium (*Plantago ovata* Forsk). In: Compendium of abstracts of the 2nd international conference on Bio-resource and Stress Management, ANGRAU & PJTSAU, Hyderabad, India, 7–10 January 2015

Singh AK (2015) Evaluation of oat (Avenasativa) germplasm for traits imparting drought tolerance in intermediate zone under J&K. In: Compendium of abstracts of the 2nd international conference on Bio-resource and Stress Management, ANGRAU & PJTSAU, Hyderabad, India, 7–10 January 2015

Singh D, Singh G (2015) Access the role of plant growth-promoting Rhizobacteria and their exopolysaccharide in drought tolerance of wheat. In: Compendium of abstracts of the 2nd international conference on Bio-resource and Stress Management, ANGRAU & PJTSAU, Hyderabad, India, 7–10 January 2015

Singh HC, Arya P, Singh SN et al (2015) Physio-morphological basis of selection in maize (Zea mays l.) for drought conditions. In: Compendium of abstracts of the 2nd international conference on Bio-resource and Stress Management, ANGRAU & PJTSAU, Hyderabad, India, 7–10 January 2015

Sinha B, Bhattacharyya K, Hansda S et al (2015) Arsenic toxicity in rice with special reference to speciation in Indian grain and its implication on human health. In: Compendium of abstracts of the 2nd international conference on Bio-resource and Stress Management, ANGRAU & PJTSAU, Hyderabad, India, 7–10 January 2015

Sivaji T, Madhavi K, Vimala B (2015) Abiotic stress management by rhizosphere bacteria. In: Compendium of abstracts of the 2nd international conference on Bio-resource and Stress Management, ANGRAU & PJTSAU, Hyderabad, India, 7–10 January 2015

Sivalingam PN, Singh D, Gurjar K et al (2015) Differential expression of transcripts during low moisture stress in *Ziziphus nummularia* (Burm.f). In: Compendium of abstracts of the 2nd international conference on Bio-resource and Stress Management, ANGRAU & PJTSAU, Hyderabad, India, 7–10 January 2015

Somasundaram G, Ambika S, Bhaskaran M (2015) Alleviation of salinity stress on rice seed germination and establishment through priming. In: Compendium of abstracts of the 2nd international conference on Bio-resource and Stress Management, ANGRAU & PJTSAU, Hyderabad, India, 7–10 January 2015

Sreenivas C, Sreedevi P,Vardhan SV et al (2015) Drainage technology: A viable tool for reclamation of saline-sodic and water logged soils of Godavari Western Delta, India. In: Compendium of abstracts of the 2nd international conference on Bio-resource and Stress Management, ANGRAU & PJTSAU, Hyderabad, India, 7–10 January 2015

Sumalini K, Pradeep T, Manjulatha G (2015) Effect of moisture stress on combining ability and gene action for polygenic traits in maize. In: Compendium of abstracts of the 2nd international conference on Bio-resource and Stress Management, ANGRAU & PJTSAU, Hyderabad, India, 7–10 January 2015

Syvarna, Patil RP, Wali V et al (2015) Assessment of salt stress impact on the germination of finger millet varieties. In: Compendium of abstracts of the 2nd international conference on Bio-resource and Stress Management, ANGRAU & PJTSAU, Hyderabad, India, 7–10 January 2015

Tal A, Kotoula-Syka E, Rubin B (2000) Seed-bioassay to detect grass weeds resistant to acetyl coenzyme A carboxylase inhibiting herbicides. Crop Prot 19(7):467–472

Tomar AS, Dalal GV, Bohra B (2015) Alleviation of salt stress in rice (*Oryza sativa* L.) by foliar spray of homobrassinolide. In: Compendium of abstracts of the 2nd international conference on Bio-resource and Stress Management, ANGRAU & PJTSAU, Hyderabad, India, 7–10 January 2015

Tripathy SK, Dash S (2015) Genotyping of drought tolerant upland races of rice using SDS-PAGE of total seed storage protein. In: Compendium of abstracts of the 2nd international conference on Bio-resource and Stress Management, ANGRAU & PJTSAU, Hyderabad, India, 7–10 January 2015

Vaidya S, Vanaja M, Lakshmi NJ et al (2015) High temperature impact on biomass production, partitioning and yield of groundnut (*Arachis hypogaea* L.) genotypes. In: Compendium of abstracts of the 2nd international conference on Bio-resource and Stress Management, ANGRAU & PJTSAU, Hyderabad, India, 7–10 January 2015

Varalaxmi Y, Sushma A, Vijayalakshmi T et al (2015) Expression analysis of sorghum SnRK2 family of genes in response to multiple abiotic stresses. In: Compendium of abstracts of the 2nd international conference on Bio-resource and Stress Management, ANGRAU & PJTSAU, Hyderabad, India, 7–10 January 2015

Varthini NV, Robin S, Raveendran M et al (2015) Identification of drought tolerant genotypes in rice based on root screening. In: Compendium of abstracts of the 2nd international conference on Bio-resource and Stress Management, ANGRAU & PJTSAU, Hyderabad, India, 7–10 January 2015

Xue ZY, Zhi DY, Xue GP et al (2004) Enhanced salt tolerance of transgenic wheat (*Tritivum aestivum* L.) expressing a vacuolar $Na+/H+$ antiporter gene with improved grain yields in saline soils in the field and a reduced level of leaf $Na+$. Plant Sci 167(4):849–859

Essence of Plants or Crops for Adaptation: Learning Lessons for Sustainable Use

10

Ratikanta Maiti

Abstract

During the process of evolution, plants and crops develop capacity of adaptation to various biotic and abiotic stresses. The plants possess many adaptive features that permit plants to thrive in diverse environments. The chapter discusses concepts, mechanisms, research advances attained on morpho-anatomical, physiological and molecular mechanisms of plants and crops for adaptation to adverse abiotic stresses such as drought, salinity, high, low temperature, abiotic stress resistance and several biotic stress resistance mechanisms. It also discusses these adaptive features as well as the evolutionary trends of plants starting from a minute algae to an angiosperm. A few of the signalling pathway for drought, salinity stress, etc. are also discussed.

10.1 Background

Starting from the origin of plants from Palaeozoic era to the present era, plants undergo a series of evolutionary sequences for adaptation to the changing environment from aquatic to terrestrial habits. Each species possesses specific morph-anatomical and physiological traits for adaptation to a particular environment. We need to know the secret of plant for its adaptive capacity to adverse environments. During the process of evolution, plant species have to struggle for existence and modify characteristics necessary for adaptation to the changing environments. Bennici (2008) discussed origin and early evolution of land plants.

During this ongoing evolutionary process, the species which can adapt to the changing environment can survive while others disappear and remain as fossil representatives of earlier plant. The plants possess many adaptive features that permit plants to thrive in diverse environments during the course of evolution. Significant research advances have been attained on morph-anatomical, physiological and molecular mechanisms of plants for adaptation to adverse abiotic stresses. These are the essence of plants for adaptation to adverse environments. Before discussing these adaptive features, I want to narrate the evolutionary trends of plants starting from algae to bryophyte, pteridophyte, gymnosperm and finally to angiosperms. It may be mentioned

R.K. Maiti (✉)
Facultad de Ciencias Forestales, Universidad Autónoma de Nuevo León, Carr. No 85, km 145, NL 67700 Linares, Mexico
e-mail: ratikanta.maiti@gmail.com

R.K. Maiti et al. (eds.), *Bioresource and Stress Management*, DOI 10.1007/978-981-10-0995-2_10

here that algae living in aquatic habitat does not require any special adaptive features except the presence of photosynthetic pigments required for photosynthesis for the elaboration of food. Structurally evolution takes place in this group from unicellular to colonial, filamentous and heterotrichous forms having root like rhizoids. Algae are gametophyte (haploid, n) representing the major part of the life cycle except the phase of zygote representing short-lived sporophytes. Next advance group is bryophytes which are adapted to terrestrial habitat with good moisture condition. They are mostly thallus, unicellular, to multicellular (*Riccia*, *Marchantia*) and finally liverworts having tree like structures, all possessing rhizoidal root like structures for absorption of moisture and nutrients from the soil. The life cycle of bryophytes are mostly gametophytic, except the sporophyte, the zygote that is short-lived, but in the case of *Anthoceros*, the sporophyte is little long lived with special elaborated structure showing advanced feature. The next group pteridophyte followed by gymnosperm represents sporophyte for a long period. In these two groups of plants, spores represent gametophytic phase which is short lived. In Gymnosperm, *Gnetum* represent highly advanced plant having ovule like structure. It stands as evolutionary link between gymnosperm and angiosperm. Finally angiosperm represents highly advanced group with highly elaborated structures such as roots, leaves and stems. In this group, the leaves of some species possess dense trichomes, thick cuticle, wax and compact palisade cells for adaptation to xerichabitats. Among the groups, pteridophytes and gymnosperm possess vascular structures viz., tracheids for transport of water and nutrients from the soils to the above ground plant organs. In the case of *Gnetum*, it has vessel-like structure for the transport of water and nutrients. In angiosperm, highly elaborated xylem vessels showing different grades of evolution starting from long xylem vessels with spiral thickening, gradually to shorter vessels with scalaform to simple opposite to alternate vessels. The highly advanced angiosperms possess short truncated and broad vessels with alternate pits. Therefore, during the course of evolution, plants undergo structural changes associated with physiological traits for adaptation to the changing environments and habitat.

Wikipedia explained nicely the process of evolution of plants: the evolution of plants has originated in widely varying levels starting from bryophytes, lycopods and ferns to the complex gymnosperms and finally to angiosperm. During the course of the evolutionary story, plants appeared relatively late in earth history. The first fossils of plants are supposed to be 475 million years old. The first land plants appeared during the Ordovician era. Gradually these land, plants diversified in the Late Silurian, around 420 million years ago, and later in Devonian era. In late Devonian era seeds evolved. The flowering plants appeared in the Triassic era ($\sim$200 million years ago). The latest major group of plants evolved were the grasses, which became important in the mid-Palaeogene, from round 40 million years ago. The grasses, as well as many other groups, evolved new mechanisms of metabolism to survive in the low CO_2 and warm, dry conditions of the tropics over the last 10 million years.

10.2 Adaptation of Plants: Logics of Evolution

It is claimed that plants descended from algae as they colonized the empty landmass. Plants couldn't have evolved until there was a formation of sufficient ozone layer to block radiation. The many structures that plants require in order to overcome gravity protect plants from drying out and reproduce, thereby would have developed gradually over millions of years. Molecular and fossil evidence explains how plants evolved from simple to complex forms. Plants must have evolved simultaneously in different ways. Seed plants are alleged to have attained an advantage over other ancient plants and finally dominated the landscape. The ancient forests are the source for the different coal beds in evolutionary theory. Flowering plants were the most recent to appear since leaves were remodelled to attract pollinators and accomplish reproductive functions. Lepidodendron, gigantic tree which dominated more than two million years ago with vascular bundle, xylem could not survive for thin weak xylem vessels and remain as fossil representative of great interest for palaeobotanists. The presence of "living fossil" plants like the *Gingko biloba* tree makes it clear that evolution is a plastic theory that can accommodate both rapid change and stability for hundreds of millions of years. The creation of plants has attained many diverse forms to form the intricate design of plant systems and the symbiotic relationships that could not have evolved in a gradual process. It is agreed by evolutionists that plants evolved from algae that slowly began to colonize the land. This could not have occurred until after there was an intact ozone layer to protect the plants from some of the atmosphere's harmful UV rays. After that alien land, plants had to adapt to the existing environments of the new landscape. The plants developed several key characteristics to prevent drying out (desiccation) in the air, to absorb nutrients from the soil, to grow upright without the support of water and to reproduce on land.

With respect to the ability to survive in a given environment in evolutionary biology, adaptations are often acquired by modifying existing structures to accomplish their function. In the case of algae turning into plants, the pre-existing structures are mostly absent. The first plants to evolve were the small mosses and liverworts (bryophytes), but these evolutionary dead ends did not lead to the vascular plants that are common today. The three groups of bryophytes are found in an unexpected sequence in the fossil record, so evolutionists must accept that they evolved separately from one another. The most likely ancestor is a mobile alga known as a "chlorophyte".

The adaptations of plants to the changing environments has allegedly developed over millions of years as plants had to acquire new features to be able to survive on land. The vascular plants are supposed to have evolved separately from other algae. The ferns, club mosses and horsetails (pteridophytes) are considered as common ancestors to the vascular plants that are present in large quantities today. The vast forests of the Carboniferous are the source of many coal and oil reserves in the evolutionary model. The next step in the evolutionary story is the development of seed plants. The first to evolve were the gymnosperms, cone-bearing plants like conifers, ginkgo and cycads. These plants had adaptive traits over the pteridophytes and dominated the forests of the late Palaeozoic era. The evolution of flowers and seed protected in a fruit led to the origin of the angiosperms with the production of flowers and fruits and many adaptations have led to diverse types of flowering plants. The presence of living trees that are virtually identical to fossil species, like the ginkgo and the Wollemi pine, demonstrates the failure of evolution to make useful predictions. Kenrick and Crane (2008) consider that the origin and early evolution of land plants occurred in the mid-Palaeozoic era, between about 480 and 360 million years ago with far-reaching consequences for the evolution of terrestrial organisms and global environments. With respect to the origin and early diversification of land plants, they mention that from a simple plant body consisting of only a few cells, land plants evolved an elaborate two phase life cycle and an extraordinary array of complex organs and tissue systems. The Late Silurian and Devonian periods were critical interval during which the initial diversification of vascular plants occurred.

In the context of the literatures on the origin of plants and their evolutionary sequences different plants starting from lower plants to the angiosperms, it is concluded that different plants originate starting from early Palaeozoic to modern age and many of them disappear for their inability to adaptation to the changing adverse environments during geological era and remain as fossils. The living *Gingo biloba* represents fossil still now owing to its special adaptive capacity still not explored.

10.3 Plant Traits and Its Relation to Adaptation to Abiotic and Biotic Stress

Solbrig (1994) discussed plant traits, adaptive strategies and their role in ecosystem function. Phenotypic plasticity can be defined as the ability of one genotype to produce more than one phenotype when exposed to different environments. The phenotypic plasticity contributes to adaptation of plants to changing environments. Sultan (1995) considered phenotypic plasticity as a major mode of adaptation in plants. Studies have been undertaken on adaptive response by genotypes to physical aspects of the environment, as well as to biotic factors such as neighbour density and the presence of bacterial symbionts. Alterations of offspring traits by parental plants of *Polygonum persicaria* are considered as a cross-generational aspect of plastic response to environment. Individual plasticity and local ecotypes are also considered as alternative bases of species ecological breadth and methodological problems in distinguishing these alternatives.

10.4 Adaptation for Stress Tolerance

It is necessary to have understanding of plant structural defense, particularly within the context of plant defense syndromes, which would not only improve our understanding of plant defense theory, but enable us to predict how plant morphological responses to climate change might influence interactions at the individual.

10.5 Biotic Stress Tolerance

10.5.1 Insect Tolerance or Resistance

Density of leaf trichomes is related to biotic and abiotic stresses in plants and crops. The role of leaf trichome density as a component of resistance to herbivores has been demonstrated in six populations of *Datura stramonium* (Valverde et al. 2001). Leaf trichome density was strongly positively correlated to resistance across populations. Trichome density was related to resistance, and positive directional selection on resistance to herbivores was observed. The results revealed the adaptive role of leaf trichomes as a component of defense to herbivores and variable selection among populations.

It has been well documented by Maiti and his team that glossy sorghums with shining waxy surface and dense pointed non-glandular trichomes confer tolerance to shoot fly. An increase in trichome density increases tolerance to shoot fly. It has been further demonstrated that glossy sorghum lines contributes to tolerance to drought, salinity and even to disease showing multiple resistance. It has been observed by Maiti that trichome density on leaves contribute to sucking pest tolerance in cotton, tolerance to aphids and virus in tomato thereby demonstrating the role of trichomes in biotic and abiotic stress mentioned above. It was further observed that some morpho-anatomical traits are related to drought resistance in few crops such as small leaves with dense trichomes. Waxy leaf, thick cuticle, compact palisade cells, strong thick collenchyma in stout long petiole in cotton, sunflower and castor (not published). Tobacco plants have short and long glandular trichomes that play several roles in the defense against biotic and abiotic stresses.

10.5.2 Herbivory Tolerance

The structural traits, such as spinescence, pubescence, sclerophylly and raphides, play important roles in protecting plants from herbivore attack. In spite of many of these morphological characteristics evolved as responses to other environmental stimuli, each of these traits can provide an important defense against herbivore attack in both terrestrial and aquatic ecosystems. It is concluded that leaf-mass–area is a robust index of sclerophylly as a surrogate for more rigorous mechanical properties used in herbivory studies (Hanley et al. 2007).

10.6 Abiotic Stress Tolerance

Abiotic stress affects the crop productivity of cereal crops at various growth stages which ultimately resulted loss in the yield. Physiological process such as flowering, grain filling and maturation was highly affected by abiotic stresses.

Salinity reduces the nutrient and water uptake and reduced plant growth under salinity stress and decreased seed reserve under drought and salinity stress (Ashraf and Hafeez 2004). Salinity affects several functions of photosynthesis such as PSII efficiency and the P-N rate, thereby allowing a good discrimination between tolerant and non-tolerant cultivars (Katerji et al. 1994). Higher salinity tolerance cultivars of winter wheat could relieve senescence at the reproductive stage (Yanhai et al. 2008).

The increasing salinity of irrigation water caused significant adverse effects on yield attributes and yield of wheat. Domestic line of wheat could be grown at a salinity concentration between 4 and 8 g l^{-1} with least reduction in biological and grain yield (Norman and Schierenbeck 2004). Salt tolerant rice cultivars increase root elongation and root number functioning as osmotic adjustment in salt tolerant rice cultivars (Maiti et al. 2005, 2009a, b, c, d, e, f). The chlorophyll a, chlorophyll b, total chlorophyll and total carotenoids contents in the stressed seedlings considerably decreased under both acidic and alkaline stresses, especially in the salt-sensitive genotype (Suriyan et al. 2009). Decrease in photosynthesis in the salinized plants depended not only on a reduction of available CO_2 by stomatal closure, but also on the increasing effects of leaf water and osmotic potential, stomatal conductance, transpiration rate, relative leaf water content and biochemical constituents such as soluble carbohydrates, photosynthetic pigments and protein. Plant is able to escape stress when the duration of salinity is short (Sultana et al. 1999).

Salinity affects reproductive growth affecting viability of pollen in sensitive cultivar and inhibits starch synthase activity in pollen more considerably in sensitive than in tolerant cultivars. The affect on stigma receptivity is more negative for seed set compared to pollen viability and other physiological parameters The net photosynthetic rate considerably decreased in all day under saline-sodic conditions compared with that under non-saline sodic conditions (Fu et al. 2008).

Salt stress leads to over-expression of reactive oxygen species-scavenging enzymes, e.g. glutathione-S-transferases and L-ascorbate peroxidase. This is associated with the physiological functions of other identified proteins such as putative sialin, putative inorganic pyrophosphatase, RNA binding protein and serine or threonine-protein kinase. For adaptation to salt stress, several biochemical changes occur such as nitrate reductase (NR) (Rao and Gnanam 1990). Accumulation of betaine and betaine aldehyde dehydrogenase (BADH) mRNA associated with the presence of ABA (Hirofumi et al. 2001), proline accumulationin response to salt stress associated with salt tolerance (Claudivan et al. 2003).

Salinity reduces sorghum productivity drastically but genotypic variability occurs in sorghum hybrids and their parents for NaCl-salinity tolerance at the seedling stage. Glossy sorghum showed variability in salt tolerance at the seedling stage and as well as contents of total chlorophyll, hydrocyanic acids and epicuticular wax (De la Rosa-Ibarra and Maiti 1995).

Some prospective strategies for improving crop salt tolerance have been put forward by Ashraf and Hafeez (2004). Salt stress in pearl millet may be alleviated through seed treatment (Ashraf and Hafeez 2004).

Salinity affect on seedling growth and genotypic variability was found in maize for tolerance to salinity (Katerji et al. 1994; Fernandez et al. 2000; Maiti et al. 2004, 2009a, b, c, d, e, f; Maiti 2010; Humberto and Maiti 2010). High emergence (%): profuse root system, high seedling vigour is considered as selection criteria for salinity tolerance in maize (2009a, b, c, d, e, f). Higher root growth under saline stress of some genotypes might be due to a mechanism of resistance in maintaining osmo-regulation (Jian-jun et al. 2010). The production of profuse lateral roots near soil surface in response to salinity function as osmotic adjustment under salt stress (Maiti 1994a, Maiti and Saucedo 1986, Maiti et al. 1981a, b, 1994c, d, 2002, 2009a, b, c, d, e, f, 2012). Salt- and mannitol-induced osmotic stresses in terms of xylem pressure change was observed when the transpiration rate of the plant was not significantly changed (Jian-jun et al. 2010).

Maize inbred lines showed high Na^+ exclusion at the root surface and the level of xylem parenchyma (Christian et al. 2005) and the application of calcium postponed the depolarization and decreased the degree of depolarization caused by NaCl. High NaCl concentration leads to depolarization of maize root cell membrane, which can partly be counteracted by calcium (Jia-Min et al. 2008).

Salinity tolerance comes from genes that limit the rate of salt uptake from the soil, the transport of salt throughout the plant, adjusts the ionic and osmotic balance of cells in roots and shoots, and regulates leaf development and the onset of senescence. Compartmentation of Na^+ in vacuole and accumulation of high salt in root system contribute to salinity tolerance (Kawasaki et al. 2001). The abundance of ascorbate peroxidase in roots was found to be much higher in salt-tolerant rice cultivar than in salt-sensitive cultivar in the absence of stress (Salekdeh et al. 2002). Over expression with the bacterial genes for trehalose synthesis enhanced salt tolerance in rice; there was fourfold greater dry weight after 4 weeks in 100 mM NaCl in transformed plants than in untransformed plants (Garg et al. 2002). Pearl millet shows variability in salinity tolerance at different (Manga and Saxena 1981; Agrawal et al. 1985; Maiti et al. 2009a, b). Salt tolerant pearl millet cultivars show increase in root elongation as well as adventitious root numbers with an increase in saline stress functioning as osmotic adjustment for salt tolerance (Maiti et al. 2009a, b).

10.6.1 Drought Stress

Drought is the major problem among all the abiotic stresses. Selection of drought resistant lines is very important for good crop production and better yield under sustainable agriculture. It is essential to identify genotypic variability for drought resistance and selecting drought resistant lines and effective use in breeding programme. Bhargav and Paranjpe (2004) studied genotypic variation in the photosynthetic competence of *Sorghum bicolor* seedlings subjected to polyethylene glycol-mediated drought stress. Simple technique is adapted to screen the rice varieties and germplasm for drought resistance. Several physiological and biochemical changes occur for resistance to drought (Wright et al. 1983) such as higher leaf water-potentials (Matthews et al. 1990), great diversity for physiological and yield traits such as chlorophyll content, leaf temperature, grain number and grain weight, Harvest index and yield (Mutava et al. 2011). Leaf-rolling may alter the leaf surface micro-climate so that stomata may remain open and growth continues without associated high rates of water loss (Matthews et al. 1990). Mobilization of pre-anthesis assimilates to grain occurred in sorghum and millet but not in maize (Muchow 1989). Osmotic adjustment (OA) in pollen grains may be used as a measure in adaptation of sorghum genotypes to drought (Patil and Ravikumar 2011). Susan (1996) identified plant physiological traits in response to environmental water availability. They used phenotypic selection analysis to test the prediction from functional and comparative studies of plants that smaller leaves and more efficient water use are adaptive in drier environments. Water-use efficiency was selected to be higher and leaf area was selected toward a small intermediate optimum in the dry environment. The optimum leaf size in the dry environment is greater for plants with higher water-use efficiency.

Several selection criteria are utilized for large-scale screening for drought resistance in maize such as difference vegetation index (NDVI): other selection criteria including leaf senescence, chlorophyll content, anthesis-silking interval, root capacitance, final grain yield and grain yield components. The most stable trait under drought stress was kernel weight. Temperate lines with a wide adaptability can be used in drought resistance breeding for both temperate and tropical environments (Yanli et al. 2011).

Drought stress leads to reduction of leaf area, stomatal closure and reduction in assimilate supply, reduction in silk growth and reduction in growth. Similar to other cereal crops the following mechanisms operate for drought resistance in maize.

Drought escape: Drought tolerant lines give early flowering and grain filling to avoid drought escape.

Drought tolerance: Drought tolerant lines give yield under drought owing to some biochemical mechanisms.

Dehydration avoidance: Dehydration avoidance is the ability of the plant to retain a relatively higher level of hydration under drought stress. Apart from physiological and biochemical mechanisms, this is achieved by reducing the stomatal number and size.

Reynolds et al. (2007) reported drought-adaptive traits derived from wheat wild relatives and landraces. They used exotic parents to increase allelic diversity in bread wheat breeding through (i) interspecific hybridization of the ancestral genomes to produce so-called synthetic derived (SYN-DER) wheat and (ii) crossing with landrace accessions, originating in abiotically stressed environments, which have become isolated from mainstream gene pools. They evaluated the inherent genetic diversity encompassed by drought-adapted landraces compared with checks using DNA fingerprinting and confirmed that some landraces were not only distant from checks but also showed significant diversity among each other. Improvement in performance of SYN-DER lines compared with recurrent parents was not related with a larger overall investment in root dry weight, but rather an increased partitioning of root mass to deeper soil and enhanced ability to extract moisture from those depths. The best Mexican landraces had shown superior ability in terms of water extraction from soil depth and increased concentration of soluble carbohydrates in the stem shortly after anthesis.

Various researches undertaken team have confirmed that glossy sorghum lines were more resistant to drought at the seedling stage compared to those in non-glossy ones (Maiti 1994b; Maiti et al. 1994a). Glossy lines showed better growth and higher water use efficiency compared to the non-glossy lines (Maiti et al. 1994b), although showing variability among genotypes at the seedling stage (Maiti et al. 1994a). Variability in physiological and biochemical characteristics of glossy or non-glossy sorghums was also observed under different abiotic stresses at the seedling stage (De la Rosa-Ibarra et al. 2000a, b). The glossy lines have greater capacity in the uptake of phosphate than that of non-glossy sorghum lines (Raju et al. 1987). During water stress, decreases of GSH, AsA, chlorophyll and relative water contents in more drought resistant hybrids were less than those in susceptible ones. Changes of H_2O_2 (Hydrogen peroxide), malionaldehyde (MDA), reduced glutathione (GSH) and ascorbic acid (AsA) contents and anti-oxidative enzyme activities correlated considerably to drought resistance of rice hybrids and more drought resistant hybrids possessed high anti-oxidation capacity (He-zheng et al. 2010). Deeper root length may allow tolerant NILs to extract more water at deeper soil layers. Enhanced rooting depth is an important approach for dehydration avoidance and rice adaptation to drought stress (Venuprasad et al. 2011). Bidinger and Rao (1987) made assessment of drought effect on pearl millet [*Pennisetum americanum* (L.) Leeke]. Bidinger et al. (2005) studied QTL effects on phenotype and adaptation in pearl millet [*Pennisetum glaucum* (L.) R. Br.] top cross hybrid.

Osmotic adjustment regulation is the main component for physiological machinery of wheat drought resistance is associated with higher K^+ content, greater soluble sugar content and greater proline (Pro) content. Osmotic adjustment of different wheat genotypes in terms of different content of osmotic solutes (Hong-Bo et al. 2008). Hydroxymethylglutathione (hmGSH) and GSH are considered to the improvement of tolerance against osmotic stress in wheat and that the 5A chromosome influences the stress-induced changes in glutathione and hmGSH levels (Gábor et al. 2004). Water stress effect reduces the growth of seminal root in durum wheat (Ahmed et al. 2005).

Water stress induces proline accumulation in wheat (Hong-Bo et al. 2007). Apart from stress-induced synthesis of proline, the tolerance to water deficit found in transgenic plants was mainly due to protection mechanisms against oxidative stress and not caused by osmotic adjustment (Eliane et al. 2007). Glycinebetaine (GB) may protect the PSII complex from damage through accelerating D1 protein turnover and maintaining anti-oxidative enzyme activities at higher level to alleviate photo damage. Diethyldithiocarbamate as well as streptomycin treatment can impair the protective effect of GB on PSII (Qian-Quan et al. 2006).

VIn maize proline content increased and the relative water content, leaf greenness, 100-kernel weight and grain yield decreased under conditions of water deficit (WD). Under severe drought stress in maize, anti-oxidant enzyme activities decreased significantly ($p<0.01$) in later stages, namely for superoxide dismutase (SOD) during the tasseling and blister stages, for peroxidase (POD) during the milk stage and for catalase (CAT) during the tasseling, blister and milk stages. Meanwhile, membrane lipid peroxidation (measured as malondialdehyde content) significantly increased ($p<0.01$) in all stages (Li-Ping et al. 2006). Under water stress, the actions of catalase (CAT), superoxide dismutase (SOD) and peroxidase (POD) in leaves and roots improved sharply. Different strategies may be adopted for maize crop improvement under drought such as (1) cultural practices such as time and method of sowing, plant density and reduced tillage; (2) early planting reduces risk of terminal drought at grain filling stage; (3) selection of cultivars having robust root system, faster rate of root growth and deeper root penetration is highly efficient in utilizing soil water; (4) alternate partial root zone irrigation; (5) response forming matching the date of planting with the rainfall distribution; (6) several management practices such as tillage, water harvesting and mulching, etc.; (7) adoption of cropping practices to promote AM and the making of AM-colonized, drought-tolerant maize cultivars through conventional breeding as well as molecular and genomic techniques (Christopher and Tony 2008).

Stress during early grain development cut off the kernel sink potential by reducing the number of endosperm cells and amyloplasts formed in pearl millet. A water deficit during any stage of grain development causes the premature cessation of grain filling (Saini and Westgate 1999). Rapid control of leaf area by senescence is the predominant mechanism at late season drought, inducing long-term avoidance of dehydration of the upper leaves on eared shoots (Doa et al. 1996). Drought stress during early phase can delay or completely inhibit flowering, both through an inhibition of floral induction and development. Several drought-resistance mechanisms, which include drought escape via suitable phenology, root characteristics, specific dehydration escaping and tolerance mechanisms, drought recovery and a deep root system with high root length density at depth, are useful in extracting water thoroughly in upland conditions (Fukai and Cooper 1995). The variation in rice root response to drought from a physiological perspective in terms of morphology and function with respect to the different growth environments (upland and lowland) commonly used by farmers (Veeresh et al. 2011). The control of reactive oxygen species (ROS) and the stability of photosynthetic pigments under stress conditions are considered to contribute to drought tolerance. These included the changes in ascorbic peroxidase (APX), superoxide dismutase (SOD), catalase (CAT) isozyme activities, chlorophyll a, b (Chl a, b) and carotenoids contents in response to water stress.

Sergei et al. (2002) investigated adaptive traits of wild barley plants of Mediterranean and desert origin. They used reciprocal introduction of seeds and seedlings to test for local adaptation and to identify a set of co-adapted traits of Mediterranean and desert ecotypes of wild barley *Hordeum spontaneum*. Evidence for local adaptation was observed in seedling introductions into intact environments and from ecotype colonization success in the first generation after seed dispersal. They found genetically determined differences between Mediterranean and desert ecotypes can be observed as the following: reproductive output was higher in desert plants, with smaller seeds than in Mediterranean plants. There was a higher competitive ability of Mediterranean plants than desert plants. Plants of desert origin showed significant reductions in yield when grown in mixed stands with Mediterranean plants; no such effect was observed for plants of Mediterranean origin. Seed germination and seedling survival was lower in seeds of desert origin. This was attributed to both genetically determined higher dormancy of desert seeds and a trade off between number of seeds and their size (directly related to seed or seedling vigour).

10.6.2 Heat Stress

Rice heat tolerant line showed tightly arranged mesophyll cells in flag leaves, fully developed vascular bundles and some closed stomata. The mesophyll cells in flag leaves of the sensitive line were severely damaged by the high temperature stress. In contrast, the mesophyll cells in flag leaves of the resistant line maintained an intact ultra structure below the high temperature stress (Gui-Lian et al. 2009).

High temperature caused a significant increase in uptake of N, P and K^+ in pearl millet but the uptake of Ca^{2+}, Mg^{2+}, Na^+ and S remained unaffected in pearl millet (Ashraf and Hafeez 2004). High temperature imposed after imbibition the germination was reduced from 50 to 45 °C and there was a small reduction in the rate of germination but not in G_m. Optimum time of sowing in the tropics when maximum daytime soil temperature at the depth of sowing is in the range of 45–50 °C (Huidobro et al. 1985).

Thus, grain fertilization and grain set were most sensitive to the maximum temperature at mid-anthesis (Rachel et al. 1998). Heat stress induced ethylene production in developing wheat grains leading to kernel abortion and increased maturation in a susceptible cultivar (Dirk et al. 2007). Physiological traits of high yielding genotypes assessed under heat stressed conditions can be used as gene pool in wheat breeding programs for tolerance to heat stress (Bilge et al. 2011).

10.6.3 Cold Stress

Cold stress cause the severe cellular dehydration that occurs with freezing in cellular damage, such as the denaturation of proteins and precipitation of various molecules injury at the membrane level (Steponkus and Webb 1992) and freeze-induced dehydration cause multiple forms of membrane lesions (Steponkus et al. 1993). Varietal differences were observed in protein phosphorylation during cold treatment of rice leaves (Abdullah and Katoa 1997).

Low temperatures (Tw) (below 20 °C) during the reproductive period in rice caused low spikelet fertility and decreased crop growth rate (CGR) with reduced radiation use efficiency (RUE) (Hiroyuki et al. 2002). Rice seedlings treated with cold-tolerant seed-coating agents under chilling stress maintained considerably higher root vigour, POD, CAT and SOD activities and chlorophyll content, had low MDA content and electrolyte leakage and accumulated more soluble sugar and free proline. The cold-tolerant seed-coating agent improved the ability of rice seedlings in resisting to chilling stress (Hai-Qing et al. 2007).

Low temperature increased spikelet sterility and the number of enlarged pollen grains per anther. It was concluded that these flowering traits were facultative in nature (Farrell et al. 2006). Low temperatures (5–10 °C) induced ethanolic fermentation in the roots and shoots of the seedlings. Grain yield was most strictly reduced by low temperature (below 20 °C) during the reproductive period, as a result of low spikelet fertility. Rice (*Oryza sativa* L.) plants are sensitive to low temperature during the young microspore stage. Priming of sorghum [*Sorghum bicolour* (L.) Moench] seeds in 300 g l^{-1} PEG for 2 days at 25 °C could be used to enhance sorghum germination at low temperature, while the inclusion of plant hormones *per se* into priming media could be more effective than double hormone combinations (Tiryaki 2009).

Cold stress during the reproductive development of spring wheat (*Triticum aestivum* L.) cause grain-set failure in the high altitudes (>1500 m) of the world (Subedi et al. 1998b). Frost killed spikelets, restricted internode extension (stem growth) and reduced yield (Whaley et al. 2004). Cold temperatures and boron deficiency caused grain set failure in spring wheat (*Triticum aestivum* L.). Genotypic variability was found in response to boron and cold temperature (Subedi et al. 1998a). The accumulation of ice in the intercellular spaces can potentially cause the physical disruption of cells and tissues caused in part by the formation of adhesions between the intercellular ice and the cell walls and membranes (Levitt 1980). Freeze-induced dehydration results in cellular damage, such as the denaturation of proteins and precipitation of various molecules. However, the best documented injury occurs at the membrane level (Steponkus and Webb 1992). Cold acclimation involves the stabilization of membranes against freeze-induced damage (Steponkus and Webb 1992). Increase in membrane-freezing tolerance that occurs with cold acclimation involves changes in membrane lipid composition (Steponkus et al. 1993). Increased freezing tolerance in an ABA-hypersensitive mutant of common wheat is observed. Elevated ABA sensitivity contributes to the improved freezing tolerance through increased expression of the ABA-regulated low-temperature signal path way. It is suggested that both positive and negative regulation of ABA response is involved in the basic mechanism of freezing tolerance in wheat (Fuminori et al. 2008).

In maize, selection of lines in stressed environments is to adopt indices based on morphological, biochemical and physiological traits. Enough genetic variation is present in well-adapted germplasm from high altitude. Field emergence of maize under early planting was improved by recurrent selection at the rate of 84 %. Seedlings germinated at 7.2 °C were transferred to field, for agronomic traits and recombined to form a population for next selection cycle. By using screening techniques, a little improvement could be achieved in field emergence and seedling vigour.

10.6.4 Adaptation to Ionizing Radiations

Adaptation is a complex process by which populations of organisms respond to long-term environmental stresses by permanent genetic change. Kovalchuk et al. (2003) presented data from the natural "open-field" radiation adaptation experiment after the Chernobyl accident and provided the first evidence of the involvement of epigenetic changes in adaptation of a eukaryote Scots pine (*Pinus silvestris*) to chronic radiation exposure. They analyzed global genome methylation of control and radiation-exposed pine trees using a method based on cleavage by a methylation-sensitive *Hpa*II restriction endonuclease that leaves a 5′ guanine overhang and subsequent single nucleotide extension with labelled [H^3]dCTP. They observed that genomic DNA of exposed pine trees was considerably hypermethylated. Moreover, hypermethylation appeared to be dependent upon the radiation dose absorbed by the trees. Such hypermethylation may be taken as a defense strategy of plants that prevents genome instability and reshuffling of the hereditary material, allowing survival in an extreme environment. Further studies are clearly needed to analyze in detail the involvement of DNA methylation and other epigenetic mechanisms in the complex process of radiation stress and adaptive response. Subsequently, Igor et al. (2004) investigated molecular aspects of plant adaptation to life in the Chernobyl zone. They analyzed the adaptability of native *Arabidopsis* plants collected from areas with different levels of contamination around the Chernobyl nuclear power plant from 1986 to 1992. Notably, progeny of Chernobyl plants resisted higher concentrations of the mutagens Rose Bengal and methyl methane sulfonate. They analyzed the possible molecular mechanisms of their resistance to mutagens and found a more than 10-fold lower frequency of extra chromosomal homologous recombination, significant differences in the expression of radical scavenging (*CAT1* and *FSD3*) and DNA-repair (*RAD1* and *RAD51-like*) genes upon exposure to mutagens (Rose Bengal and x-rays) and a higher level of global genome methylation. This result suggests that adaptation to ionizing radiation is a complex process involving epigenetic regulation of gene expression and genome stabilization that improves plants resistance to environmental mutagens.

References

Abdullah S, Katoa A (1997) Varietal differences in protein phosphorylation during cold treatment of rice leaves. Phytochemistry 45(7):1329–1335

Agrawal MC, Singh R, Varma SK et al (1985) Yields of bajra and wheat in saline water applied through sprinkler and surface irrigation methods. Ann Arid Zone 21:9–14

Ahmed M, Meriem KH, Othmane M (2005) Impact of water deficit intensity on durum wheat seminal roots. C R Biol 328(10–11):918–927

Ashraf M, Hafeez M (2004) Thermotolerance of pearl millet and maize at early growth stages: growth and nutrient relations. Biol Plant 48(1):81–86

Bennici A (2008) Origin and early evolution of land plants. Problems and considerations. Commun Integr Biol 1(2):212–218

Bhargav S, Paranjpe S (2004) Genotypic variation in the photosynthetic competence of *Sorghum bicolor* seedlings subjected to polyethylene glycol-mediated drought stress. J Plant Physiol 161(1):125–129

Bidinger MV, Rao GDP (1987) Assessment of drought resistance in pearl millet [*Pennisetum americanum* (L.) Leeke]: factors affecting yields under stress. Aust J Agric Res 38:37–48

Bidinger S, Rizvi H, Yadav RS et al (2005) Field evaluation of drought tolerance QTL effects on phenotype and adaptation in pearl millet [*Pennisetumglaucum* (L.) R. Br.] top cross hybrids. Field Crop Res 94(1):14–32

Bilge B, Mehmet Y, Cemal Y (2011) Heat and drought resistance criteria in spring bread wheat (*Triticum aestivum*L.): morpho-physiological parameters for heat tolerance. Sci Res Essays 6(10):2212–2220

Christian Z, Anja N, Sandra K et al (2005) Molecular characterization of Na+/H+ antiporters (ZmNHX) of maize (Zea mays L.) and their expression under salt stress. J Plant Physiol 162(1):55–66

Christopher BR, Tony JV (2008) Maize drought tolerance: potential improvements through arbuscularmycorrhizal symbiosis. Field Crop Res 108(1):14–31

Claudivan FL, José C, Marco AO et al (2003) Solute accumulation and distribution during shoot and leaf development in two sorghumgenotypes under salt stress. Environ Exp Bot 49(2):107–120

De la Rosa-Ibarra M, Maiti RK (1995) Biochemical mechemical mechanism in glossy sorghum lines for resistance to salinity stress. J Plant Physiol 146:515–519

De la Rosa-Ibarra M, Maiti RK, Ambridge Gutierez LA (2000a) Evaluation of salinity tolerance of some sorghum genotypes seedlings (*Sorghum bicolor* L. Moench). Phyton Int J Exp Bot 66:87–92

De la Rosa-Ibarra M, Maiti RK, Quezada MR (2000b) Physiological and biochemical characteristics of glossy/non-glossy sorghums developed under three stress factors. Phyton Int J Exp Bot 68:1–10

Dirk H, Hwa Do J, Richard EM et al (2007) Heat stress induced ethylene production in developing wheat grains induces kernel abortion and increased maturation in a susceptible cultivar. Plant Sci 172(6):1113–1123

Doa W, Cournac L, Louguet P (1996) Impact of late-season drought on water relations in a sparse canopy of millet [*Pennisetum glaucum* (L.) R. Br.]. Field Crop Res 48(2–3):103–113

Eliane CGV, Ivan S, Marcos P et al (2007) Stress-induced synthesis of proline confers tolerance to water deficit in transgenic wheat. J Plant Physiol 164(10):1367–1376

Farrell TC, Fox KM, Williams RL et al (2006) Genotypic variation for cold tolerance during reproductive development in rice: screening with cold air and cold water. Field Crop Res 98(2–3):178–194

Fernandez JR, Maiti RK, Foroughbakhch R et al (2000) Genotypic variability among maize hybrids (*Zea mays* L.) for resistance to drought, salinity and high temperature at the seedling stage. Res Crops 1(1):12–19

Fu Y, Zhengwei L, Zhichun W et al (2008) Relationship between diurnal changes of net photosynthetic rate and influencing factors in rice under saline sodic stress. Rice Sci 15(2):119–124

Fukai S, Cooper M (1995) Development of drought-resistant cultivars using physiomorphological traits in rice. Field Crop Res 40(2):67–86

Fuminori K, Shigeo T, Chiharu N (2008) Increased freezing tolerance in an ABA-hypersensitive mutant of common wheat. J Plant Physiol 165(2):224–232

Gábor K, Gabriella S, Gábor G (2004) Effect of osmotic stress on glutathione and hydroxyl methylglutathione accumulation in wheat. J Plant Physiol 161(7):785–794

Garg K, Owens R, Choi KLV et al (2002) Trehalose accumulation in rice plants confers high tolerance levels to different abiotic stresses. In: Compendium of abstracts of the National Academy of Sciences, USA, 2002

Gui-lian Z, Li-yun C, Shun-tang Z et al (2009) Effects of high temperature stress on microscopic and ultrastructural characteristics of mesophyll cells in flag leaves of rice. Rice Sci 16(1):65–71

Hai-qing Z, Ying-bin Z, Guo-chao X et al (2007) Effect and mechanism of cold tolerant seed-coating agents on the cold tolerance of early Indica rice seedlings. Agric Sci China 6(7):792–801

Hanley ME, Lamont BB, Fairbanks MM et al (2007) Plant structural traits and their role in anti-herbivore defense. Perspect Plant Ecol Evol Syst 8(4):157–178

He-zheng W, Lian-he Z, Jun MA et al (2010) Effects of water stress on reactive oxygen species generation and protection system in rice during grain-filling stage. Agric Sci China 9(5):633–641

Hirofumi S, Satoshi I, Reda M (2001) Effect of salinity and abscisic acid on accumulation of glycinebetaine and betaine aldehyde dehydrogenase mRNA in Sorghum leaves (*Sorghum bicolor*). J Plant Physiol 158(7):853–859

Hiroyuki S, Toshihiro H, Kazuto I (2002) Response of growth and grain yield in paddy riceto cool water at different growth stages. Field Crop Res 73(2–3):67–79

Hong-Bo S, Shui-Yuan J, Feng-Mei L et al (2007) Some advances in plant stress physiology and their implications in the systems biology era. Colloids Surf B Biointerf 54(1):33–36

Hong-Bo S, Li-Ye C, Cheruth AJ et al (2008) Water-deficit stress-induced anatomical changes in higher plants. C R Biol 331(3):215–225

Huidobro JG, Monteith JL, Squire GR (1985) Time, temperature and germination of pearl millet (*Pennisetum typhoides* S. & H.): III. Inhibition of germination by short exposure to high temperature. J Exp Bot 36(2):338–343

Humberto GR, Maiti RK (2010) Perspectives of salinity tolerance of some crops: a review. Int J Agric Environ Biotechnol 3(1):45–82

Igor K, Vladimir A, Igor P et al (2004) Molecular aspects of plant adaptation to life in the chernobyl zone. Physiology 135(1):357–363

Jia-min H, Xiao-li W, Fu-qin Z et al (2008) Effects of NaCl and Ca^{2+} on membrane potential of epidermal cells of maize roots. Agric Sci China 7(3):291–296

Jian-jun Z, Xin-fu BAI, Qing-mei BU et al (2010) An analysis to the driving forces for water and salt absorption in roots of maize seedlings under salt stress. Agric Sci China 9(6):806–812

Katerji N, Van Hoorn JW, Hamdy A et al (1994) Effect of salinity on emergence and on water stress and early seedling growth of sunflower and maize. Agric Water Manag 26(1–2):81–91

Kawasaki S, Borchert C, Deyholos M et al (2001) Gene expression profiles during the initial phase of salt stress in rice. Plant Cell 13:889–905

Kenrick P, Crane PR (2008) Origin and early evolution of land plants problems and considerations. Commun Integr Biol 1(2):212–218

Kovalchuk O, Burke P, Arkhipov A et al (2003) Genome hypermethylation in *Pinus silvestris* of Chernobyl-a mechanism for radiation adaptation. Mutat Res Fundam Mol Mech Mutagen 529(1-2):13–20

Levitt J (1980) Responses of plants to environmental stress: chilling, freezing, and high temperature stresses. Academic, New York

Li-Ping BAI, Fang-Gong SUI, Ti-Da GE et al (2006) Effect of soil drought stress on leaf water status, membrane permeability and enzymatic antioxidant system of maize. Field Crop Res 16(3):326–332

Maiti RK (1994a) Estimation of chlorophyll and epicuticular wax of some glossy and nonglossy sorghum at the seedling stage. Int Sorghum Millet Newsl 35:119

Maiti RK (1994b) The role of morpho-physiological traits in shoot fly resistance in sorghum. Int Sorghum Millet Newsl 35:107

Maiti RK (2010) A novel technique for evaluating and selecting crop cultivars for salinity tolerance: its progress. Int J Bio-resource Stress Manag 1(1):51–53

Maiti RK, Saucedo JMR (1986) Studies of some biochemical characters of 5 glossy and 5 non-glossy sorghums related to drought resistance. Sorghum Newsl 29:91

Maiti RK, Raju PS, Bidinger (1981a) Evaluation of visual scoring for seedling vigour in sorghum. In: Abstracts of 19th ISTA Congress, Vienna, 6–13 June 1981

Maiti RK, Raju PS, Bidinger et al (1981b) Evaluation of visual scoring for seedling vigour in sorghum. Seed Sci Technol 9:613–622

Maiti RK, Reddy YVR, Rao KV (1994a) Seedling growth of glossy and non-glossy sorghums [*Sorghum bicolor* (L.) Moench.] under water stress and non-stress conditions. Phyton 55:1–8

Maiti RK, Flores Campos LO, Dominguez L (1994b) Growth analysis and productivity of 15 genotypes of glossy sorghum for forage and grain production. Int Sorghum Millet Newsl 35:133–134

Maiti RK, Heredia-Rojas JA, Cazares-Aonso (1994c) Electrophoretic analysis of protein profile of some glossy and nonglossy sorghum lines: a preleminary study. Phyton Int J Exp Bot 56:105–107

Maiti RK, De La Rosa MI, Sandoval ND (1994d) Genotypic variability in glossy sorghum lines for resistance to drought, salinity and temperature stress at the seedling stage. J Plant Physiol 143:241–244

Maiti RK, Singh VP, Wesche-Ebeling P et al (2002) Research advances on sorghum for resistance to drought, high and low temperature and salinity and the mechanisms of resistance. Res Crops 3:257–281

Maiti RK, Singh VP, Wesche-Ebeling P et al (2004) Research advances on cold, drought and salinity tolerance and its mechanisms of resistance in maize (*Zea mays* L.): a review. Crop Res 27:1–30

Maiti RK, Vidyasagar P, Banerjee PP et al (2005) Research advances on seed physiology, salinity and some other factors affecting crop growth in rice (*Oryza sativa* L.). Res Crops 6:379–392

Maiti RK, Pawar R, Humberto GR et al (2009a) Salt tolerance of pipeline Bt-cotton (*Gossypiumhirsutum*) hybrids subjected to NaCl stress. Int J Agric Environ Biotechnol 2(2):125–132

Maiti RK, Rodriguez HG, Yadav VK et al (2009b) Salt tolerance of nine rice hybrids and their parents at the seedling stage. Int J Agric Environ Biotechnol 2(3):199–205

Maiti RK, Ghosh SK, Gozalez-Rodriguez H et al (2009c) Salt tolerance of pearl millet hybrids and parents of Vibha seeds at germination and seedling stage. Int J Agric Environ Biotechnol 2(3):206–210

Maiti RK, Koushik S, Rodriguez HG et al (2009d) Salt tolerance of twelve maize hybrids at seedling stage. Acta Agron 57(4):120–135

Maiti RK, Sahakar SC, Gupta A (2009e) Genotypic variability in salinity tolerance of pipe line hybrids of sunflower of Vibha seeds. Int J Agric Environ Biotechnol 1:52–56

Maiti RK, Sahapur SC, Gupta A et al (2009f) Evaluation and slection of sunflower hybrids and parents for tolerance to different levels of salinity at the seedling stage. Int J Agric Environ Biotechnol 2:57–63

Maiti RK, Rajkumar D, Vidyasagar P (2012) Screening of rice varieties for drought resistance at the seedling stage. Res Crops 13(3):791–794

Manga VK, Saxena MBL (1981) Effect of salinity on germination and early seedling growth of different millet crops. Agric Sci Dig 1:100–102

Matthews RB, Azam-Ali SN, Peacock JM (1990) Response of four sorghum lines to mid-season drought. II. Leaf characteristics. Field Crop Res 25(3–4):297–308

Muchow RC (1989) Comparative productivity of maize, sorghum and pearl millet in a semi-arid tropical environment II. Effect of water deficits. Field Crop Res 20(3):207–219

Mutava RN, Prasad PVV, Tuinstra MR et al (2011) Characterization of sorghum genotypes for traits related to drought tolerance. Field Crop Res 123(1):10–18

Norman EKA, Schierenbeck KA (2004) Hybridization as a stimulus for the evolution of invasiveness in plants. Cross Mark 97:137043–137050

Patil BS, Ravikumar RL (2011) Osmotic adjustment in pollen grains: a measure of drought adaptation in sorghum. Research communication. Curr Sci 100(3):377–382

Qian-Quan M, Wei W, Yong-Hua L et al (2006) Alleviation of photoinhibition in drought-stressed wheat (*Triticum aestivum*) by foliar-applied glycinebetaine. J Plant Physiol 163(2):165–175

Rachel F, Ellis RH, Wheeler TR et al (1998) Effect of high temperature stress at anthesis on grain yield and biomass of field-grown crops of wheat. Ann Bot 82(5):631–639

Raju PS, Clark RB, Maiti RK et al (1987) Phosphorus uptake, distribution and use by glossy and nonglossy sorghum. J Plant Nutr 10:2017–2024

Rao KR, Gnanam A (1990) Inhibition of nitrate and nitrite reductase activities by salinity stress in *Sorghum vulgare*. Phytochemistry 29(4):1047–1049

Reynolds M, Dreccer F, Trethowan R (2007) Drought-adaptive traits derived from wheat wild relatives and landraces. J Exp Bot 58(2):177–186

Saini H, Mark E, Westgate (1999) Reproductive development in grain crops during drought. Adv Agron 68:59–96

Salekdeh GH, Siopongco J, Wade LJ et al (2002) A proteomic approach to analyzing drought- and salt-responsiveness in rice. Field Crop Res 76(2–3):199–219

Sergei V, Samuel M, David W (2002) Adaptive traits of wild barley plants of Mediterranean and desert origin. Oecologia 133(2):131–138

Solbrig OT (1994) Plant traits and adaptive strategies: their role in ecosystem function. Biodivers Ecosyst Funct 99:97–116

Steponkus PL, Webb MS (1992) Freeze-induced dehydration and membrane destabilization in plants. In: Somero G, Osmond B (eds) Water and life: comparative analysis of water relationships at the organisms. Cellular and molecular level. Springer, Berlin, pp 338–362

Steponkus PL, Uemura M, Webb MS (1993) A contrast of thecryostability of the plasma membrane of winter rye and spring oat: two species that widely differ in their freezing tolerance and plasma membrane lipid composition. Adv Low Temper Biol 2:211–312

Subedi KD, Floyd CN, Budhathoki CB (1998a) Cool temperature-induced sterility in spring wheat (*Triticumaestivum* L.) at high altitudes in Nepal: variation among cultivars in response to sowing date. Field Crop Res 55(1–2):141–151

Subedi KD, Gregory PJ, Summerfield RJ et al (1998b) Cold temperatures and boron deficiency caused grain set failure in spring wheat (*Triticum aestivum* L.). Field Crop Res 57(3):277–288

Sultan SE (1995) Phenotypic plasticity and plant adaptation. Acta Bot Neerl 44(4):363–383

Sultana N, Iked T, Itoh R (1999) Effect of NaCl salinity on photosynthesis and dry matter accumulation in developing rice grains. Environ Exp Bot 42(3):211–220

Suriyan C, Kanyaratt S, Chalermpol K (2009) Comparative effects of salt stress and extreme pH stress combined on glycinebetaine accumulation, photosynthetic abilities and growth characters of two rice genotypes. Rice Sci 16(4):274–282

Susan AD (1996) Differing selection on plant physiological traits in response to environmental water availability: a test of adaptive hypotheses. Evolution 50(1):92–102

Tiryaki B (2009) Seed priming combined with plant hormones: influence on germination and seedling emergence of sorghum at low temperature. Seed Sci Technol 37(2):303–315

Valverde P, Fornoni J, Nunez F (2001) Defensive role of leaf trichomes in resistance to herbivorous insects in *Daturastramonium*. J Evol Biol 14(3):424–432

Veeresh G, Amelia H, Akira Y et al (2011) Root biology and genetics improvement for drought avoidance in rice. Field Crop Res 122(1):1–13

Venuprasad I, Veeresh G, Atlin GN et al (2011) Rice near-isogenic-lines (NILs) contrasting for grain yield under lowland drought stress. Field Crop Res 123(1):38–46

Whaley JM, Kirby EJM, Spink JH (2004) Frost damage to winterwheatin the UK: the effect of plant population density. Eur J Agron 21(1):105–115

Wright GC, Smith RCG, Morgan JM (1983) Differences between two grain sorghum genotypes in adaptation to drought stress. III. Physiological responses. Aust J Agric Res 34(6):637–651

Yanhai Z, Zhenlin W, Xuezhen S et al (2008) Higher salinity tolerance cultivars of winter wheat relieved senescence at reproductive stage. Environ Exp Bot 62(2):129–138

Yanli L, Zhuanfang H, Chuanxiao X et al (2011) Large-scale screening for maize drought resistance using multiple selection criteria evaluated under water-stressed and well-watered environments. Field Crop Res 124(1):37–45

Breeding, Genetics and Biotechnology

11

Aruna Kumari and Ratikanta Maiti

Abstract

This chapter give a concise review of research trends in crops, dealing with various aspects of breeding such as evaluation, genotypic variability, genetic advancement, character association, path coefficients, combining ability, etc., and biotechnology in various crops. Several high-yielding cultivars are evaluated for genetic divergence and stability characters across environments. It also discusses the research advances in screening of genotypes against abiotic and biotic stresses, the development of early-maturing hybrids, the expression of stability traits against environments, the interactive effect of genotype and environment, heterosis, and attributes for selection of genotypes.

In view of increasing global warming and growing populations there is a great necessity to increase crop productivity by utilizing improved breeding technology. Of the papers presented at the Second International Conference on Bio-resource and Stress Management, held in Hyderabad in 2015, a good number dealt with various aspects of breeding such as evaluation, genotypic variability, genetic advancement, character association, path coefficients, combining ability, etc. and biotechnology in various crops, revealing research trends. These research trends are highlighted below.

11.1 Sorghum

Sorghum [*Sorghum bicolor* (L.) Moench] is an important crop of dry land agriculture. In the northern part of India, sorghum is grown mainly for fodder production whereas in the central and southern parts of India, it is grown as a source of food and fodder. Forage sorghum is one of the most widely adapted forage crops in drought-prone areas because of its higher productivity, better palatability and digestibility.

A. Kumari (✉)
Department of Crop Physiology, Agricultural College, Professor Jayashankar Telangana State Agricultural University, Polasa, Jagtial, Karimnagar 505 529, Telangana, India
e-mail: arunasujanagcjgl@gmail.com

R.K. Maiti
Facultad de Ciencias Forestales, Universidad Autónoma de Nuevo León, Carr. No 85, km 145, NL 67700 Linares, Mexico
e-mail: ratikanta.maiti@gmail.com

R.K. Maiti et al. (eds.), *Bioresource and Stress Management*, DOI 10.1007/978-981-10-0995-2_11

The improvement of forage sorghum is much emphasized owing to its importance as a fodder crop to feed the increasing livestock population in India. Significant research activities have been directed towards various aspects of breeding on sorghum.

11.1.1 High-Yielding Cultivars

Several high-yielding varieties of sorghum have been generated and released for cultivation. In the present era of nutritional security, development of quality-rich genotypes is of significance. The Sorghum Research Unit, Dr. PDKV, Akola (MS) has developed one *kharif* dual-purpose sorghum genotype SPV-1786 with excellent dough and *roti*-making quality. SPV-1786 recorded grain yield (37.97 q ha^{-1}) nearly equal to that of checks SPV-1616 (38.44 q ha^{-1}), SPV-462 (40.49 q ha^{-1}) and CSV-15 (37.87 q ha^{-1}). It has further recorded higher fodder yield (130.47 q ha^{-1}) than the checks SPV-462 (129.13 q ha^{-1}) and CSV-15 (121.72 q ha^{-1}) and comparable to that of check SPV-1616 (131.88 q ha^{-1}) during the 3-year AICSIP (All India Coordinated Sorghum Improvement Project) trials during 2006, 2007 and 2008 under zone II of the AICSIP trials. The genotype SPV-1786 possessed comparable grain and stover quality to checks SPV-1616, SPV-462 and CSV-15. However, the dough and *roti*-making quality of SPV-1786 was found to be excellent over the checks SPV-1616, SPV-462 and CSV-15. Hence, SPV-1786 holds promise for further use in breeding programmes oriented towards improving quality characteristics in sorghum (Kalpande et al. 2015d).

11.1.2 Abiotic Stress

11.1.2.1 General Combining Ability

Drought or water stress is a frequently occurring phenomenon. The condition is still compounding due to global warming. *Rabi* sorghum is more prone to the effect of water stress during its growth period. In a study conducted (Ghorade et al. 2015b) during the post-rainy *rabi* season at the Sorghum Research Unit, Dr. PDKV, Akola (MS) to identify promising lines and testers for grain yield in *rabi* sorghum under drought conditions, the estimates of general combining ability effects indicated that out of five lines, the line MS 104 A was a good general combiner for grain yield per plant (0.93*) along with six yield components like plant height, panicle breadth, panicle length, panicle weight, seed setting percentage and 1000-seed weight. Among the testers, AKSV 13 R was a good general combiner for grain yield per plant (2.22**) along with six drought tolerance traits, viz plant height, days to 50 % flowering, panicle breadth, panicle weight, seed setting percentage and 1000-seed weight. Another tester, M 35-1, showed a desirable GCA for grain yield per plant (3.61**) along with five yield components, viz panicle breadth, panicle length, panicle weight, seed setting percentage and 1000-seed weight. Another tester, SPV 504, has also transmitted desirable genes for grain yield per plant (2.07**) along with four yield components like leaf plant height, panicle weight, seed setting percentage and 1000-seed weight. The tester Parbhani Moti was also a good general combiner for grain yield per plant (1.62*) as well as for seed setting percentage and 1000-seed weight. The tester CSV 216 R (2.61**) was also observed to be a good general combiner for grain yield per plant. Thus, the study specifies that there is a need to exploit all these six testers and one line for a future breeding programme to develop high-yielding hybrids in *rabi* sorghum.

11.1.2.2 Biomass Production

In indigenous or high-yielding sorghum varieties or fodder sorghum varieties or sweet sorghum varieties, production of a high amount of biomass is most desirable. Eight sweet sorghum genotypes, four exotic (Keller, Wray, Urja and BJ 248), four indigenous (SSV 84, SSV 74, CSV 19SS and CSH 22SS), two grain sorghums (CSV 15 and SPV 1616) and one fodder sorghum (CSV 21 F) were evaluated during the *kharif* season at the Directorate of Sorghum Research (DSR), Rajendranagar, Hyderabad, Andhra Pradesh, for higher biomass production during the key crop growth stages (Pancholee et al. 2015). Morphological character, viz the number of internodes and number of leaves, was associated with high biomass production in the genotypes. Higher numbers of internodes were observed in genotypes CSH 22SS, SSV84 and CSV 19SS. Further, a significantly higher biomass (g per plant) was obtained in the genotypes CSH 22SS, SSV 74 and SSV 84 at the milky stage. The juice obtained was slightly acidic in nature; the pH ranged from 4.07 to 6.01. Yield components like test weight, panicle length and per panicle weight were significantly higher in SPV 1616, SSV 74, CSV 15, CSH 22SS and SSV 84.

Indian sorghum genotypes performed well in respect of biomass as well as yield in comparison with exotic genotypes. The genotypes SSV 74 and CSH 22SS were found to be superior for biomass production as well as yield and among all stages of growth milky stage were best for biomass production. Biomass was not significantly affected by stay-green character.

11.1.3 Sorghum Screening for Abiotic and Biotic Stress Resistance

11.1.3.1 Screening for Abiotic Stress

The post-rainy season sorghums are unique to India and are important from the perspective of both food and fodder, unlike *kharif* sorghums wherein the grain losses are more in *kharif*. The commercial value of grain sorghum of *kharif* is decreased due to infestations of grain mould. Among the abiotic stress affecting yield, terminal drought and cold stress at anthesis assume importance. Terminal drought is of common occurrence, as *rabi* sorghum is grown under receding soil moisture conditions. A set of 16 sorghum genotypes were evaluated at Rahuri, Maharashtra, for three post-season seasons in two soil types (shallow and medium soil depths) (Reddy and Patil 2015). The soil depth was found to influence the agronomic and physiological traits. The genotype interacted significantly with the year for all the traits except stomatal conductance and transpiration rate. Genotype significantly interacted with the soil depth for all the traits except plant height, leaf area index and biomass yield at flowering. The variety RSV 1037 with a grain yield of 2.5 t ha^{-1} and fodder yield of 10.5 t ha^{-1} was the best performing in medium soils, while the genotype RSV 850 with a grain yield of 1.2 t ha^{-1} and fodder yield of 6.4 t ha^{-1} was the best performing in shallow soils.

Hybrid breeding for *rabi* adaptation suffers from the constraints of poor yield heterosis, poor seed set and poor grain quality. Hybrids involving post-rainy season parents lack hybrid vigour and those involving *kharif* parents exhibit good yield heterosis but have poor grain quality. The loss in grain yield (open panicle grain yield – selfed panicle grain yield) was highest in the hybrid at about 22.9 g followed by B lines, varieties and R lines. Higher grain yields can be obtained in the hybrid with improvement in the pollen viability and germinability. Hybrids involving *kharif* females were early and those involving *kharif* restorers were sensitive to sowing date for seed set percentage. The hybrids with both *kharif* parents had good panicle length. The hybrids involving post-rainy season restorers were sensitive to sowing date for grain yield and the hybrids with at least one *rabi* parent had good seed size. The grain yield under selfing was governed by additive gene action while the seed set percentage under selfing was governed by dominance gene action. The seed set was good in CSV 8R, Phule Anuradha, Phule Chitra, Phule Vasudha, PKV Kranti and SPV 1595 when used as both male and female parents, indicating their tolerance to cold stress. Among the hybrid parents, R lines showed stability for many characters across the dates of sowing while B lines were less stable indicating the need for the development of stable B lines for the *rabi* season (Reddy and Patil 2015).

It is concluded that *rabi* sorghum varieties should be developed focusing on the traits that are specific to the soil depths. Breeding *rabi* hybrids should focus on developing female parents with *rabi* adaptation and restorers with *kharif* adaptation, both improved for grain yield and seed set percentage.

11.1.3.2 Sorghum Shoot Fly Resistance Characters

A serious pest of sorghum in the early stages of development is shoot fly. Twenty-six derived lines and three checks of sorghum evaluated for variability and correlation of grain yield and shoot fly resistance characters (Ranjith and Ghorade 2015) has shown that highest genetic advance and high heritability was observed in trichome density per mm^2. Leaf glossiness showed high positive correlation with number of shoot fly eggs and dead hearts. The oviposition percentage on 14th, 21st and 28th days after seedling emergence showed significant positive correlation with dead hearts. Leaf trichome density at abaxial leaf surfaces showed significant negative correlation with shoot fly dead hearts. The leaf trichome density revealed higher level of resistance due to non-preference for oviposition. Seedling vigour showed weaker association with dead heart percentage. Genotypic correlation confirmed that the number of trichomes on surface of lamina and leaf glossiness contributed resistance to shoot fly. Thus, these characters can be used as selection criteria for breeding shoot fly resistance genotypes. High heritability and high genetic advance as well as positive associations of the characters obtained can be explored in the selection of shoot fly resistance in sorghum.

11.1.3.3 Mould Resistance

Sorghum is a major cereal crop of Maharashtra state grown in both *kharif* and *rabi* seasons and serves both food and feed requirements of rural population and cattle. In spite of higher productivity of rainy season sorghum achieved by releasing high-yielding hybrids and varieties, deterioration of grain due to grain mould is a serious problem. So efforts were made to develop high-yielding CMS-based hybrids coupled with grain mould tolerance. Twenty-four CMS-based hybrids were developed utilizing six female and four male parents following a line × tester mating design during *rabi* 2011–2012 and these along with ten parents and two checks, viz PVK 801 and CSH 25, were evaluated for heterosis of grain yield and grain mould parameters (More et al. 2015b). A highest heterotic effect for grain yield and its associated traits was observed in the cross 372 A × C43 (100.60 %) followed by PMS 98 A × KR 199 (48.94 %), DNA 10 × KR 199 (45.93 %), DNA 10 × KR 196 (42.72 %) and IMS 12 A × KR 196 (38.61 %) over hybrid check CSH 25. For grain mould parameters the crosses AKMS 85A × KR 196, AKMS 85A × C43, DNA 10 × KR 196, DNA 10 × KR 199 and PMS 98A × C43 showed highest significant negative heterosis over tolerant check PVK 801, which shows that these hybrids had significantly less incidence of grain mould as compared to the check. The hybrids PMS 98 A × C43, DNA 10 × KR 196, IMS 12A × KR 199 and DNA 10 × KR 199 recorded high mean and heterosis for grain yield and less incidence of grain mould. Therefore, these hybrids need to be tested across the environment for their stable performance.

11.1.4 Genetic Variability

The genetic variability and nature of character association in 87 genotypes (50 B lines, 25 R lines and 12 inbred lines) of *kharif* sorghum [*Sorghum bicolor* (L.) Monech] for 17 quantitative traits studied (More et al. 2015a) revealed the presence of statistically significant differences among the treatments for all the traits under study, indicating the presence of variability among the genotypes. The genotypic coefficient of variation (GCV) estimates were lower than the phenotypic coefficients of variation (PCVs), indicating the effect of environment on these traits, however the differences were of lower magnitude. High estimates of GCV and PCV were observed for plant height, days to 50 % flowering, fodder yield, number of primary branches per panicle, grains per primary branch and harvest index. These traits also showed high heritability percentage and genetic advance indicating the presence of additive gene effect. Heritability ranged from 62.78 % (threshed grade score) to 99.84 % (days to 50 % flowering). On the basis of percentage performance for grain yield and grain mould parameters, genotypes ICSB204, PMS7B, PMS46B, ICSR89058, ICSR90003 and ICSR160 were recorded as promising ones. The traits panicle length, panicle width, germination percentage, test weight, number of primary branches, per grain primary branches and harvest index exhibited significant and positive correlations with grain yield. Field grade score and threshed grade score showed negative and non-significant correlations with germination percentage, test weight, carbohydrate content and grain yield, while plant height showed significant and positive correlation with fodder yield per plant both at genotypic and phenotypic level.

The nature and magnitude of genetic variability for productivity related and grain quality traits in minicore collections during the post-rainy season of 2011–2012 at AICSIP, UAS Dharwad, was assessed (Karadi and Kajjidoni 2015). The analysis of variance exhibited the presence of highly significant differences for all characters like plant height, panicle length, panicle width, seed yield per plant, 100-seed weight, seed volume, bulk density, true density and seed size at both 5 % and 1 % levels of probability.

High estimates of the PCV and GCV were observed for most of the characters except for seed bulk density and seed true density traits. High heritability coupled with high GAM was recorded for all characters except seed true density. Highly significant positive association was observed for grain yield per plant with plant height, ear head width, 100 seed weight, seed volume and seed size at both levels. Hundred-seed weight showed highly significant positive association with seed yield per plant, seed volume and seed size (Karadi and Kajjidoni 2015).

Path analysis revealed that seed yield per plant had direct effects on plant height, ear head width and 100-seed weight; direct selection for these traits will be rewarding for improvement of grain yield per plant. Minicore collections exhibited high magnitude of variation with respect to ear head and grain quality characters. Among minicores, accession IS-19975 was superior for seed yield per plant, IS-27034 for test weight, IS-473 for bulk density, IS-29714 for true density and IS-19975 for seed size (Karadi and Kajjidoni 2015).

11.1.5 Genetic Divergence

Genetic divergence studies (Gaikwad et al. 2015) undertaken on 32 genotypes of forage sorghum [*Sorghum bicolor* (L.) Moench] for fodder yield and its contributing characters grouped the genotypes into seven clusters. Maximum inter-cluster distance was observed between clusters IV and VI (59.66). Thus, the parents involved in cluster IV (AKFG-09-4, AKFG-09-3, AKFG09-5 and Improved Ramkel) and cluster VI (IS47802) can be utilized for improvement of green fodder yield and its contributing characters. Relative contribution of characters towards total divergence realized the importance of green fodder yield followed by stem girth and leaf length.

11.1.6 Stability of Sorghum Genotypes Across Environments

Sixty genotypes of sorghum evaluated at two locations (Hyderabad and Deesa) on two dates of sowing for six traits, namely, days to flag leaf emergence, days to 50 % flowering, plant height, panicle length, 100 seed weight and grain yield per plant (Gomashe et al. 2015), has revealed that the photoperiod response index (PRI) in 15 genotypes was less than 10. This indicated their insensitivity to changes in photoperiod. Genotypes CSV 17, IS 35251, IS 35255, IS 35803, AKMS 14B, 2219B and AKR 150 were highly photoperiod insensitive. Stability analysis was carried out using Eberhart and Russell and GGE biplot analysis. In stability analysis mean squares due to genotypes, environments (linear) and pooled deviations were highly significant ($p \leq 0.01$) for all the traits. This indicates that genetic diversity existed among genotypes and environments. Mean squares due to genotypes × environments were significant ($p \leq 0.05$) for grain yield and highly significant ($p \leq 0.01$) for days to flag leaf emergence and days to 50 % flowering, revealing the differential response of varieties under different environments only for these three traits. Eberhart and Russell method gave useful information on stable genotypes but it failed to predict a broader picture considering high yields. GGE biplot analysis gave a simple graphical picture of the genotypes based on their mean performance as well as stability across the environment.

11.1.7 Stability in Expression of Traits

Climate change poses a major challenge to feed the ever-growing population. Sorghum is one of the important staple food grains supporting more than 500 million lives in the arid and semi-arid tropics of Africa and Asia, which are predicted to be most adversely affected by climate change. A study undertaken on the stability in expression of 12 sorghum descriptors in three popular hybrids (CSH 9, CSH 14 and CSH 16), their parental lines (296A, AKMS 14A, 27A, CS 3541, AKR 150 and C 43) and two other popular male parents (RS 29 and Indore 12) over four to six seasons, including rainy and post-rainy seasons (Hariprasanna et al. 2015b), has shown that expression of traits like time of panicle emergence, natural height of plant up to base of flag leaf, total height, leaf length and width, stem diameter, panicle length and 1000-grain weight varied slightly between rainy and post-rainy seasons. However, variation was minimal among the years under the same season except in the case of time of panicle emergence (CSH 14 and CSH 16; very early to medium or medium to very late) and 1000-grain weight (CSH 14; low to high).

11.1.8 Genotype × Environment Interaction

Genotype × environment (linear) interaction was significant for all the studied characters in 13 genotypes of forage sorghum along with check variety evaluated for stability at three locations of Maharashtra state (Akola, Washim and Amravati) during *kharif* 2012 (Deshmukh and Zarea 2015) except stem weight per plant. This indicated the linear response of the genotypes to environmental changes. The non-linear component (pooled deviation) was highly significant for all the characters except leaf length, leaf area per plant, stem girth and plant moisture percentage, indicating that the major components for differences in stability were due to deviation from the linear function. The results illustrated that the genotypes AKFG-09-6, SSG-59-3 and Imp Ramkel were stable genotypes for the majority of characters. Two genotypes, viz AKFG-09-6 and AKFG-09-8, exhibited below average stability. These genotypes are suited to favourable environmental conditions.

11.1.9 Combining Ability and Heterosis

In a study, three male sterile lines were crossed with 22 testers in a line × tester design and produced 54 hybrids to estimate the estimates of specific combining ability effects in *rabi* sorghum hybrids in order to identify the potential cross combinations for yield and its components in *rabi* sorghum. Among the 66 hybrids studied, a total of four crosses showed positive and significant SCA effects for the character grain yield per plant. AKRMS 80A × Rb 307–11 hybrid exhibited highest significant SCA effects for grain yield per plant (18.23**) along with the component traits like panicle breadth, number of primaries per panicle and fodder yield per plant. Second cross AKRMS 80A × RS 585 exhibited higher significant SCA effects for grain yield per plant (12.08**) along with three component characters like plant height, number of primaries per panicle and 1000 grain weight. Third cross AKRMS 47A × AKSV 70R showed higher significant SCA effects for grain yield per plant (11.99**) along with other seven component characters, i.e. days to 50 % flowering, days to maturity, plant height, number of primaries per panicle, number of grains per panicle, 1000 seed weight and fodder yield per plant. These three crosses exhibited positive significant SCA effects along with positive significant standard heterosis for grain yield per plant. So these three crosses can be utilized well using the heterosis breeding. It is very well known that if SCA variance, which is a measure of non-additive genetic variance, is high for characters and also observed heterosis is also high, such crosses can be utilized for commercial exploitation of heterosis. Fourth cross AKRMS 80-1-1-1A × Rb 307–11 showed positive significant SCA effects for grain yield per plant (11.73**) but in this cross high SCA is not associated with heterosis (Sakhare et al. 2015a).

Five male sterile lines were crossed with eight testers in line × tester design and produced 40 hybrids. Studies (Ghorade et al. 2015d) were undertaken with 40 hybrids and 13 parents and widely used standard check CSG-9 at Sorghum Research Unit, Dr. Panjabrao Deshmukh Krishi Vidyapeeth, Akola during *kharif* 2010–2011 to identify the potential lines and testers suitable for development of dual-purpose hybrids in rainy season sorghum based on general combining ability effects. Dual-purpose hybrids have good grain yield along with good fodder yield. Line MS-27A was the best general combiner for yield per plant (6.18**) as well as for fodder yield per plant (15.27**). Among testers, AKR-456 showed significant and desirable GCA effects for grain yield per plant (16.19**) along with fodder yield per plant (10.85**). Another tester, AKR 422, also showed significant positive GCA effects for grain yield per plant (9.44**) along with the positive and significant GCA effects for fodder yield per plant (16.93). Thus line MS-27A along with testers AKR 456 and AKR 422 need to be exploited extensively in development of promising dual-purpose rainy season sorghum hybrids.

11.1.10 Development of Early-Maturing Hybrids for the Post-rainy Season

In post-rainy season sorghum, the early-maturing genotypes escape the terminal drought conditions. The adverse effects of moisture stress on the yield are less in such genotypes in the post-rainy season. Three female lines (AKRMS-80 A, AKRMS-80-1-1A and AKRMS-47A) were studied (Sakhare et al. 2015b) to identify a line suitable for developing high-yielding and early-maturing hybrids in *rabi* sorghum. The line AKRMS 47A was found to be suitable for developing high-yielding and early-maturing hybrids in *rabi* sorghum due to its positive significant GCA effects for grain yield (2.91**) along with negative significant GCA effects for days to 50 % flowering (−1.47**) and days to maturity (−0.23**) Similarly, among the 22 testers, tester RL 5–1 gave highest significant and desirable GCA effects for grain yield per plant (7.95**) along with desirable component characters like days to 50 % flowering (−0.83**) and days to maturity (−1.47**). Similarly, another tester, Rb 397–2, exhibited significant positive GCA effects for grain yield (6.36**) along with negative significant GCA effects for days to 50 % flowering (−1.02**) and days to maturity (−0.3**). Another tester, AKSV 70R, with positive significant GCA effects for grain yield (5.35**) along with negative significant GCA effects for days to 50 % flowering (−0.85**) was also promising. A fourth tester, G 45-3-1-1, with positive significant GCA effects for grain yield (4.70**) along with negative significant GCA effects for days to maturity (−0.45**) was also found to be desirable. This unique combination of potential of a line or tester for grain yield along with early maturity can be exploited in the development of high-yielding and early-maturing hybrids in post-rainy sorghum, which is the need of present day *rabi* sorghum for overcoming terminal drought stress.

11.1.11 Heterosis

Fourteen *kharif*-based sorghum hybrids along with 2 checks and 13 parental lines (5 cytoplasmic male sterility-based lines and 8 restorers) were evaluated to estimate the heterobeltiosis, standard heterosis and mid-parent heterosis (More et al. 2015b). Heterosis was estimated for grain yield per plant, fodder yield per plant, days to 50 % flowering, plant height, third leaf area, panicle length, panicle breadth, 1000-seed weight, ear head weight, chlorophyll content, relative water content, free sugar and protein content. The mean squares in respect of genotypes were significant for all the traits. Four hybrids yielded higher grain weight per plant and were on a par with the highest yielding check, CSH – 14 (90.20 g per plant), and six hybrids gave significantly highest fodder yield over the standard checks. These hybrids also performed well for other agronomical and physiological traits. High and significant heterosis over mid-parent and better parent was observed in crosses PMS 28A × C –43 and PMS 28 A × KR 200 for grain yield per plant and most of the yield components. A total of four crosses exhibited positive and higher heterosis and heterobeltiosis for free sugar content, seven crosses for protein content and seven crosses for chlorophyll content. The cross PMS 28 A × C – 43 exhibited high and significant standard heterosis for grain yield per plant.

11.1.12 Dual-Purpose Sorghum

In post-rainy sorghum, a dual-purpose sorghum hybrid is the one with high grain yield along with high fodder yield. Among the 66 post-rainy sorghum hybrids studied (Sakhare et al. 2015c), 2 hybrids, viz AKRMS 80A × Rb 307–11 and AKRMS 47A × AKSV 70R, exhibited significant standard heterosis for grain yield as well as fodder yield. In the first cross, i.e. AKRMS 80A × Rb 307–11, the standard heterosis for grain yield per plant over the check CSH 19 R was 22.99 % while for fodder yield per plant it was 21.90 %. Besides standard heterosis, this cross also exhibited positive significant SCA effects for grain yield per plant (18.23**) as well as for fodder yield per plant (18.54**). In another cross, AKRMS 47A × AKSV 70R, the standard heterosis for grain yield per plant was 15.76 % while for fodder yield per plant it was 28.32 %. Along with high standard heterosis, this cross also showed positive significant SCA effects for both grain yield per plant (11.99**) and fodder yield per plant (10.79**). In these two crosses, high standard heterosis for grain yield and fodder yield was accompanied with positive and significant SCA effects for grain yield as well as fodder yield. This clearly indicated that these two crosses can be very well exploited using heterosis breeding for development of dual-purpose *rabi* sorghum hybrids. These hybrids need to be tested in multi-location trials on a large scale for their evaluation and commercial exploitation.

11.1.13 High-Yielding Early-Maturing Hybrids

High-yielding and early-maturing hybrids are popular among farmers as these fit well in cropping systems. The field is vacated early, which facilitates timely sowing of the following *rabi* crop. In order to identify such early-maturing and high-yielding *kharif* sorghum hybrids, promising hybrids were sorted out (Ghorade et al. 2015c) based on positive significant standard heterosis for grain yield along with negative significant standard heterosis for either days to 50 % flowering or days to maturity. Negative values for the traits like days to 50 % flowering or days to maturity indicate earliness. In total, three hybrids showed positive significant standard heterosis for grain yield along with negative significant standard heterosis for either days to 50 % flowering or days to maturity. AKMS- 89A × AKR-422 exhibited the highest standard heterosis (20.56 %) over the check CSH-9 for grain yield along with negative significant standard heterosis for days to maturity (−10.69 %). The cross AKMS 30A × AKR-456 with positive significant standard heterosis for grain yield per plant (9.39 %) and negative significant standard heterosis for days to maturity (−6.60**) was also found to be promising. A third cross, AKMS 89A × AKR-456, also showed positive significant standard heterosis for grain yield per plant (4.15 %) along with negative significant standard heterosis for both days to 50 % flowering (−8.13*) and days to maturity (−12.57**). All these cross combinations also showed positive significant SCA effects for grain yield per plant and negative significant SCA effects for either days to 50 % flowering or days to maturity. Thus, these three cross combinations need to be exploited for development of high-yielding and early-maturing hybrids in *kharif* sorghum.

11.1.14 Attributes for an Efficient Breeding Programme

In one study (Ghorade et al. 2015a), days to 50 % flowering showed positive and significant correlation with grain yield per plant ($r = 0.71^{**}$) and exerted positive and direct effect on grain yield (0.86). It showed maximum positive indirect effect on grain yield per plant through plant height (0.17). Plant height showed positive significant correlation with grain yield per plant ($r = 0.399^{*}$) and exhibited positive and direct effect on grain yield (0.394), and the maximum positive indirect effect on grain yield per plant was through days to 50 % flowering (0.381). Panicle breadth showed negative non-significant correlation with grain yield per plant ($r = -0.179$) and it showed negative direct effect on grain yield per plant (−0.354), thereby revealing that the character panicle breadth has no role to play in improving the grain yield. Thousand-seed weight showed positive significant correlation with grain yield per plant ($r = 0.430^{*}$). It also exhibited positive direct effect on grain yield per plant (0.251). It also showed highest positive indirect effect on grain yield through plant height (0.1419). Fodder yield per plant exhibited positive significant correlation with grain yield per plant ($r = 0.576$) but it exhibited highest positive indirect effect on grain yield per plant through panicle breadth (0.084). From the study, it was concluded that the characters days to 50 % flowering, plant height and 1000-seed weight exhibited positive significant correlation with grain yield per plant along with positive direct effect on grain yield per plant. Selection based on these attributes may be helpful in planning an efficient breeding programme for high grain yield in *kharif* sorghum.

11.1.15 Selection for Reduction of Dead Heart

The number of eggs per plant at 14 DAE showed positive but no significant correlation with dead heart count at 28 DAE ($R = 0.30$). It had a negative direct effect (−0.26) on dead heart count at 28 DAE. Number of eggs per plant at 21 DAE showed positive significant correlation with dead heart count at 28 DAE ($r = 0.58^{**}$); it had positive direct effect (0.288) with dead heart count at 28 DAE. Number of eggs per plant at 28 DAE showed positive significant correlation with dead heart count at 28 DAE ($R = 0.432^{*}$). It exhibited positive direct effect (0.032) with dead heart count at 28 DAE. Chlorophyll content index showed negative non-significant correlation with dead heart count at 28 DAE ($r = -0.116$) and had negative direct effect (−0.328) on dead heart count at 28 DAE. Trichome density per plant showed negative but significant correlation with dead heart count at 28 DAE ($r = -0.495^{**}$) and had negative direct effect (−0.401) on dead heart count at 28 DAE. Seedling vigour showed positive and significant correlation with dead heart count at 28 DAE ($r = 0.379$); it had negative direct effect (−0.350) on dead heart count at 28 DAE. Leaf glossiness showed negative but significant correlation with dead heart count at 28 DAE ($r = -0.492^{**}$); it exhibited direct negative effect on dead heart count at 28 DAE (−0.194). Dead heart count at 14 DAE showed positive significant correlation with dead heart count at 28 DAE ($r = 0.646^{**}$) and it registered direct positive effect on dead heart count at 28 DAE (0.419). Dead heart count at 21 DAE showed positive significant correlation with dead heart count at 28 DAE ($r = 0.756^{**}$) and it registered direct positive effect on dead heart count at 28 DAE (0.543). The study revealed that selection based on characters like trichome density and leaf glossiness may be helpful in reducing the dead heart count at 28 DAE, as these two characters exhibited negative significant correlation with dead heart count at 28 DAE along with the negative direct effect on dead heart count at 28 DAE (Ghorade et al. 2015e).

11.2 Pearl Millet

11.2.1 Character Association for Terminal Drought

Correlation studies (Patil and Jadeja 2015a, b) under terminal drought conditions indicated that grain yield per plant was highly correlated with panicle harvest index, drought response index and harvest index, whereas under irrigated conditions it was highly correlated with ear head weight per plant and test weight. Drought response index was correlated to grain yield per plant under drought environments only. Under both conditions, number of productive tillers per plant exhibited positive significant association with maximum number of characters, and ear head girth showed negative significant association with maximum number of characters. Drought-responsive traits had mutual association with each other under terminal drought conditions, but not under irrigated conditions. Hence, while selecting drought-responsive genotype, emphasis should be given to drought stress-responsive traits, viz root length per plant, proline content and drought response index. Thus, correlation results reveal that alleles controlling the grain yield in stress and non-stress environments are partially different and this underlines the need for selection in target environments to improve the performance of genotypes.

11.2.2 Genetic Variability for Micronutrients

In two released and commercial open-pollinated varieties (ICTP 8203 and ICMV 221) of pearl millet studied for intra-population variability using S1 progenies for grain iron (Fe) and zinc (Zn) densities and 1000-grain weight (Kanatti et al. 2015), analysis of variance exhibited highly significant variability for all the traits in both populations. In ICTP 8203, Fe density varied from 44 to 76 mg kg^{-1} and Zn density from 40 to 60 mg kg^{-1}, while in ICMV 221, Fe density varied from 40 to 78 mg kg^{-1} and Zn density from 31 to 52 mg kg^{-1}. Higher broad-sense heritability was observed for both micronutrients in both populations (77 % in ICTP 8203 and 86 % in ICMV 221 for Fe; and 71 % in ICTP 8203 and 72 % in ICMV 221 for Zn). This substantial genetic variability coupled with high heritability implies good prospects for improvement of both populations for these micronutrients. The highly significant and high positive correlation observed between Fe and Zn densities ($r = 0.66$ in ICTP 8203 and $r = 0.72$ in ICMV 221, $p < 0.01$) suggested that both micronutrients can be effectively improved simultaneously. Both micronutrients had non-significant or small positive correlations with 1000-grain weight, implying these can be improved in large-seeded genetic backgrounds.

11.2.3 Line × Tester Analysis

Line × tester analysis was conducted in pearl millet (*Pennisetum glaucum* L.) for estimating the general and specific combining ability effects (Kathale et al. 2015b). Analysis of variance for means revealed significant differences for seven characters. Close agreement between GCA and per se performance of parents was detected for most of the characters studied. Combinations having high per se performance also exhibited high SCA effects and involved at least one good general combining parent. Among females, MS 88004 A was the best general combiner for grain yield and seven other characters. Among restorers, AIB-214 was the best general combiner for grain yield, plant height, ear head girth, ear head length and fodder yield per plant. The combination MS 88004 A × AIB-214 was the best specific combination for grain yield per plant, whereas the combination MS 841 A × IC-1179 was the best specific combination for grain yield per plant, height, total number of tillers per plant and ear head girth, ear head length. The combinations MS 841 A × IC-223-1, MS 863 A × IC-1153, MS863A × IC-14, MS 99111 A × IC 1153 and MS 98222A × R-1665-3 produced significant and desirable SCA effects for most of the traits studied, indicating potential for exploiting hybrid vigour in a breeding programme.

11.2.4 Heterosis

The extent of heterosis present and its effects on yield and yield components in 50 F_1s (five male sterile lines crossed with ten restorers) in pearl millet evaluated during *kharif* 2010

(Kathale et al. 2015c) revealed that analysis of variance for means exhibited significant differences for all the 11 characters studied, viz days to 50 % flowering, days to maturity, plant height (cm), total number of tillers per plant, number of effective tillers per plant, 1000 grain weight (g): number of grains per cm^2, ear head girth (cm), ear head length (cm), grain yield per plant (g) and fodder yield per plant (g). The interaction mean square for parents vs hybrids was significant for seven characters, demonstrating the existence of sufficient variability in parents and hybrids. The magnitude of heterosis and heterobeltosis for all the characters was highly acceptable. Among all the characters, the magnitude of heterosis was highest for fodder yield per plant, being 26.27 % and 21.24 % over mid- and better parent, respectively, in the cross MS 841 A × IC-223-1. This was followed by the character 1000 grain weight (22.67 % and 21.23 %) in the cross MS 863 A × IC-1179 and grain yield per plant (18.34 % and 17.79 %) in the cross for mid-parent and better parent heterosis, respectively.

11.3 Foxtail Millet

Foxtail millet is an important cereal crop grown as a staple food grain in China, India and Japan and for silage and hay in North and South America, Australia and North Africa. It is one of the oldest domesticated crops and is an ideal climate-resilient crop, highly drought tolerant and usually grown where climatic hazards do not permit the cultivation of other cereals.

11.3.1 Morpho-agronomic Descriptors

The genotypic and phenotypic variance, PCV and GCV, heritability and genetic advance for 12 morpho-agronomic descriptors was studied (Hariprasanna et al. 2015a) in 40 foxtail millet genotypes including elite breeding lines and released varieties as checks during the rainy season of 2014. The genotypes showed highly significant differences for all the descriptors studied indicating sufficient genetic variability. High coefficients of variation were observed for flag leaf width, panicle exertion, number of basal tillers, grain and fodder yields per plant and 1000-grain weight. The estimates of PCV and GCV for almost all the parameters except grain and fodder yields were nearly equal. This revealed the existence of immense inherent variability that remained unaltered by environmental conditions. Very high PCV and GCV values were observed for number of basal tillers and grain yield per plant, which indicated the existence of comparatively high variability for these traits and hence could be exploited for improvement of the traits through selection in advanced generations. Estimation of heritability along with GCV gave an idea of expected gain from selection. Heritability (broad-sense) estimates ranged from 0.33 to as high as 0.96. There was high heritability for all the traits studied except fodder yield. This indicated that genotype plays a more important role than environment in determining the phenotype, suggesting the predominance of additive gene action in the inheritance. Very high heritability accompanied by high genetic advance as a percentage mean was observed for peduncle length (0.8 %, 27 %), panicle exertion (0.94 %, 114 %), bristle length (0.86 %, 50 %), plant height (0.84 %, 20 %) and number of tillers (0.96 %, 122 %), and hence these traits could be improved by simple selection methods. The D^2 analysis grouped the genotypes into ten clusters based on the Torcher method with seven clusters having a single member. Four genotypes (GPU 27, KOPFM 15, GPU 28 and KOPFM 18) had significantly higher grain yield than the check varieties RAU 2 and Si A 326. These genotypes can be further evaluated and utilized to develop high-yielding varieties.

11.4 Finger Millet

Finger millet is an important cereal crop among the small millets and third in importance among millets in the country, in terms of area and production after sorghum and pearl millet.

It is cultivated mostly as a rain-fed crop in India for its valued food grains and its adaptability to a wide range of geographical areas and agro-ecological diversity, mostly countries in Africa and Asia. India is the major producer of finger millet in Asia.

11.4.1 Protein and Calcium Content

Finger millet is highly nutritious as its grains contain 65–75 % carbohydrates, 5–8 % protein, 15–20 % dietary fibre and 2.5–3.5 % minerals. Considering its importance in food, adequate information on its nutritious quality is meagre in finger millet. The protein and calcium content was estimated in 48 finger millet genotypes. The estimated protein content among the genotypes was minimum 5.95 g 100 g^{-1} and maximum 9.33 g 100 g^{-1}, and the average was 7.49 g 100 g^{-1}. The estimated calcium content among the genotypes was maximum 324.47 mg 100 g^{-1} and minimum 188.57 mg 100 g^{-1}, with an average value of 235.74 mg 100 g^{-1}. SNVD-11-88 contained maximum protein and SRS-13435 contained minimum protein. The maximum calcium content was in the genotype SNVD-11-15, while it was minimum in the CMLS-01336-SEL-11 genotype. The protein content of 20 genotypes out of 48 was more than the average value, and content of calcium in 25 genotypes was more than the average value. This huge variation in protein and calcium content among the genotypes can be useful in improving the protein and calcium content in other genotypes or varieties. Genotypes with protein and calcium content more than average values can be used in bio-fortification of finger millet (Das et al. 2015b).

11.5 Wheat

11.5.1 Character Association: Bread Wheat

Knowledge of correlation coefficients among various desirable characters is one of the most important aspects for any breeding to elucidate the degree of favourable and unfavourable associations between different useful characters. An investigation undertaken (Diyali et al. 2015) to assess genotypic and phenotypic correlations between yield and its components and also between the contributing components for simultaneous improvement of character through selection on the basis of one of the characters, including path coefficient analysis, showed that yield was positively and significantly correlated with number of spikes per plant, number of spikelets per spike, number of grains per spike, weight of grain per spike, flag leaf area, 1000 grain weight and amylose content at both genotypic and phenotypic levels. Weight of grain per spike, number of tillers per plant, number of spikes per plant and flag leaf area showed positive direct effects along with highly positive significant correlations with grain yield. Therefore, direct selection for any of the characters would be effective for yield improvement in wheat with simultaneous improvement of the rest of the character.

11.5.2 Genetic Variability

Thirty-one genotypes of wheat were evaluated for variability (Bhalerao et al. 2015). High GCV and PCV values were recorded for number of tillers, grains per panicle, yield per plant, yield per hectare and number of grains per spikelet, indicating that these characters are governed by additive gene effects, and directional selection for these traits would be more effective.

Correlation studies revealed positive and significant genotypic and phenotypic correlations for plant height, number of tillers per plant, number of grains per spikelet, number of grains per panicle, yield per plant and harvest index, indicating the importance of these characters for improving grain yield of wheat per hectare. The path coefficient analysis revealed that the characters plant height, number of tillers per plant, grains per panicle, yield per plant, yields per plot, test weight and gluten content had high positive direct effects on grain yield per hectare.

Therefore, while imparting selection in wheat characters, plant height, number of tillers per plant, number of grains per spikelet, number of grains per panicle, yield per plant and harvest index must be given preference.

11.5.3 Wheat Selection Criteria

The yield-contributing characters in 40 genotypes of wheat were studied through range, coefficient of variation (CV), heritability, genetic advance, correlation coefficient and path analysis during 2011 (Muundiyara and Kerkhi 2015), with analysis of variance showing significant differences among all genotypes for all the characters, viz days to 50 % flowering, days to maturity, plant height, number of spikelets per spike, productive tillers per plant, ear length with awn, number of grains per spike, 1000 grain weight, biological yield per plant, grain yield per plant, harvest index and gluten content. A broad range of variation was observed for plant height and biological yield per plant, number of grain per spike and test weight. The phenotypic variance value for most characters was closer than the corresponding genotypic variance value, showing little environment effect on the expression of these characters. The estimated value of broad-sense heritability was found to be between 45 % (productive tillers per plant) and 95 % (gluten content). Heritability values were determined as 92 %, 92 %, 89 %, 85 %, 84 %, 71 %, 68 % and 67 % for plant height, days to maturity, test weight, days to 50 % flowering, ear length with awn, harvest index, number of grains per spike and grain yield per plant, respectively. High heritability coupled with high genetic advance as a percentage over mean was observed for plant height, test weight, number of grain per spike and days to maturity, suggesting that selection for these traits would give good responses. The grain yield per plant exhibited highly significant and positive correlation with biological yield per plant, test weight, productive tillers per plant and number of grains per spike, number of spikelets per spike, spike length with awn and plant height showing significant and positive correlation with harvest index. It was suggested that these characters could be selected for wheat improvement programmes.

11.6 Maize

11.6.1 Agroclimatic Indices

The crop growth response is influenced largely by the microclimate environment in the crop; in particular, the temperature plays a vital role in the potentiality of crop in terms of growth, development and yield. The growth and yield of maize hybrids of different duration with agroclimatic indices was predicted based on growing day degree (GDD): heliothermal units (HTU): phenothermal index (PTI) and also heat use efficiency (HUE) during the 2011 and 2012 rainy season at the Maize Research Centre, Hyderabad. Results revealed that on a pooled basis, significantly higher grain yield (6416 kg ha^{-1}) was obtained with a normal date of sowing (25 June) and was on par with 15 days early and 15 days late (10 June and 10 July) sowings. However, further delay in sowing decreased the grain yield significantly. Among the hybrids the medium- and late-maturity hybrids gave significantly higher grain yields (7351 and 7162 kg ha^{-1}, respectively) due to accumulation of higher GDD and HTU compared to early- and extra early-maturity hybrids (5645 and 5231 kg ha^{-1}, respectively). The phenothermal index (PTI) for consecutive phenophases indicated that the PTI gradually decreased from emergence to maturity in all the dates of sowing during both the years and the highest PTI was observed with early sowing dates (up to 10 July) in comparison with later dates of sowing. Higher HUE was observed with the early sowing date than with the later sowing dates. This trend was observed with all the hybrids of different maturity groups. However, the early-maturity hybrids were found to be relatively stable with respect to yield compared to the medium- and late-maturity hybrids, as the HTU did not vary much across the dates of sowing with their short maturity period. The study indicated that the sowing time has a significant influence on the yield of maize under rain-fed conditions. It may therefore be necessary to manipulate sowing of hybrids of different maturity to avoid drought during the critical periods of crop development; furthermore, it would allow farmers to grow the variety that best meets their needs (Sreelatha et al. 2015).

11.6.2 Moisture Stress in Maize

The seasonal analysis tool DSSAT v 4.6 was used to assess the risk of soil moisture stress on grain yield of rain-fed maize using 30 years of weather data from 1981 to 2010 in the southern Telangana agroclimatic zone of Telangana (Sreenivas et al. 2015a). A higher mean grain yield was observed when the crop was sown on 20 June and grain yield was comparable with crops sown on 30 June and 10 July. Significantly lower grain yield was noticed under late-sown conditions. Scenario analysis showed that less risk was associated with the 20 June sown crop as the probability of exceeding 50 % moisture stress was very low during silking as well as the grain-filling stage. Beyond 10 July there was a more than 20 % chance of exceeding 50 % moisture stress during these critical phases of the crop. Based on strategic analysis, the moisture stress impact was reduced by providing either one critical irrigation at the silking stage (55 DAS) or two irrigations, one at silking and another at the grain-filling stage (75 DAS), increasing the yield by 13.8 % and 19.1 % with decreases in the standard deviation from 2054 to 1360 and 1162, respectively, over the rain-fed conditions.

11.6.3 Genotypic Variability

The genetic variability, broad-sense heritability and genetic advance were estimated in 60 inbreds lines of maize during *kharif* 2013 (Sandeep et al. 2015) for 12 quantitative characters, viz days to 50 % tasselling, days to 50 % silking, days to maturity, plant height, ear height, ear length, ear girth, number of kernel rows per ear, number of kernels per row, 100-seed weight, grain yield per plant and shelling percentage, exhibiting highly significant mean sum of squares values for all the characters indicating greater diversity among the genotypes. The GCVs for all the characters studied were lower than the PCVs, indicating the effect of the environment. High GCV and PCV values were observed for grain yield per plant and ear height. The difference between the estimates of PCV and GCV were low for all the characters except shelling percentage, thereby a lesser role was played by the environment in the expression of this character. The difference between the estimates of PCV and GCV was high for shelling percentage, indicating the significant role played by the environment in the expression of these characters. All the characters showed high estimates of heritability except shelling percentage. High heritability coupled with high genetic advance as a percentage of the mean was observed for grain yield per plant, ear height, plant height, number of kernels per row, 100 seed weight and ear length, indicating the role of additive genes in governing the inheritance of these traits, which could be improved through simple selection.

11.6.4 Combining Ability

The combining ability and gene action for grain yield and yield components in maize was estimated in 60 crosses of maize (Pole et al. 2015) obtained by crossing of six lines and ten testers. GPM-210 showed good GCA effects for grain yield, days to 50 % tasselling, days to 50 % silking, days to maturity, plant height, and number of grains per row and fodder yield. EC-558714 and GPM-213 were the best combiners for grain yield and yield components. Among the testers, EC-558719 exhibited good GCA effects for days to 50 % tasselling, days to maturity, plant height, ear length, ear girth, kernel rows and number of grains per row, 100 grain weight, fodder yield and grain yield. The testers IC-558620, IC-541068 and EC-558715 were the best combiners for one or other characters including grain yield. The hybrids IC-32118 × IC-558620, GPM-35 × IC-541067, GPM-154 × GPM-405, GPM-210 × EC-558711 and GPM-213 × EC-558719 showed desirable SCA effects for most of the traits studied, revealing potential for exploiting hybrid vigour in a breeding programme. The estimates of variance indicated predominance of additive gene action for plant height, number of grains per row, 100 grain weight and grain yield.

In maize, the parent GPM-19 was found to be a good general combiner for eight characters, while IC-541068 and IC-541060 were for seven characters and IC-541066 were for six characters. The above four parents could be utilized in a breeding programme of maize. There was close agreement between pre-SCA performance and GCA as well as SCA effects for most of the characters. Observation of various characters indicated that the cross showing high heterosis and high SCA effects had high per se performance and they involved at least one high combining parent (Kathale et al. 2015d).

Seven inbred lines were crossed in full diallel fashion. The resulting 42 crosses along with seven parents were employed for studying the variance. Analysis of variance revealed significant differences for all the 11 characters studied. The parent versus hybrid interaction mean square was significant for all characters, indicating variability in them. The highest magnitude of heterosis over mid-parent EC-558715 × IC-541060 (53.23 %) and better parent EC-558715 × IC-541060 (50.91 %) was observed for grain yield per plant. The cross GPM-230 × IC-541068 (8.296 %) showed the highest magnitude of standard heterosis, followed by IC-541058 × GPM-230 (10.32 %) (Kathale et al. 2015a).

The correlation and path analysis of yield and yield-contributing characters in seven inbred lines crossed in a diallel mating design (method II, model I) resulting in 21 crosses was evaluated during *rabi* 2012–2013 at the Agricultural Research Station, Karimnagar, Telangana, for studying the association of yield components with popping ability and path analysis of these characters on yield and popping quality for enabling the selection of inbred lines for the ultimate usage in popcorn hybrid seed production. Correlation studies revealed that grain yield per plant was positively associated with plant height, ear height and ear length and ear girth, number of kernel rows per ear, number of kernels per row, 100-kernel weight and popping expansion ratio. The path coefficient analysis at genotypic level revealed that the character ear height exhibited the largest direct effect on grain yield per plant followed by days to 50 % silking, number of kernels per row, number of kernel rows per ear and ear length (Sridhar et al. 2015).

Seven promising inbred lines were used for making crosses in full diallel fashion to produce 42 crosses. These were evaluated along with seven parent and three standard checks during *kharif* 2010 (Patil et al. 2015a) for 11 characters, viz days to 50 % tasselling, days to 50 % silking, days to maturity, plant height, ear length, ear girth, number of kernel rows, number of grains per row, 100 grain weight, fodder yield (g) and grain yield (g). Analysis of variance for means revealed significant differences for all the 11 characters studied. The parent versus hybrid interaction mean squares were also significant for all the characters, indicating variability in parents and hybrids. The highest magnitude of heterosis over mid-parent (53.23 %) EC-558715 × IC-541060 and over better parent EC-558715 × IC-541060 (50.91 %) was observed for grain yield. There was close agreement between per se performance and GCA as well as SCA effects for most of the characters. Observation of various characters indicated that the cross showing high heterosis and high SCA effects had high per se performance and involved at least one high combining parent.

Studies were undertaken on heterosis and combining ability analysis in maize (*Zea mays* L.) to assess the possibility of estimating the extent of heterosis for yield and yield-contributing characters in 30 crosses to identify superior parents and crosses. Variability existed among the genotypes for all the characters studied, i.e. days to 50 % tasselling, days to 50 % silking, days to maturity, plant height, cob length, cob girth, number of grains per cob, 100 grain weight and grain yield per plant (g). The crosses NM-0913 × NM-0973, NM-099 × NM-0973 and NM-0984 × NM-0914 had high mean performance for yield per plant and other yield-contributing traits. The mean squares due to the interaction effects of parents versus crosses were found to be significant for all the characters, indicating the choice of exploitation of heterosis. The mean squares for lines were significant for days to 50 % tasselling, days to 50 % silking, plant height, cob length, cob girth, number of grains per cob, 100 grain weight and grain yield per plant, while testers were significant for plant height, cob girth, number of grains per cob, 100-grain weight and grain yield per plant. The higher magnitude of variance in line × tester interaction suggested the presence of greater variability among the crosses than among the parents. The GCA estimates of lines and testers emphasized the importance of five parents (NM-0935, NM-0913, NM-099, NM-0969 and NM-0973) for their use as good general combiners for yield and important yield-contributing characters. These parents could be utilized for the development of either a synthetic variety or an elite breeding population. Out of 30 crosses studied, 4 crosses (namely NM-099 × NM-0973, NM-0984 × NM-0914, NM-0969 × NM-0967 and NM-0981 × NM-0914) were identified as the highest potential crosses for hybrid production (Patil et al. 2015b).

11.6.5 Genetic Divergence

Thirty-three genotypes were evaluated for genetic divergence to identify the likely desirable and potential parents for a maize breeding programme aimed at yield and earliness improvement. Genetic divergence was determined for nine characters: days to 50 % tasselling, days to 50 % silking, days to maturity, plant height (cm), cob length, cob girth, number of grains per cob, 100-grain weight (g) and grain yield. Mahalanobis generalized distance values for characters were used in this study for computing genetic divergence. The analysis of dispersion for nine correlated variables, using Wilk's criterion, revealed highly significant differences between genotypes for aggregate of the nine characters. The 33 genotypes were grouped into six clusters by Tocher's method. The maximum inter-cluster distance was recorded between cluster IV and cluster VI (62.538). The canonical analysis indicated that number of grains per cob, days to 50 % tasselling, grain yield^{-1}, 100-grain weight, cob length and days to 50 % silking were the important sources of variation in vector I. In vector II, plant height, days to maturity, days to 50 % tasselling and number of grains per cob were important sources of variation. In vector III, cob length, number of grains per cob and days to 50 % tasselling were important. In vector IV, 100 grain weight, cob girth, days to 50 % tasselling and plant height were important. In vector V, days to 50 % tasselling, number of grains per cob and days to 50 % silking were important. The genotypes belonging to a distant cluster and exhibiting high performance in the desirable direction for days to 50 % tasselling, number of grains per cob, days to 50 % silking and plant height were identified as the potential parents for the hybridization programme. Genotypes NM-098, NM-093, NM-0918, NM-0991, NM-0911, GP-87, NM-0923, NM-0924, NM-0955, NM-0912, NM-0921, NM-097, NM-0915(W): NM-0959, GP-93, NM-0992 and GP-92 were identified as diverse parents to be crossed with either PKVM-Shatak or Maharaja (Solanke et al. 2015).

11.7 Rice

11.7.1 Compartmentation of Photosynthetic Enzymes

The high photosynthetic capacity of C_4 plants is due to their exclusive mode of CO_2 incorporation, featuring stringent compartmentation of photosynthetic enzymes into two distinctive cell types, mesophyll and bundle sheath. Firstly, CO_2 assimilation is carried out in mesophyll cells. The chief carboxylating enzyme, phosphenol pyruvate carboxylase (PEPC), in concert with carbonic anhydrase, is vital to create rapid equilibrium between CO_2 and HCO_3. This step is responsible for the hydration and fixation of CO_2 to produce the C_4 acid oxaloacetate. C_3 photosynthesis is an inefficient process, because the enzyme that lies at the heart of the Calvin cycle, ribulose-1,5-bisphosphate carboxylase-oxygenase (RuBisCO) is itself a very inefficient enzyme. The oxygenase activity of RuBisCO is an unavoidable side reaction, which is a consequence of its reaction mechanism. The product of oxygenation, glycolate 2-P, has to be retrieved by photorespiration, a process that results in the loss of a quarter of the carbon that was originally present in glycolate 2-P. Photorespiration therefore reduces carbon gain to plants and increases the carbon level in the atmosphere, which leads to increases in global warming. Purely in terms of the carbon economy, there is, therefore, a strong selection pressure on plants to reduce the rate of photorespiration so as to increase carbon gain, but it also improves water and nitrogen-use efficiency. So, it is necessary to manipulate the C_3 pathways into C_4 mode. Possibilities for the manipulation of plants to decrease the amount of photorespiration include the introduction of improved RuBisCO from other species, reconfiguring photorespiration, or introducing carbon-concentrating mechanisms, such as inorganic carbon transporters, carboxysomes or pyrenoids, or engineering a full C_4 Kranz pathway using the existing evolutionary progression in C_3–C_4 intermediates as a blueprint. One such attempt is possible in rice, which is belongs to the Poaceae family, as the members of Poaceae, like foxtail millet, sugarcane, maize, etc., have a Kranz type of leaf anatomy (Gangurde et al. 2015).

11.7.2 Multiple Gall Midge Biotype-Resistant Rice Genotype

JGL 13595 (MTU 4870 × JGL 418) is a short-duration multiple gall midge biotype-resistant high-yielding fine-grain rice genotype developed as a substitute for BPT 5204 (a popular cultivated variety in India). This culture was developed by a pedigree selection method of breeding by evaluation in station trials, MLT and minikit testing in farmers' fields. It is suitable for sowing from 1 June to 31 July. The seed-to-seed duration during the rainy season and post-rainy season is 120–125 days. It has an erect plant type and attains a height of 95–105 cm with medium tillering ability. It has a medium slender grain type nearer to BPT 5204 with high yield with a non-lodging habit and multiple gall midge biotype resistance. The parents of this variety are MTU 4870 (a long-duration high-yielding BPH-resistant culture with a medium slender grain) and JGL 418 (a medium-duration culture with high yield and resistance to gall midge). This new culture is readily accepted by farmers due to its high yield and resistance to multiple gall midge biotypes, and is accepted by traders as the grain type is nearer to BPT 5204 (Raju et al. 2015).

11.7.3 Aromatic Rice

The association between yield traits was worked out in the F_1 generation in 28 crosses of aromatic rice during *kharif* 2012 to study the nature and extent of correlation among yield and yield-attributing characters. Grain yield per plant was positively and significantly associated with plant height, panicle length and panicle weight, indicating the importance of these traits as selection criteria in yield improvement programmes in aromatic rice (Krishna et al. 2015).

11.7.4 Genotypic Variability

Field experiments were conducted in 31 rice genotypes during *kharif* 2012 and *kharif* 2013 at the Agricultural Research Station, Kampasagar, Telangana State in South India, to estimate the genetic variability parameters for yield and yield-contributing characters. Pooled analysis over the environments, genotype and environment interactions were significant for traits like days to 50 % flowering, plant height, productive tillers per plant, filled grains per panicle, 1000 grain weight and grain yield, implying differential behaviour of genotypes under the two environments for these characters. The estimates of the PCV were slightly higher than those of the GCV for all the traits studied across the seasons. Slight differences were observed for all the characters with regard to variability coefficients, heritability and genetic advance percentage of the mean in different environments. Number of grains per panicle, number of productive tillers per plant, 1000 grain weight and grain yield were recorded as having moderate to high GCV, PCV, heritability and genetic advance percentage of the mean. These characters could be transmitted to the progeny and phenotypic selection based on these characters would be effective (Bhadru et al. 2015).

11.7.5 Sources of Resistance to Brown Plant Hopper

Host plant resistance in paddy to brown plant hopper *Nilaparvata lugens* was undertaken by screening the germplasm accessions using a standard seed box technique. Among 400 germplasm accessions, four accessions (IC-578672, 578151, 464186 and 463837) were found to be resistant and 17 accessions (IC-577478, 578145, 463851, 578665, 577482, 463828, 465106, 577741, 578144, 577663, 465109, 466428, 463887, 578137, 464128, 578017, 578413) were found to be moderately resistant. The mechanisms of resistance of selected rice germplasm accessions were examined using an antixenosis mechanism judged by honeydew production by BPH and antibiosis mechanism judged by nymphal survival, ovicidal test and gain in body weight of BPH. Tolerance was judged by days to wilting of BPH-infested plants and analysed as the biochemical aspects for resistance, viz estimation of phenols, reducing sugars, ascorbic acid, NPK and proteins. The results revealed that the resistant and moderately resistant varieties showed low amounts of honeydew excretion, lower numbers of nymphal survival, higher numbers of unhatched eggs and a greater time to wilting. Survival percentage of nymphs was lower in resistant varieties and they took the maximum time to wilt in comparison with susceptible check TN1. Rice varieties with lower quantities of honeydew excreted usually had lower survival rates (Ramulamma et al. 2015).

11.7.6 Genetic Divergence

Genetic divergence among genotypes plays an important role in the selection of parents having wider variability for different characters and ultimately for rational use of genetic resources. The landraces maintained by farmers are endowed with tremendous genetic variability and diversity, as they are not subjected to subtle selection over a long period.

The genetic diversity was estimated among 70 landraces of rice using Mahalanobis D^2 analysis for 16 agro-morphological and grain quality traits. The mean data after computing for each character were subjected to standard methods of analyses of variance and diversity analysis. Based on the relative magnitude of D^2 values, the experimental material was distributed into nine clusters by using Tocher's method. Out of nine clusters, cluster I was the largest, comprising 50 landraces, followed by cluster III with eight genotypes. The clustering pattern revealed that the geographical diversity and genetic diversity were not related. Total number of grains per panicle, decorticated grain length, 100-grain weight, amylose content, time of heading and stem length together contributed the maximum (94.16 %) to the total genetic divergence. A wide range of variation was found in cluster means for all the traits studied. The promising landraces from these clusters with high mean values for different traits may be directly used for adaptation or may be used as parents in future hybridization. Maximum inter-cluster distance was exhibited between clusters VIII and IX, followed by clusters V and IX. The greater the distance between two clusters, the wider the genetic diversity among the genotypes of these clusters. Hence, hybridization between the landraces from these clusters would produce high heterotic recombinants (Umarani et al. 2015).

A set of 76 rice genotypes were subjected to Mahalanobis D^2 analysis to assess genetic diversity among them. All these genotypes were grouped into nine clusters with maximum inter-cluster distance between cluster VI and cluster IX (5805.13) and minimum inter-cluster distance was observed between cluster I and cluster II

(374.05). Among the ten characters studied, days to 50 % flowering contributed the maximum towards genetic divergence (41.58 %) followed by number of unfilled grains per panicle (12.49 %) and number of filled grains per panicle (11.16 %) (Srinivas et al. 2015).

11.7.7 Line x Tester Analysis

Line × tester analysis using a set of four females (lines) and ten males (testers) was carried out in 40 hybrids to estimate the extent of heterosis and combining ability for yield and its components in rice (*Oryza sativa* L.) (Patil et al. 2015d). Analysis of variance for all the characters revealed that parents and hybrids showed significant differences for all the characters, indicating a considerable amount of genetic variability in the material studied. Parent versus hybrid comparison was found to be significant for all the characters except panicle length, plant height and grain yield per plant, indicating substantial amount of heterosis among hybrids. Among the parents, NAUR-1, Pusa Sughandh-5 and Indrayani were the best-performing parents for grain yield per plant and some of its related attributes. High magnitude of heterobeltiosis was observed for test weight, grain yield per plant, grains per panicle, panicle length, days to 50 % flowering and 1000-grain weight. Crosses NAUR-1 × Indrayani, NAUR-1 × Pusa Basmati-370 and NVSR-6029 × Pusa Sughandha-5 were found to be most heterotic for grain yield per plant. Simultaneous increase in grains per panicle and panicles per plant had positive effects towards higher grain yield. The per se performance of hybrids were, in general, related to the heterotic response in the majority of characters. This indicated that the selection of crosses on the basis of per se performance or heterotic response would be equally important. Combining ability analysis revealed that both GCA and SCA variances were important for inheritance of various traits. However, SCA variances were higher than GCA variances for all the characters, which indicated that non-additive gene action was predominant in the expression of the traits. The estimation of GCA effects for parents revealed that female NAUR-1 was observed to be a good general combiner for yield contributing traits, whereas in males, Pusa Sughandh-5, Indrayani and Pusa Basmati-370 were good general combiners for yield and yield contributing characters. The best specific cross combinations were average × average, average × average and poor × good, respectively, for grain yield per plant.

An experiment consisting of 15 lines, 3 testers and their 45 crosses on heterosis was conducted at a research farm of Birsa Agriculture University, Kanke, Ranchi, Jharkhand during the *kharif* season, showing that the cross combination BAU-274-92 × IR-36 gave significantly positive heterosis to the tune of 31.39 %, followed by BR-8 × IR-36 and BR-8 × BD-202 (27 %) and BAU-211-90 × IR-36, BAU-211 × BD-202 and BAU-269-92 × IR-36 (25.83, 24.67 and 25.97 %, respectively) for grain yield per plant. These crosses also showed positive and significant SCA effects for the character. The GCA status of the parents involved revealed high × high, high × low and low × low combinations, meaning involvement of both additive and non-additive gene effects. The combinations can further be improved through simple selection procedures (additive × additive) or after advancing the generations through transgression effects (Madhuri et al. 2015a).

11.8 Cowpea

Legumes are considered to be an important component of a subsistence cropping system because of their ability to form nodules in symbiotic association with *Rhizobium* bacteria, which convert atmospheric nitrogen (N_2) into an assimilable form of ammonia, to add a substantial amount of organic matter to the soil and to grow better than many other crops with low inputs under harsh climatic and edaphic conditions. Cowpea is one such legume, with 12 characters related to nodulation investigated at the Instructional Farm UBKV, Pundibari, West Bengal, in 2014, showing that the indigenous breeding lines had medium-sized spherical nodules distributed mostly on tap roots and secondary roots. Highest GCV and PCV were observed for nodule dry weight and nodule fresh weight. High heritability coupled with high genetic advance was recorded in nodule dry weight followed by root dry weight and nodule fresh weight. Seed yield per plant was found to be strongly and positively correlated with nodules per plant, nodule size and nodule fresh weight. The correlations between nodules per plant and nodule fresh weight was highly significant and positive. Highest direct effect on seed yield per plant was exerted by shoot fresh weight followed by nodules per plant. Thus, the native rhizobia, when present in the soil rhizosphere, could be exploited by selection of correct genotypes. An avenue for the development of appropriate new breeding lines or varieties through breeding, considering biological nitrogen fixation as one of the essential components for improving the seed yield, could be provided (Das and Sarkar 2015).

11.9 Black Gram

Black gram (*Vigna mungo* L. Hepper) is an important food legume with high nutritive value and consists of up to 26 % protein. Lack of suitable genotypes with adaptation to local environmental conditions is one of the factors affecting its production. Efforts to genetically improve the crop productivity of local cultivars involve identification of important morphological and yield parameters followed by development of advanced breeding lines. It is a thermosensitive crop. Eighteen genotypes of black gram were evaluated (Yerra et al. 2015) for the effect of environmental conditions on yield. The maximum indirect positive effect on seed yield was exhibited by number of branches per plant through number of effective capitula per plant (0.499), followed by days to 50 % flowering (0.474). Direct negative influences were observed for days to 50 % flowering (−0.425) and oil content (%) (−0.055), indicating the importance of depict selection for identification of a genotype tolerant of high temperature. The temperatures during the crop growth period were more than 35 °C and the reproductive phase was exposed to high temperature of 41.7 °C. High temperature affected pod setting and seed filling and influenced the seed yield. The seed yield of the genotypes ranged from 1.36 to 6.73 g per plant with an average of 3.50 g per plant. Among 18 genotypes, the highest seed yield was recorded with PU-19 (6.73 g per plant) and it also produced highest number of clusters (25), pods (40) and seeds (161). It was very pertinent that seed yield during the season was directly influenced by seed number rather than cluster number or pod number. These findings clearly indicated that the seed set and seed filling in black gram crop are crucial for realizing better yield at high temperatures.

11.9.1 Mutagens in Black Gram

Seeds of black gram [*Vingo mungo* (L.) Hepper] variety VBN 4 and ACM 07002 were exposed to gamma ray doses at 200, 250, 300 and 350 Gy and EMS (ethyl methane sulphonate) doses at 20, 30 and 40 mM. In variety VBN 4, a total of 40 mutants were observed in the gamma ray-treated plants and 17 mutants were observed in the EMS-treated plants. A greater number of mutants was observed with gamma rays. Among the four doses of gamma ray treatment and three doses of EMS, the dose 350 Gy in gamma ray mutagen and 40 mM in EMS mutagen registered the highest frequency. In ACM 07002, a total of 28 mutants were observed in the gamma ray-treated plants and 16 mutants were observed in the EMS-treated plants. The number of mutants was higher with gamma rays than with EMS. Among the four doses of gamma ray treatment and three doses of EMS, the dose 300 Gy in gamma ray mutagen and 40 mM in EMS mutagen registered the highest frequency (Ramchander et al. 2015).

The magnitude of genetic variability and correlation coefficient among different yield components in mung bean was assessed in 17 genotypes (Sayyad et al. 2015). Significant variations were observed among the genotypes for all the characters. The highest genetic coefficient of variation was observed for number of pods per plants, followed by yield per plant, number of primary branches and plant height The yield per plant showed highly significant and positive associations with plant height, number of primary branches, number of secondary branches and number of pod per plant, 100 seed weight and days to maturity. The number of pods per plant had a high positive direct effect on yield per plant, followed by days to maturity, number of primary branches, number of seeds per pod and plant height. Therefore, days to maturity, number of primary branches, number of secondary branches, number of pods per plant and 100 seed weight were identified to be the important characters that could be used in selection for yield of mung bean.

11.10 Soybean

11.10.1 Genetic Variability

Genetic parameters of variability, correlation and path coefficients were studied in 30 genotypes along with 3 checks of soybean for 9 characters (Bodkhe et al. 2015). Analysis of variance revealed the presence of variability for number of pods per plant, plant height, days to maturity, seed yield per plant, days to 50 % flowering, days to first flowering, 100 seed weight, oil content and number of branches per plant. The characters number of pods per plant, plant height, number of branches per plant and seed yield per plant showed higher GCV with high heritability estimates and genetic advance. Significant and positive correlations were found between seed yield and plant height, number of pods per plant, days to 50 % flowering and days to maturity, and the highest association was between number of pods per plant and seed yield. Days to first flowering followed by 100 seed weight, plant height and number of branches per plant exhibited the highest direct effects on seed yield. The study revealed that it would be rewarding to lay emphasis in a selection programme of soybean for characters such as plant height, number of pods per plant, days to 50 % flowering and days to maturity, number of pods per plant, early flowering, tallness, number of branches per plant and 100-seed weight.

11.10.2 Resistance to Mechanical Damage

Soybean viability and vigour are affected due to the mechanical damage caused to seeds during processing. Soybean seed germplasm was screened for resistance to mechanical damage (Shelar et al. 2015) by the use of a pendulum test device that struck the seeds from heights of 11.00, 13.00 and 14.50 cm. The percentage of cracked and split seeds increased with an increase in the pendulum height, whereas the percentage of sound seed was decreased as the pendulum height was increased, irrespective of soybean line. Among the 256 soybean lines screened, the lines DS-59, DS-61, DS-108, DS-138, DS-139, DS-154, DS-94, DS-95, DS-120, DS-130 and DS-143 were found to be resistant to mechanical damage whereas 48 lines were susceptible to mechanical damage.

11.11 Pea

Quality parameters were evaluated in 45 F_1s and 45 F_2s derived from ten parents in diallel in table pea (Singh et al. 2015). The results showed high estimates of heritability in narrow sense for number of developed ovules per pod and shelling percentage based on F_2 generation, while green pod yield per plant, harvest index and protein content showed high heritability estimates based on both the generations. The genetic advance was high only for green pod yield per plant in both the generations, while it was moderate for remaining characters at $K = 2.06$, which indicated that both additive and non-additive genes were effective for controlling these traits, while yield was under control of additive genes. An advancement of 53 g based on F_1 and 46 g based on F_2 in green pod yield per plant per cycle of selection was observed. It was about 10 % for harvested index and 8 % for shelling percentage at the same intensity while protein content showed about 4 % from a single cycle of selection. The probable genetic gain was high for all the characters. Protein content showed positively significant correlation with number of developed ovules per pod and harvest index. A simple pedigree method followed by recurrent selection may be effective for improving these quality parameters.

11.12 Green Gram

Characterization of 84 germplasm accessions of green gram (Sowmya et al. 2015b) for morphological characters revealed the presence of variability among these accessions. Based on plant growth pattern, 9 were erect, 11 were spreading and the remaining accessions were semi-spreading type, while on the basis of plant habit pattern, only 8 had determinate growth and the remaining were indeterminate type. Based on leaf size at the fifth node of the plant, 11 had small-sized leaves, 39 had medium-sized leaves and remaining genotypes possessed large-sized leaves. Therefore, combination of these qualitative characters in association with pod and seed (both qualitative and quantitative) characters could be used to differentiate the existing genotypes from one another, and superior genotypes could be further used in breeding programmes.

11.13 Lentil

Lentil (*Lens culinaris* Medik) is among the oldest domesticated crops of mankind, which with time has become important as a food, feed and cropping system in many countries of the world. In India, wilt and rust are the most devastating diseases, which together cause yield losses up to 80 %. To date, no cultivar or land races of *L. culinaris* have been identified possessing resistance to both the diseases. Some of the accessions of wild species, viz *Lens orientalis*, *Lens ervoides* and *Lens nigricans*, possess resistance to either or both the diseases. During the post-rainy season 2012–2013, accessions of wild species were evaluated for their resistance or tolerance against wilt and rust. In the same year, interspecific crosses comprising seven good agronomic bases of domesticated lentil were made with three accessions of *Lens orientalis*. Since *L. culinaris* and *L. orientalis* are the primary gene pool, the crosses were successful between the two and 50–65 % seed setting with viable hybrid seeds was obtained. The same good agronomic bases of domesticated lentils were also crossed with three accessions of *Lens ervoides* (secondary gene pool). In these cases initially after pollination, the embryo development started but at a later stage, the hybrid embryo started aborting. As soon as zygote abortion was observed, growth hormone GA_3 (10 ppm) was applied, which resulted in checking of embryo abortion to some extent. The cultivars of domesticated lentil were also crossed with three accessions of wild species *Lens nigricans*, a secondary gene pool. After pollination, the hybrid embryo collapsed in all the cases, with the result that no seed setting in any of the crosses was obtained. The crossed seed of interspecific crosses *L. culinaris* × *L. orientalis* and *L. culinaris* × *L. ervoides,* when in a wilt sick plot along with susceptible check and in a normal field along with susceptible check for rust, has shown resistance to wilt as well as rust. These interspecific crosses were back crossed to their respective recurrent parent and sufficient numbers of crossed seeds were obtained. The BC_1 generations were grown in a wilt sick nursery, and the resistant plants were picked up to identify genotypes possessing high yield along with resistance to wilt and rust (Katiyar et al. 2015).

11.14 Pigeon Pea

The response of red gram varieties Asha, PRG-158, Maruthi, Laxmi and MRG-1004 to planting geometry under square planting revealed that square planting of red gram at 100 × 100 cm recorded the highest seed yield (2338 kg ha^{-1}) at 90 × 90 cm (2120 kg ha^{-1}) (Rani et al. 2015a).

11.14.1 CMS-Based Pigeon Pea Hybrids

CMS-based pigeon pea hybrids (16 crosses) obtained from crosses of two females (lines), viz AKCMS-92A and AKCMS-87A, and eight males (testers), viz AKPR-364, AKPR-324, AKPR-372, AKPR-303, AKPR-057, AKPR-359, AKPR-215 and AKPR-277, along with the standard checks PKV-TARA and ASHA, were evaluated (Pawar et al. 2015) to estimate the per se performance, extent of useful heterosis, average heterosis, heterobeltiosis and combining ability effects to find out superior cross combinations for their further exploitation, showing that the mean squares due to genotypes were highly significant for all the traits studied. This indicated the presence of substantial genetic variability among genotypes for all the traits studied. The highest significant useful heterosis for grain yield per plant was recorded in AKCMS-87A × AKPR-324 (27.63 % over check PKV- TARA and 37.39 % over check ASHA). The highest significant average heterosis was observed in cross AKCMS-87A × AKPR-303 (97.95 %). The same cross recorded the highest significant heterobeltiosis (87.51 %) for grain yield per plant. Among female parents, AKCMS-87A recorded a significant GCA effect for maximum of the three characters number of clusters, number of pods and grain yield per plant; also the same female parent, AKCMS-87A, was found to possess highest GCA effect for days to 50 % flowering, days to maturity, number of branches, number of seeds per pod and 100 seed weight. The male parent AKPR-324 was found to possess the highest significant GCA effect for plant height, number of branches, number of clusters, number of pods, number of seeds per pods, 100-seed weight and grain yield per plant. The male parent AKPR-372 recorded the highest significant GCA effect for days to 50 % flowering and days to maturity. The cross AKCMS-87A × AKPR-324 (3.05) recorded the highest significant positive SCA effect for grain yield per plant, followed by AKCMS-87A × AKPR-359 (2.88). The cross AKCMS-87A × AKPR-324 depicted high mean performance (38.62), high magnitude of useful heterosis (27.62 % over check PKV-TARA and 37.39 % over check ASHA) and positive SCA effect, and both parents involved revealed high GCA effects. This cross could be successfully utilized to obtain superior segregants in further segregating generations. Another cross, AKCMS-87A × AKPR-359, revealed high mean performance (36.17 g), high magnitude of useful heterosis (19.53 % over check PKV-TARA and 28.67 % over check ASHA) and positive SCA effect, and both parent involved revealed high GCA effects. One of the parents (AKPR-372), though a low combiner for grain yield, was a good combiner for days to 50 % flowering and days to maturity. This cross may be employed to exploit a non-additive component along with high heterotic response. However, the performance of this cross has to be evaluated in large-scale trials.

11.15 Lathyrus

11.15.1 Genetic Divergence

Fifty-seven genotypes were evaluated for genetic divergence to identify potential parents for a *Lathyrus* breeding programme aimed at yield and earliness improvement for eight characters, i.e. days to 50 % flowering, days to maturity, plant height (cm), number of primary branches, seeds per pod, pods per plant, 100 seed weight (g) and grain yield. The genetic divergence was computed by Mahalanobis generalized distance for eight characters. The analysis of dispersion for eight correlated variables, using Wilk's criterion, revealed a highly significant difference between genotypes for aggregate of the eight characters. The 57 genotypes were grouped into eight clusters by Tocher's method. The maximum inter-cluster distance was observed between cluster IV and cluster VIII (110.67). The canonical analysis revealed that the plant height at maturity, pods per plant, yield per plant, days to 50 % flowering, seeds per pod, number of primary branches and days to maturity were significant and important sources of variation in vector I; in vector II, primary branches, days to maturity and days to

50 % flowering were important sources of variation. In vector III, days to 50 % flowering, days to maturity, number of seeds per pod and 100 seed weight were important sources of variation. The genotypes belonging to a distant cluster and exhibiting high performance in the desirable direction for plant height, yield per plant, days to 50 % flowering and pods per plant were identified as the potential parents for a hybridization programme. Twenty genotypes, viz L-3, L-31, L-33, L-25, L-32, JRL-16, RLK-279, L-37, RLK-1045, L-44, L-14, L-11, L-07, L-08, RLK-240, L-39, L-05, RLK-602, L-16 and BioR-208, were identified as potential and diverse parents for use in a crossing programme (Khillari et al. 2015).

11.16 Breeding for Stress Tolerance

11.16.1 Garden Pea

The performance of eight varieties of garden pea studied showed that all the varieties had considerable variation in their performance for most parameters. Better yield parameters in terms of number of green pod per plant, green pod weight, green pod length, green pod yield per plant and yield per hectare were noticed in all varieties (Khichi et al. 2015).

11.16.2 Sunflower

The correlation studies of seed and seedling characters with yield of eight sunflower hybrids were studied at the Oilseeds Research Unit, Dr. PDKV, Akola, and at the Seed Testing Research Unit, Dr. PDKV, Akola, during *kharif* 2012. The character association revealed that the seed characters kernel to hull ratio (0.986), seed kernel percentage (0.984), volume weight (0.783) and 100-seed weight (0.741) showed highly significant positive correlations with yield per plant, indicating that yield of sunflower hybrids could be increased by increase in these seed characters. Seed hull percentage (−0.949) and electrical conductivity of seed leachate (−0.407) recorded negative correlations with yield per plant. This indicates that a decrease in these characters will result in an increase in yield. All the seedling characters studied, viz germination percentage (0.976), root length (0.986), shoot length (0.729), seedling vigour index (0.966) and seedling dry weight (0.841), showed highly significant positive correlations with seed yield per plant, indicating the vital role of these characters towards corresponding gain in yield of sunflower hybrids. The sunflower hybrids showing high value of desirable seed and seedling characters also yielded higher; this may be due to high germination percentage and early vigour of these hybrids. Thus, more emphasis should be given to seed and seedling characters as an early indicator of yield performance, and seed and seedling characters may be used as selection criteria for identification of high-yielding sunflower hybrids (Nichal et al. 2015).

11.16.3 Mustard

11.16.3.1 Combining Ability

Thirty crosses obtained by crossing two testers (drought-tolerant donors) with 15 lines along with their parents were evaluated to analyse the combining ability of available genotypes with drought-tolerant donors and to identify superior crosses for their exploitation in developing drought-tolerant varieties of mustard (Puttawar et al. 2015). Variation among the genotypes was highly significant for all the characters studied, except days to maturity, thereby indicating a substantial amount of genetic variability. The significance of parents versus crosses noticed for days to 50 % flowering, days to maturity, plant height, number of branches per plant, number of siliqua per plant and seed yield per plant satisfied the basic prerequisites for comparing the expression of parents and crosses for different characters; per se performance revealed the superiority of the parents GM-6, Bio-902, Seeta and ACN-9 for seed yield per plant and number of siliqua per plant. Among the crosses, GM6 × Geeta, Aashirvad × RH 819, SEJ 2 × Geeta and Seeta × Geeta exhibited superiority for seed yield per plant and number of siliqua per plant.

Significant variations among lines were observed for most characters like days to 50 % flowering, plant height, number of siliqua per plant, 1000 seed weight and seed yield per plant, and also for line × tester interaction, except days to 50 % flowering, which allowed estimation of general combining ability and specific ability effects for the above characters, while the mean squares for testers were significant only for number of siliqua per plant. The parents Urvashi, JD –6 and Vardan were identified as good general combiners. The crosses SEJ 2 × Geeta and Ashirvad × RH-819 were identified as promising crosses based on mean and SCA effects for yield and number of siliqua per plant, which were identified for forwarding to the next generation for making improvement in the seed yield along with drought tolerance, followed by a simple selection method.

11.16.3.2 Variability in F_3 and F_4 Generations

Thirty experimental strains of *Brassica rapa* L. var yellow sarson for F_3 and F_4 generations were evaluated for variability (Das et al. 2015a). The F_3 generation bilocular groups showed significant variation for siliqua on the main raceme and seeds per siliqua. The F_4 generation bilocular groups showed significant variations for most of the characters except plant height, height up to first fruiting branch, length of main raceme, siliqua on main raceme and seed yield. GCV for all the traits were higher than the corresponding PCV in F_4 generation. Heritability (bs) estimates recorded high for siliqua on main raceme and 100-seed weight, and genetic advance as a percentage of the mean was recorded higher for siliqua on primary branches, length of main raceme, siliqua on main raceme and 100-seed weight in the bilocular group. In F_3 generation, seed yield showed significantly positive phenotypic and genotypic correlations with plant height, siliqua on primary branches, length of main raceme, siliqua on main raceme, seeds per siliqua and 100-seed weight, and showed significantly negative correlation with height up to first fruiting branch. The F_4 generation showed significantly positive correlation of seed yield with number of primary branches. In the F_3 and F_4 generations the siliqua on primary branches showed highest positive direct effect on seed yield. In the F_4 generation, while siliqua on primary branches exerted high positive direct effect on seed yield in both the generations, length of main raceme exerted highly negative direct effect on seed yield.

11.16.4 Safflower

In 150 safflower germplasm lines and five checks evaluated for yield and yield-related components and their direct and indirect contribution towards yields, analysis of variance and mean performance for seed yield and its components revealed significant differences among all the germplasm lines for all characters there by indicating presence of variability in genotypes. Seed yield showed highly positive significant correlations with number of effective capitula per plant and number of secondary branches per plant. Path analysis studies revealed that the highest direct effect on seed yield per plant was observed for the character number of effective capitula per plant (0.413), followed by days to maturity (0.353) and number of secondary branches per plant (0.096). The maximum indirect positive effect on seed yield was exhibited by days to 50 % flowering through days to maturity (0.325), with number of effective capitula per plant (0.189) indicating the importance of depict selection for these characters (Gopal et al. 2015).

A random mating population of safflower was evaluated for yield improvement (Tayade et al. 2015) of recurrent selection and selection of promising half-sib families over checks AKS-207, Bhima and PKV-Pink. The result showed that there was remarkable variability in the material used. The additive genetic variance and phenotypic variance among half-sib families were high for plant height (269.32 and 66.33), followed by seed yield per plant (160.68 and 40.17). Heritability estimates in narrow sense were high for seed yield per plant (70.20 %), followed by number of capitulums per plant (42.96 %) over population mean and over check varieties, i.e. AKS-207 (51.40 %), Bhima (41.79 %) and PKV-Pink (44.40 %) at 5 % selection intensity for the same character. The seed yield per plant was positively correlated with plant height (0.21*), number of primary branches per plant (0.45*) and oil content (0.27*). Out of 135 families, 20 half-sib families exhibited significant increases for yield-contributing characters besides seed yield over check AKS-207, Bhima and PKV-Pink. These families should be utilized for the next recombination cycle to improve seed yield and yield-contributing characters in a future population improvement programme.

Path analysis studies in 40 safflower germplasm including 2 checks (*Carthamus tinctorius* L.) showed that the highest direct effect on seed yield per plant was exhibited by the character number of effective capitula per plant (0.723), followed by 100 seed weight (0.548) and days to maturity (Ingole et al. 2015).

11.16.5 Sesamum

The general and specific combining ability evaluated (Dhuppe et al. 2015) in 25 hybrids of sesamum (*Sesamum indicum* L.) obtained by crossing 5 × 5 genotypes in line L × T fashion with their ten parents and two standard checks for 11 characters, viz days to first flower, days to 50 % flowering, days to maturity, plant height, number of branches per plant, number of capsules per plant, capsule length, number of seed per capsule, 1000 seed weight, oil content and seed yield per plant, showed that OSC-560 was observed to be a good general combiner for seed yield and days to maturity. RT-54 was found to be good general combiner for plant height, number of capsules per plant, number of seeds per capsule and oil content. Among males, IS-200 was a good general combiner for seed yield per plant and number of seed per capsule. The cross combinations RT-54 × NIC-16194, OSC-560 × KMR-11, OSC-207 × IS-200, MT-10-13-01 × IS-200, MT-10-13-01 × NIC-8600-A, MT-10-13-01 × SI-982, SSD-01 × NIC-16194 and RT-54 × KMR-11 showed positive and significant SCA effects. Hence, these are recommended for yield improvement.

11.16.6 Groundnut

11.16.6.1 Adaptability and Stability of Groundnut Genotypes

Groundnut (*Arachis hypogaea* L.) is one of the most important oilseed crops, which is commercially cultivated in rainy, post-rainy and summer seasons in India. The adaptability and stability of some groundnut genotypes over diverse growing environments in a total of 14 environments with simulation was assessed (Ladole et al. 2015). Pooled analysis of variance over diverse environments revealed the presence of genotype × environment interactions and significant differences among the genotypes and environments for pod yield and other component traits. The partitioning of G × E interaction revealed that the linear component was significant against pooled deviation for pod yield and seed oil content. However, the linear component was non-significant for the other component traits, indicating that the response of genotypes to environments was not controlled genetically. The non-linear component (pooled deviation) was also significant when tested against the pooled error, revealing presence of non-linear responses of the genotypes to the changing environments.

The genotype AK 320 showed specific adaptability for favourable environments, and the genotype AK 342 showed adaptability for unfavourable environments with respect to pod yield. However, both genotypes were not very stable across environments for the seed oil content, but their seed oil content improved proportionally to the improving environment. For the trait pods per plant, the genotype AK 327 was found to have average stability. The genotype TAG 24 exhibited average stability for harvest index. In general, the genotypes that showed a high level of stability were not high yielders, and the high yielders were found to be unstable in respect of pod yield and its component traits.

11.16.6.2 Genetic Variability and Heritability

The knowledge of variability of genetic material would be of immense importance in any breeding programmes. Continuous breeding will lead to narrowing down the variability among the genotypes, which will further hinder the yield in any crop. One of the ways to increase the variability is by introduction of genetically diverse new genotypes. Hence, the success of planned hybridization in self-pollinated crops like groundnut (*Arachis hypogaea* L.) to recombine the desirable traits from different plants is influenced by the amount of the variability existing in the gene pool.

Variability and heritability were evaluated in 25 genotypes of groundnut (Kahate et al. 2015) including five checks, viz JL-24, TAG-24, LGN-1, JL-220 and AK-159, for 17 characters, viz pod yield per plant, kernel yield per plant, test weight, shelling percentage, oil content, harvest index, late leaf spot severity, stomatal frequency per mm^2 (adaxial and abaxial), stomata length (adaxial and abaxial), stomata breadth (adaxial and abaxial), non-reducing sugar, reducing sugar, SCMR and phenol content. The estimates of GCV and PCV indicated that maximum GCV and PCV were observed for LLS severity, followed by pod yield, kernel yield per plant, reducing sugar and non-reducing sugar. The estimates of heritability was appreciably high for LLS severity, stomatal frequency per mm^2 (adaxial and abaxial), test weight, stomata length (adaxial and abaxial), reducing sugar, non-reducing sugar, phenol content, oil content, stomata breadth (adaxial and abaxial), pod yield per plant, kernel yield per plant and SCMR.

Genetic variability was studied in a set of 22 groundnut genotypes under conditions favouring normal growth and expression of all the characters (Khan et al. 2015). The analysis of variance showed significant differences among the genotypes tested for all the eight characters, justifying the selection of genotypes for the study. The genotypes showed considerable amounts of genetic variation for all the characters and indicated the good scope for selection of suitable basic material for further improvement. PCV values were of higher magnitude than GCV values for all parameters under study. The estimates of PCV and GCV were high for kernel yield (38.11 and 33.76) and dry pod yield (36.30 and 32.26). High heritability (h^2) and high genetic advance as a percentage of the mean (GAM) was observed for kernel yield (78.49 % and 61.62 %) and dry pod yield (78.96 % and 59.05 %), while low h^2 was noticed for 100-kernel weight (28.51 %) and sound mature kernels (8.14 %). However, the GAM values observed were moderate for 100-kernel weight (11.43) and low for other important traits like days to 50 % flowering, day to maturity, shelling (%) and sound mature kernels (%).

High estimates of GCV, PCV, heritability and genetic advance as a percentage of the mean were observed for kernel yield per plant, pod yield per plant, number of pods per plant, harvest index and test weight under both stress and non-stress situations. It indicated the role of additive gene action and hence the usefulness of phenotypic selection for bringing improvement. Pod yield showed positive significant association with number of pods per plant, test weight, kernel yield per plant, number of seeds and harvest index at genotypic and phenotypic levels under non-stress conditions, whereas pod yield per plant also showed positive significant associated with kernel yield, number of pods per plant, test weight, harvest index and number of seeds per pod under stress conditions. The path analysis revealed that the number of pods per plant exerted the highest positive direct effect on pod yield, whereas harvest index and kernel yield exhibited the highest positive direct effect on pod yield under stress conditions. Therefore, it would be rewarding to lay due emphasis on the selection of these characters for rapid improvement in pod yield (Abubakker et al. 2015).

The genetic variability and character association for traits related to yield and water use efficiency in 225 genotypes consisting of 184 groundnut minicore accessions and 41 advanced breeding lines studied (Sowmya et al. 2015a) revealed that there was high GCV, PCV, heritability and GAM for pods per plant, pod yield per plant, kernel yield per plant, SMK and 100-kernel weight, suggesting the presence of considerable variability for these traits. Low GCV, PCV and high heritability coupled with low GAM were noticed for SLA and SCMR. Pod yield per plant had significant correlation with all the traits except SMK. Path analysis revealed the maximum direct effect of kernel yield per plant on pod yield per plant. The accessions ICG 13942, ICG 12000, ICG 532, JL24, ICG 5745, ICG 5662, ICG 14482, GBFD5272, M282 and ICG 13099 exhibited higher SCMR values, and the accessions ICG 13942, JL24, GKVK13, ICG 5745, ICG 2772, ICG 5016, ICG 10036, ICG 9777, ICG 875 and ICG 532 had lower SLA values. These accessions were identified as high water use efficient based on their high SCMR and low SLA values, and need to be tested under both water stress and non-stress conditions in order to confirm their drought tolerance. High pod yield per plant was recorded by genotypes ICG 13942, JL24, ICG 930121, GKVK13, ICG 6057, ICG 99003, ICG 10036, ICG 10556, ICG 11426 and ICG 10479. The identified superior accessions can be further utilized in breeding programmes to develop high-yielding and drought-tolerant genotypes.

11.16.7 Mustard

Thirty crosses of mustard obtained by crossing two testers (drought-tolerant donors) with 15 lines along with their parents were evaluated for analysis of combining ability of these available genotypes with drought-tolerant donors and to identify superior crosses for their exploitation in developing drought-tolerant varieties. Variation among the genotypes was highly significant for all the characters studied except days to maturity, thereby indicating a substantial amount of genetic variability. The significance of parents versus crosses noticed for days to 50 % flowering, days to maturity, plant height, number of branches per plant, number of siliqua per plant and seed yield per plant satisfied the basic prerequisite for comparing the expression of parents and crosses for different characters; per se performance revealed the superiority of the parents GM-6, Bio-902, Seeta and ACN-9 for seed yield per plant and number of siliqua per plant. Among the crosses, GM6 × Geeta, Aashirvad × RH 819, SEJ 2 × Geeta and Seeta × Geeta exhibited superiority for seed yield per plant and number of siliqua per plant. Significant variations among lines were observed for most characters like days to 50 % flowering, plant height, number of siliqua per plant, 1000 seed weight and seed yield per plant, and also for line × tester interaction except days to 50 % flowering, which allowed the estimation of general combining ability and specific ability effects for the above characters, while mean squares for testers were significant only for number of siliqua per plant. The parents Urvashi, JD −6 and Vardan were identified as good general combiners. The crosses SEJ 2 × Geeta and Ashirvad × RH-819 were identified as promising crosses based on mean and SCA effects for yield and number of siliqua per plant, which were identified for forwarding to the next generation for making improvement in the seed yield along with drought tolerance followed by a simple selection method (Puttawar et al. 2015).

11.16.8 Okra

11.16.8.1 Okra Mutagens

Mutation frequency, mutagenic effectiveness and efficiency of gamma rays (at 15, 30, 45 and 60 kR) and ethyl methane sulphonate (at 0.2 %, 0.4 %, 0.6 %, 0.8 % and 1.0 %) were assessed in M_1 and M_2 generations with mutagenic treatments in okra genotype Parbhani Kranti. The gamma ray dose was found to be more effective to induce a wide range of mutation than ethyl methane sulphonate. Among gamma rays doses, 30 kR doses were found to be more effective to induce mutation. A number of mutants for one or more traits, viz plant height, maturity, branching and fruit size, were isolated in the M_2 generation. Breeding behaviour of some selected mutants was studied in the M_3 generation. The means for various characters increased at lower doses of mutagens in the M_2 and M_3 generations. The magnitude of the CV was significantly higher than control. Among a large number of mutants in the M_2 generation, 9 in Parbhani Kranti were selected and may prove useful for an okra improvement programme (Jadhav et al. 2015a).

Gamma rays were more effective and efficient than ethyl methane sulphonate in increasing the production of number of mutants. The frequency for chlorophyll mutations ranged from 0.42 to 1.54 % and for viable mutations it ranged from 0.34 to 1.98 % in Parbhani Kranti. The efficiency of the mutagen was measured in terms of percentage of lethality and sterility induced. The efficiency and effectiveness for chlorophyll and viable mutations in Parbhani Kranti was highest at 0.6 % ethyl methane sulphonate and 30 kR gamma rays, respectively, on a lethality basis. There was a progressive increase in mutation frequency of chlorophyll and viable mutations with increases in gamma radiation and ethyl methane sulphonate doses (Jadhav et al. 2015b).

11.16.8.2 Path Coefficient Analysis

Correlation and path coefficient analysis for yield components in 50 genotypes of okra [*Abelmoschus esculentus* (L.) Moench] (Burse et al. 2015) showed that the yield per plant was significantly and positively correlated with plant height, number of primary branches per plant, number of fruiting nodes per plant, number of fruits per plant, average fruit weight and length of fruit. Leaf area, days to first harvest, diameter of fruit and moisture content were significantly and negatively correlated with yield per plant.

Path coefficient analysis indicated that average fruit weight, diameter of fruit, plant height, and length of fruit, leaf area and number of fruits per plant showed high positive direct effects as the major yield-contributing traits for enhancing the yield of okra. Based on direct and indirect effects of different yield components on yield, it would be rewarding to give stress on the number of fruits per plant, number of fruiting nodes per plant, average fruit weight, length of fruit and plant height, while formulating selection indices for improvement of yield in okra (Burse et al. 2015).

11.16.9 Tomato: Genetic Variability

Genetic variability, heritability and genetic advances in tomato (*Solanum lycoperscicum* L.) were evaluated in 38 genotypes of tomato for yield and various yield-contributing characters. All the genotypes exhibited a wide range of variability for traits under consideration. These traits were also found to be highly heritable. The PCV was found to be greater than the GCV for all the characters studied. High heritability was observed for fruit yield per plant followed by average fruit weight, while primary branches gave lowest heritability results. High heritability combined with high genetic advance was shown by average fruit weight, suggesting that additive gene action plays a major role in governing these traits and these traits can be improved by simple selection (Thapa et al. 2015b).

11.16.10 Tomato: Character Association

Correlations and path coefficients were studied in 38 tomato (*Solanum lycopersicum* L.) genotypes for 13 yield-contributing characters (Thapa et al. 2015a), revealing that yield per plant was highly significant and positively correlated with average fruit weight, fruits per plant, locule number per fruit, pericarp thickness, fruits per cluster and number of clusters per plant. Strong association of these traits indicated that selection based on these traits would ultimately improve the fruit yield.

Path coefficient analysis is used to obtain a clear picture of the inter-relationship between yield per plant and its components, direct and indirect effects. Analysis revealed that highest positive direct effects on fruit yield per plant was exhibited by average fruit weight followed by fruits per plant, clusters per plant, fruits per cluster and pericarp thickness; they showed positive and significant associations with fruit yield per plant in both correlation and path coefficient analyses (Thapa et al. 2015a).

11.16.11 Cassava

The nature and magnitude of genetic diversity of 18 genotypes of cassava was studied (Rao et al. 2015). The analysis of variance showed significant differences among the genotypes for all the traits. Based on Mahalanobis D^2 analysis, all the 18 genotypes were grouped into five clusters. Cluster IV had the maximum number of genotypes (7) followed by cluster I (5). Maximum inter-cluster distance was observed between cluster IV and V (62.40): while the intra-cluster distance was maximum in cluster IV (17.2). Tuber dry matter content (64.05 %) contributed the maximum percentage followed by plant dry matter content (20.26 %), height of first branching (9.15 %), starch content (5.23 %), number of leaves per plant (0.65 %) and stem diameter (0.65 %), suggesting that selection of one or two elite genotypes from divergent (IV and V) and (II and V) clusters based on the above characters and crossing would result in more heterosis and novel hybrids.

11.16.12 Sweet Potato

Sweet potato, *Ipomoea batatas* (L.) Lam., is the seventh most important food crop worldwide. It is grown in more than 100 countries, and among the world's root and tuber crops, it is second in production to the white potato, *Solanum tuberosum* L. Sweet potato is an important staple food in several parts of the world. It is also grown for human consumption as a vegetable (both fleshy roots and leaves), for processing as a snack food, for animal feed, for industrial starch extraction and fermentation, and for various other processed products. The crop produces adequate to high yields in both low-technology and high-technology systems.

The most important biotic constraints on sweet potato production worldwide are the sweet potato weevils *Cylas formicarius elegantulus* (summers) and *C. f. formicarius (Fabricius)*. *C. f. elegantulus* is found in the Western Hemisphere (North, South and Central America, the Caribbean Basin and parts of the Pacific Basin). These pests feed on roots, inducing the plant to produce bitter-tasting terpene compounds, which make even slightly damaged roots unfit for consumption. Therefore, it is not uncommon for these weevils to reduce yield by as much as 60–100 % in the absence of adequate control measures (Mohanty et al. 2015c).

Thirty genotypes of sweet potato were evaluated to elicit information on the magnitude of genetic variability, heritability and diversity, and to predict the gains realized through selection, character association, cause-and-effect relationship and divergence for the quantitative characters in sweet potato (*Ipomoea batatas* L.) (Mohanty et al. 2015a). Among all the genotypes studied, genotype Accession-22 had the highest root yield per hectare and was found to be suitable for the local agroclimatic conditions. The genotypes 440127, SWA-2 and ST-14 were also found to be elite for different characters. High PCV and GCV were observed for characters like vine length, vine internodal length, number of branches per plant, number of leaves per plant, total leaf area, number of roots per plant, root yield per plant, β-carotene content, starch content, total sugars, reducing sugars, non-reducing sugars and total root yield per hectare, indicating high variability available in the germplasm for these characters for further improvement. High heritability (h^2) coupled with high genetic advance as a percentage of the mean was observed for the characters vine length, vine internodal length, number of branches per plant, length of leaf lobe, number of leaves per plant, total leaf area, root girth, root yield per plant, β-carotene content, starch content, total sugars, reducing sugars, non-reducing sugars and total root yield per hectare, indicating that these characters were least influenced by the environmental effects and were governed by additive genes, and selection will be rewarding for improvement of such traits. The total root yield per hectare (t ha^{-1}) had significant positive correlation with traits like number of branches per plant, number of roots per plant, root girth, root yield per plant and β-carotene content, and also showed direct effects on number of branches per plant, number of roots per plant, root length, root yield per plant, starch content and reducing sugars. The high direct effect of these traits appeared to be the main factor for their strong association with total root yield per hectare. Analysis for divergence using a D^2 statistic revealed highly significant differences for different traits, grouping the 30 genotypes into six clusters. Cluster II had the maximum number of genotypes (8) followed by cluster I (7). The highest percentage contribution to divergence came from β-carotene content and starch content, suggesting that selection of one or two elite genotypes from divergent (II and VI) and (III and VI) clusters based on the above characters and crossing would result in more heterosis and novel hybrids.

For crop improvement in sweet potato, knowledge of genetic diversity helps the breeder in choosing desirable parents for use in the breeding programme. The diverse genotypes or accessions can be crossed to produce superior high-yielding hybrids possessing resistance to various abiotic and biotic stresses. The genotypes were characterized based on response for 10 morphological and 18 quantitative characters. Correlation and path coefficient analyses were carried out for 18 characters of yield and its components. Character association indicated that tuber yield per hectare was positively and significantly associated with number of tubers per plant, tuber yield per plant and β-carotene content at phenotypic and genotypic correlation levels, and tuber yield per plant was positively and significantly associated with vine length, vine internodal length, leaf area and tuber girth at both phenotypic and genotypic levels. Path analysis indicated that number of branches per plant, root length, root yield per plant and starch had direct effects on tuber yield per hectare; the remaining characters had negligible to low and moderate indirect effects through other component characters. Number of tubers per plant, tuber yield per plant and β-carotene content are reliable characters to use for improvement of sweet potato (Mohanty et al. 2015b).

11.16.13 Bitter Gourd

Bitter gourd (*Momordica charantia* L.) is an important tropical and subtropical cucurbit vegetable grown throughout India for its tender fruits. It is considered a prized vegetable because of its pharmaceutical and high nutritive value, especially ascorbic acid and iron. Bitter gourd has been used for centuries in the ancient traditional medicine of India, China, Africa and Latin America, as its extract possess antioxidant, antimicrobial, antiviral, antihepatotoxic and antiulcerogenic properties, and also has the ability to lower blood sugar. In spite of the potential economic and medicinal importance of the crop, due attention has not been given to a need-based crop improvement programme. There is a prime need for its improvement and to develop varieties or hybrids suited to specific agro-ecological conditions. Being monoecious in nature, its sex expression is influenced by temperature.

Evaluation of the performance of parents and their F_1 hybrids for yield and earliness during the summer season revealed significant variation in mean performance among the parents and hybrids for all the characters studied. Among the hybrids, the best performing hybrids were IC-044438 × IC-045339, IC-044417 × IC-470558 and IC-033227 × IC-044438 for most of the characters like yield, vine length, number of laterals per vine, number of fruits per vine and fruit length, while for earliness in terms of days to first female flowering and node number at which the first female flower appeared, the hybrids IC-044438 × IC-045339 and IC-045339 × IC-470560 were found to be superior. Among the parents, genotype IC-044438 followed by IC-033227 and IC-470560 recorded highly significant yield and yield-related traits like number of fruits per vine, average fruit weight and pulp thickness, while IC-470560 and IC-044438 were found to be early as they recorded minimum days to the first male flower, the first female flower and sex ratio (male to female). These superior hybrids and parents may be further tested in different locations during the summer season for their stable performance and may be recommended for commercial cultivation and also can be used in a future breeding programme (Rani et al. 2015b).

11.16.14 Chilli

The genetic diversity among 63 genotypes using Mahalanobis D^2 statistic for ten quantitative and six qualitative characters in chilli was assessed (Janaki et al. 2015). The fruit diameter followed by yellow carotenoids, red carotenoids, ascorbic acid and capsaicin contributed the maximum towards genetic divergence. The 63 genotypes were grouped into eight clusters. Among all the clusters, clusters III and V were the largest, containing 17 genotypes, followed by cluster IV (11), cluster I (8) and cluster II (7). The clusters VI, VII and VIII were monogenotypic. The mutual relationships between the clusters revealed that inter-cluster distance values were greater than intra-cluster values. The intra-cluster D^2 values ranged from 0.00 (VI, VII and VIII) to 434.43 (V). The inter-cluster D^2 values varied from 117.25 (between cluster I and III) to 4139.41 (between cluster IV and VIII). The high intra- and inter-cluster distances indicated the presence of wide genetic diversity among the genotypes present within and between clusters. The genotypes from clusters IV, V and VIII may be utilized in breeding programmes for yield improvement.

11.16.15 Cotton

Cotton, or 'white gold', is a premier cash crop of the rainy season and one of the prominent eco-industrial crops of India, generating sizeable employment. Cultivated cotton genus *Gossypium* bears spinnable seed coat fibres. It plays a triple role by producing lint, oil and protein. Cotton stalk are also used as fuel and for making particle and paper board. India has the largest area in the world. China and India are the major cotton-consuming countries in the world (around 58 % of world cotton consumption). As regards export, the USA and India export around 55 % of the world's cotton, with the USA's share being 3.1 million tons and India's share being 1.2 million tons (Deshmukh et al. 2015).

The genetic variability was analysed in 28 hybrids in a line × tester design, obtained by crossing of seven lines with four testers along with parents and two checks. There were significant positive associations of number of monopodia per plant, number of sympodia per plant, number of bolls per plant and ginning percentage with seed cotton yield per plant. Boll weight had significant and positive associations with ginning percentage and uniformity ratio, and a positive and non-significant association with micronaire value. There was a significant negative association of uniformity ratio and micronaire with fibre strength.

Path coefficient analysis revealed that number of monopodia, number of sympodia, number of bolls per plant and ginning percentage had the maximum direct positive effect on seed cotton yield per plant and number of bolls per plant, exhibiting an indirect effect on seed cotton yield, viz the number of sympodia, fibre length, number of monopodia per plant and plant height. Among the fibre quality traits, fibre length showed a negative indirect effect on yield via the ginning percentage, micronaire value and uniformity ratio, while the ginning percentage exhibited indirect positive effects via number of sympodia, boll weight, number of bolls per plant, micronaire value and uniformity ratio on seed cotton yield per plant.

Path coefficient analysis revealed that bolls per plant and fibre strength should be given greater emphasis in a cotton yield improvement programme as well as for modern ginning and spinning mills, as they contribute maximum direct effects, while the number of sympodia per plant contributes maximum indirect effects through the trait bolls per plant (Patil et al. 2015c).

Combining ability analysis is an important tool for identifying superior parents suitable for developing both new cultivars and hybrids. The general combining ability of the parents and specific combining ability and heterosis of hybrids obtained by crossing three female lines with six testers in a line × tester fashion for seed cotton yield and fibre quality traits were evaluated.

Analysis of variance for combining ability indicated the predominance of non-additive gene action for all the characters under study except plant height, 2.5 % span length and oil content. The parents AK032 and AK053 among the lines and DR-7R among the testers were found to possess significant GCA effects for most of the yield-contributing characters. Maximum heterosis for seed cotton yield per plant was recorded by AK023 × DR-7R over the standard check. The highest significant heterosis in desirable direction for fibre strength was recorded by AK023 × DR-7R followed by AK023 × AKH-976 over the check (Madhuri et al. 2015b).

The combining ability effects in parents and their hybrids in diploid cotton, obtained by cross combinations using a line × tester mating design, were evaluated (Solanke et al. 2015) in a set of ten parents—two females (GMS line), viz GAK-423 and GAK-8615, and eight males (tester), viz AKA-9703, AKA-0110, AKA-9620, AKA-9009-1, AKA-0209, AKA-5, AKA-7 and HD-162—and their 16 crosses along with check PKV Suvarna. The mean squares due to genotypes were highly significant for all the traits studied expect for number of monopodia per plant, ginning percentage and oil content. This indicated the presence of substantial genetic variability among genotypes for all the traits studied. Among female parents, GAK-8615 recorded significant GCA effects for seed cotton yield per plant, number of sympodia per plant and number of bolls per plant. Among males, AKA-9703 was the best general combiner, with high GCA effects for plant height, number of bolls per plant and 2.5 % span length. The tester AKA-5 exhibited significant GCA effects for seed cotton yield per plant. The tester AKA-7 exhibited significant GCA effects for seed cotton yield per plant and 2.5 % span length. The cross GAK-423 × AKA-0110 (6.396) recorded the highest positive SCA effect for seed cotton yield per plant. It also exhibited SCA effect in a desirable direction for fibre strength, ginning percentage, number of sympodia per plant and days to 50 % flowering.

Analysis of variance for combining ability in 28 hybrids obtained (Deosarkar et al. 2015) by a line × tester design by crossing seven lines and four testers revealed that both additive and non-additive variances were important in the inheritance of various traits studied. Combining ability revealed significant differences within parents, crosses, line (except for days to 50 % flowering, number of sympodia per plant, days to maturity, boll weight, ginning percentage, fibre length, uniformity ratio and fibre strength), testers (except for number of monopodia per plant, number of sympodia per plant, plant height, number of bolls per plant and micronaire value) and line × tester (except for number of monopodia per plant and ginning percentage). However, significant differences were also observed for parents and crosses for all the traits studied, except for number of monopodia per plant, boll weight, micronaire value, seed cotton yield per plot and seed cotton yield qt ha^{-1}. General combining ability is generally associated with additive gene action while specific combining ability is genetically due to dominance and epistasis. The variance due to SCA was higher in magnitude than the variance due to GCA for the characters days to 50 % flowering, number of monopodia per plant, number of sympodia per plant, day to maturity, plant height, number of bolls per plant, boll weight, ginning percentage, uniformity ratio, fibre length, micronaire value, bundle strength and seed cotton yield per plant, seed cotton yield per plot and seed cotton yield qt ha^{-1}, which indicated the predominance of non-additive gene action.

The general combining ability of the parents and specific combining ability and heterosis of hybrids was estimated for yield and fibre quality traits. The four females (lines), viz CAK-8660, CAK-3456, CAK-2160 and CAK-32, and five males (testers), viz AKH-5120, AKH-073, AKH-545, AKH-976 and AKH-32NS, and their 20 crosses were evaluated. The parents CAK-2160 (10.417) among the lines and AKH-976 (24.665) among the testers were found to possess significant GCA effects for most of the yield-contributing characters. The cross CAK-2160 × AKH-976 recorded the highest significant positive SCA effects (22.667 **) for seed cotton yield per plant. Maximum useful heterosis for seed cotton yield per plant was recorded by CAK-2160 × AKH-976 (71.00 % over check PKV HY-5 and

95.43 % over check PKV HY-2). GCA effects of parents and SCA effects of two crosses, viz CAK-2160 × AKH-976 and CAK-2160 × AKH-545, were identified as the best crosses for a further crop improvement programme (Gedam et al. 2015).

The 24 crosses through a line × tester design involving six diverse female lines and four male parents were evaluated to study the extent of heterosis and to estimate general and specific combining ability effects (Kalpande et al. 2015a). Analysis of variance for means revealed significant differences for all the 13 characters studied. The line × tester interaction mean square was significant for all the characters except number of boll and days to maturity, indicating sufficient genetic diversity among them. The variance due to GCA and SCA indicated that the non-additive type of gene action was predominant for all the characters studied. The magnitudes of heterosis, heterobeltosis and standard or economic heterosis for all the characters were highly appreciable. Among all the characters, the magnitude of heterosis was highest for plant height, measuring to the extent of 117.89 % over standard check NH-615 in the cross PA-532 × AKA-7, followed by seed cotton yield per plant (65.32 %) in the cross PA-528 × AKA-7 over the mid-parent. Among females, PA-08 was found to be the best general combiner for nine characters, viz days to 50 % flowering, day to boll bursting, number of bolls per plant, plant height, lint yield, span length, fibre strength, uniformity ratio and short fibre index, with significant GCA effects. The females PA-720 and PA-532 were the best general combiners for five characters including seed cotton yield per plant. Among males, AKA-7 was found to be best general combiner for seed cotton yield and its contributing characters, which had significant GCA effects for five characters, viz number of sympodia, bolls per plant, boll weight, plant height and seed cotton yield per plant. The male parent Dwd-arb-10-1 was also found to be best general combiner for five seed cotton yield and fibre quality characters. There was close agreement between per se performance and GCA as well as SCA effects for most of the characters. Observations on various characters indicated that the crosses showing high heterosis and high SCA effects had high per se performance and they involved at least one high combining parent. The combinations PA-720 × JLA-802, PA-528 × AKA-7, PA-402 × JLA-802, PA-532 × JLA-802 and PA-08 × JLA-802 showed significant and desirable SCA effects for most of the yield and fibre quality traits studied, indicating potential for exploiting hybrid vigour in a breeding programme.

11.16.16 Roselle

Genetic variability, correlation and path analysis studied in a set of 60 diverse genotypes of roselle (*Hibiscus sabdariffa* L.) for 11 traits showed that the PCVs as well as the GCVs were high for important productivity characters like fibre yield per plant, dry stick weight per plant, green fresh weight per plant and green dry weight per plant (for PCV) and medium for other characters like plant height and bark thickness. High heritability coupled with high expected genetic advance was observed for fibre yield per plant, dry stick weight per plant, green fresh weight per plant and green dry weight per plant. Fibre yield per plant was found to be significantly and positively correlated with plant height, base diameter, bark thickness, green fresh weight per plant, green dry weight per plant and dry stick weight per plant. Plant height, base diameter, green fresh weight per plant, green dry weight per plant and dry stick weight per plant also exhibited significant positive inter-correlation among themselves. Partitioning of correlation coefficients of various components upon fibre yield per plant into direct and indirect contributions revealed that green fresh weight per plant had the maximum direct effect upon fibre yield per plant, followed by plant height, bark thickness and dry stick weight per plant (Satyanarayana et al. 2015).

11.16.17 Cashew Nut

Cashew (*Anacardium occidentale* L.) is one of the most important export earning crops of India, occupying an area of 9.91 lakh hectares, with a production of 6.25 lakh tonnes. Although India is the largest producer and exporter of cashew in the world, the productivity is very low (only 722 kg ha^{-1}) as compared to other countries, which is primarily due to seedling progeny of poor genetic stock. Being a highly cross-pollinated crop, development of hybrids exhibiting heterosis for important economic characters is one of the ways to enhance the productivity. Despite the importance of cashew as a commercial export-oriented crop, the crop is challenged with a number of problems like low and variable nut yield, low % kernel turnover and susceptibility to insect pests as a result of establishing cashew farms with unselected seeds.

11.16.17.1 Fruit Set

Variability studies in cashew nut on flowering parameters, fruit set, apple weight, nut weight and kernel weight revealed the existence of wide variability. Total number of flowers varied from 270.85 (H 117) to 748.57 (H 94). The fruit set under natural conditions was found to be 7.83 to 28.62. Sex ratio varied from 0.08 to 0.32 and apple weight was recorded to be maximum in H 116, and nut weight, kernel weight and nut yield per tree were maximum in H 94 (Sreenivas et al. 2015b).

11.16.17.2 Heterosis

The extent of heterosis determined in 20 F_1 cashew hybrids (Sethi et al. 2015) has indicated that the hybrids A-71, B-27, C-30 and C-41 exhibited better heterosis in terms of nut weight (g), kernel weight (g) and overall nut yield (kg per tree) than the rest of the tested hybrids. The relative heterosis, heterobeltiosis and standard heterosis among these superior hybrids varied from 85.71 % to 94.88 %, 57.33 % to 65.85 % and 15.51 % to 30.70 %, respectively. Hence, these hybrids may be recommended for cultivation to increase production and productivity of cashew under agroclimatic condition of Odisha.

11.16.17.3 Evaluation of Cashew Nut

Twenty-five released varieties of cashew from different parts of India were evaluated (Tripathy et al. 2015) for yield-attributing and yield parameters, showing that cashew varieties such as Amrutha, BPP-4, Madakathara-2, K-22-1, VRI-3, Ullal-4, UN-50 and Goa-1 were identified for dwarf architecture, which need further evaluation. Similarly, both Priyanka and Dhana can be exploited for cashew apple processing purposes. The out turn % demonstrated by the majority of the varieties was within the acceptable international standard. The cashew varieties BPP-8, Vengurla-7, Bhaskara, NRCC Sel-2, Ullal-3, Kanaka, Vengurla-4, Bhubaneswar-1, VRI-3 and Dhana combined two or three traits that were significantly different from the rest of the varieties. However, cashew varieties such as BPP-8, Vengurla-7, Bhaskara, NRCC Sel-2 and Ullal-3 seem outstanding for most of the parameters.

11.16.18 Conservation of Citrus Species

The northeastern region of India is considered one of the 'mega-biodiversity hotspots' in the world because of its diverse presence of flora and fauna. This region stretches from 21° 57′ N to 29° 28′ N and from 89° 40′ E to 97° 25′ E, and is considered as one of the natural homes of *Citrus*. In a naturally cross-pollinated genus like *Citrus*, nature has eventually created different forms of *Citrus* and the region has the conducive environment, suitable soil and topography for perpetuation of these various forms. It can be regarded as 'a live museum of *Citrus*'. The wide adaptability of *Citrus* is also reflected in its general distribution of topographical situations. It also holds a unique position in the world map of *Citrus* occurrence and diversity and possesses wide diversity of *Citrus* species either in cultivated or in wild forms. About 17 species are found in the northeastern region of India and 8 species are indigenous to this region. There is every possibility of occurrence of these *Citrus* species as the states are a contiguity of the Himalayan belt. The Khasi mandarin of the region is unique in its quality. Other than the commercial species, some of other species of *Citrus*, namely rough lemon, Kamala Australia, Samphola, Citron, Singkin, various limes and lemons, pummelos, grapefruit, etc., are available in various types either in homesteads or in forests. Due to changes in climatic condition, the productivity of *Citrus* has shown a decreasing trend over the last few years. Many of the underutilized *Citrus* species are on the verge of extinction due to various interventions. Considering the importance of the valuable *Citrus* wealth of this region, emphasis has been laid on collection and maintenance of *Citrus* species (Hazarika 2015).

11.16.19 Guava

Guava (*Psidium guajava* L.) is one of the richest natural sources of vitamin A and pro-vitamin A. Red-coloured varieties contains lycopene twice that of tomato. Nectar and fruit juices are a convenient and refreshing way to provide nutrition to the people. Except for jam and jelly, the consumption of other guava processed products in India is limited due to problems like grittiness and browning associated with its products. So far, red-coloured varieties are not exploited for nectar preparation in India.

11.16.19.1 Standardization of Nectar Recipe

Five red pulped varieties, namely Kohir red, Kohir long, red-fleshed, hybrid 1–6 (SRD) and Lalit, available within germplasm collections, were analysed for physicochemical characteristics (Kumar et al. 2015b). The hybrid 1–6 and red-fleshed varieties were found to be the best varieties for processing in terms of the colour of the pulp, chemical constituents such as TSS, acidity, ascorbic acid content and total sugars. Lycopene plays an important role in masking the browning of the product. Nine recipes of nectar standardized and evaluated for organoleptic qualities for two selected varieties showed that the nectar prepared with the composition of 20 % pulp, 17 °Brix TSS and 0.3 % acidity was rated superior for all quality attributes such as appearance, aroma and flavour, taste and overall acceptability.

11.16.20 Papaya

Papaya (*Carica papaya* L.) is one of the important delicious fruit crops grown in the tropical and subtropical parts of the world. Regular consumption of papaya can ensure a good supply of vitamins A and C, which are essential for good health. Papaya is rich in the enzyme papain, which helps in the digestion of proteins. Biofertilizers play a very significant role in improving soil fertility by fixing atmospheric nitrogen both in association with plant roots and without it.

11.16.20.1 Integrated Nutrient Management

Integrated nutrient management studies (Supriya et al. 2015) in Papaya revealed that treatment with biofertilizers with 100 % NPK + AZO + AZS + PSB exhibited maximum results in plant height (69.72 cm), number of functional leaves (23.8), leaf area (708 cm^2), stem girth (5.81 cm), petiole length (38.2 cm) and petiole girth (2.94 cm), number of fruits per plant (20.6), unit fruit weight (1.35 kg), fruit length (26.1 cm), fruit circumference (36 cm), fruit width (16.9 cm), fruit yield (38,955 kg ha^{-1}) and dry matter yield (4500 kg ha^{-1}).

11.17 Miscellaneous

11.17.1 Tuberose Dye

The time of immersion for tinting the spikes of tuberose cultivar for use in various food dyes is important for value addition. A study of the spikes of tuberose cultivar Suvasini treated with different food dyes and times of immersion showed that 5 % concentration of food dyes with 2 h of immersion was the optimum one, resulting in the best colour shades without edging at petals, with the maximum bud opening percentage (26.43) and the highest mean floret size (3.95).

11.17.2 Marigold

Marigold cultivar Pusa Narangi Gainda exposed to different doses of ^{60}Co gamma rays (5, 10,

15, 20, 25, 30, 35, 40 Gy) (Majumder et al. 2015) revealed that 20 Gy and 15 Gy doses were identified as LD_{50} values for in vivo and in vitro conditions (53.24 % and 52.55 % plant survival). Morphological variations existed between irradiated and non-irradiated plants. Treatment with 20 Gy induced early flower bud (49.18 days) compared to control (55.56 days) under in vivo study. Under in vitro conditions, the flower colour was changed to yellow (yellow 10 YR) compared to the control (orange 5 YR) at 15 Gy. In the M_1 generation, two mutants in flower form (M_1 and M_3), one in colour (M_4) and another one in flower earliness (M_2), were isolated under in vivo mutation, whereas five mutants in flower colour (υm1, υm2, υm5, υm6 and υm7) and two in flower form (υm3 and υm4) were isolated in irradiated plants, which were quite distinct.

Variability caused by induced mutations need not be essentially different from variability caused by spontaneous mutation during evolution. Therefore, it is necessary to carry on the M_2 generation. The mutants selected from the M_1 generation were selfed to raise the six putative mutants (pm1, pm2, pm3, pm4, pm5 and pm6) of the M_2 generation. These could successfully maintain their distinct traits. Among them, pm3 produced early flowering in 47.89 days and pm4 gave light orange-coloured flowers (orange 10 YR) (Majumder et al. 2015).

11.17.3 Breeding for Micronutrients

Micronutrient malnutrition in the recent past has emerged as a serious threat to the world's nutritional security, predominantly in developing countries. In the process of bio-fortification, breeding of essential micronutrients into widely consumed staple food crops is one of the promising strategies for combating hidden hunger. At the same time, it aids by improving availability of micronutrients in edible plant parts, in higher concentrations. Success of a bio-fortified crop relies on various important aspects, viz an efficient crop improvement programme, significantly high nutrient availability along with increased retention capacity, wider adoptability by farmers and greater acceptance by the target audience—malnourished populations (Joshi et al. 2015).

11.18 Innovative Technology for Bio-resource Management

11.18.1 Molecular Characterization Using DNA Markers

Variability exists among genotypes for several characters. Molecular markers are utilized in identifying the genetic variability for traits as disease or pest tolerance or susceptibility, evaluation of traits for yield, for abiotic or biotic responses, etc. Several molecular markers are available with wide utility. The dendrograms obtained by usage of these markers exhibit the variable degrees of relationship existing among the genotypes.

Molecular markers are utilized in identifying the similarity between the two genotypes from different locations. In cotton, the inter-simple sequence repeat DNA (ISSR) marker was used for molecular evaluation of 12 elite cotton genotypes to *Gossypium hirsutum*. The dendrogram generated by the ISSR marker showed two clusters of the genotypes of *Gossypium hirsutum* that were brought from two locations, viz Dharwad, Nanded and Parbhani. The similarity coefficient value obtained by the ISSR marker dendrogram was in the range of 0.32–0.80. Among the 12 elite genotypes evaluated, only 2 cotton genotypes, NDH-58 and NDH-88 showed the highest similarity. The genotype PH-1009 showed the lowest similarity coefficient of 0.428. ISSR primers IS-12 and IS-13 were effective in generating the maximum number of amplicons of eight in number, while ISSR primers IS-2 and IS-14could generate only four amplicons (Kalpande et al. 2015b).

11.18.2 DNA Finger Printing of Rice Cultivars

DNA finger printing is employed to determine the allelic diversity and relationship among the organisms. Closely related cultivars could be identified with the technique of fingerprinting. The sequence repeat (SSR) markers analysis that was done to determine the allelic diversity and relationship among 23 rice varieties developed by Dr B.S.K.K.V. Dapoli revealed a total of 35 SSR primer pairs to be polymorphic and generated 184 alleles with an average of 5.26 alleles per primer pair. The number of alleles amplified for each primer pair ranged from 2 to 8. The polymorphic information content (PIC) values of each primer pair ranged from 0.23 to 0.78 with an average of 0.58. The UPGMA grouped 23 rice varieties into two main clusters, which were further divided into two sub-clusters. The first major cluster consisted of 7 released rice varieties, and 16 varieties formed the second major cluster. Thus, most of the closely related cultivars of rice that were released from Dapoli were identified with the fingerprinting of the polymorphic SSR primer pairs. These results could be useful for resolving the problem that arises in the seed certification programme, as well as determination of genetic diversity of the rice varieties very quickly. Similarly, it will protect the varieties from commercial exploitation by others because of its uniqueness in amplification pattern (Nikam et al. 2015).

11.18.3 Genetic Strategies for Bio-fortification

One of the major stresses limiting crop productivity is nutrient deficiency stress. Several elements are reported to be below critical levels in the soil around the globe. Among the micronutrients, it is reported that 25–30 % of cropped area is deficient in Zn globally. This not only limits the productivity of crops but also means it is in short supply for human nutrition, where approximately one third of the world's population is suffering from zinc deficiency. In view of this, there is an urgent need to develop Zn-dense crops and therefore bio-fortification is considered a cost-effective and long-lasting solution to overcome zinc malnutrition. Micronutrient-dense crops can be obtained through consistent efforts through genetic strategies to obtain long-lasting outcomes.

Among various genetic strategies that are available, genetic transformation and molecular breeding approaches have been attempted in tomato and pigeon pea, respectively, for bio-fortification with zinc.

Tomato was transformed with the high-affinity zinc transporter gene through a standardized tissue culture protocol. A set of 250 pigeon pea accessions were screened to identify the high grain zinc donor lines. A biparental mapping population was developed and an SSR marker system was used to screen for parental polymorphism. Estimation of zinc content in the F_3 seeds resulted in identification of several promising transgressive segregants (Mathapati et al. 2015).

11.18.4 Genetic Stability of In Vitro Multiplication

Medicinal plants, because of their excessive usage and deforestation, are being lost and are becoming endangered. An endangered medicinal plant was efficiently regenerated by in vitro plant regeneration via an indirect organogenesis leaf-derived callus for the endangered medicinal plant. An optimal level of callus was developed from young and fresh leaves cultured on Murashige and Skoog (MS) medium supplemented with 4.44 μM 6-benzyladenine (BA), 5.37 μM 1-naphthalene acetic acid (NAA). The highest frequency of shoot multiplication was observed on MS medium supplemented with 13.62 μM thidiazuron (TDZ) by gradual sub-culturing of the proliferated callus mass, at which 89 % shoot multiplication was obtained, with an average number of 23.8 shoots per callus. SCoT (start codon targeted polymorphism) markers were employed to evaluate the genetic stability of in vitro regenerants with that of the donor plant. The analysis of 18 selected in vitro regenerants that was done using 10 responded SCoT primers revealed 44 scorable bands (330–1690 bp). The banding pattern of each primer confirmed the genetic uniformity of the regenerants with its donor plant and substantiated the efficacy and suitability of this protocol for in vitro propagation of this potent medicinal sub-shrub.

A protocol that was developed would be helpful in reducing the pressure on natural populations for secondary metabolite production, especially for extraction of essential oils and alkaloids (Seth and Panigrahi 2015).

11.18.5 Genetic Variability in Wheat Using RAPD Marker

Genetic variability among 25 wheat genotypes was analysed by a random amplified polymorphic DNA (RAPD) marker method for assessing the link between the different accessions of wheat to exploit genetic diversity for a future breeding programme. The results revealed the existence of high genetic diversity among the wheat genotypes studied. Out of 25 primers, 11 selected primers could amplify 23 clear and identifiable bands, of which 15 bands were polymorphic, accounting for 65.22 % of genetic polymorphism. The remaining primers were not amplified for the set of genotypes that were studied. All the wheat genotypes studied could be divided into two major groups. Jaccard's coefficient of similarity ranged from 0.54 to 0.95. Several polymorphic bands were also observed in different genotypes, which helped in molecular diversity analysis of these genotypes. The wheat genotypes were found to be clustered into two major groups; the first group includes 15 genotypes and the second included 10 genotype. The highest similarity was observed between AKAW-3717 and WSM-1472 genotypes. The most distant genotypes were AKAW-2344 and AKAW-4073. RAPD technology is a potentially simple, rapid, reliable and effective method of detecting polymorphism for assessing genetic diversity between genotypes, and these help in the selection of parents for hybridization (Ninghot et al. 2015).

11.18.6 Genetic Diversity Under Temperature Stress

Wheat production in many environments around the world is restricted due to heat stress, which has a serious impact on productivity. The effects of heat stress are seen during both germination and grain-filling stages. Terminal heat stress refers to the rise in temperature at the time of grain filling. This terminal heat stress is more responsible for the decline in wheat production and productivity. Among the wheat-cultivated six agroclimatic zones, four zones experience high temperature stress and cover an area of 13.5 m ha that is under heat stress. The previous 11 QTLs associated and identified with the stress traits, when validated in an Indian genetic background, indicated that among the QTL linked markers, 11 markers produced distinct and sharp scorable bands for all the 38 genotypes grown in two different environments.

Further, out of these 11, 6 microsatellites produced monomorphic bands without any differentiation to the QTL linked population parents. Five microsatellite markers that were polymorphic were selected for validation of the Indian wheat genetic backgrounds for different heat stress adaptive traits. These markers lay in different genomic regions on the 1B, 2D, 5A, 5D and 7D chromosomes. Two microsatellite markers (gwm131 and gwm304) showed significant marker-trait associations particularly with the canopy temperature depression and total chlorophyll content under heat stress; besides producing genotypic data with QTL linked SSR markers the phenotypic variances explained by these markers can be further confirmed in mapping populations or backcross populations and thus can be used subsequently in future breeding programmes as donors of the traits (Gajghate et al. 2015).

Genetic diversity and association analysis were carried out in 35 wheat genotypes using 17 morpho-physiological characters under high temperatures during grain-filling stage conditions. Based on genetic distance, 35 genotypes were grouped into 5 multi-genotypic clusters. Clusters III and IV had a maximum ten genotypes each, while cluster II had three genotypes only. Both cluster I and V each consisted of six genotypes. Maximum intra-cluster distance (3.826) was exhibited by the cluster III followed by cluster I (3. 125) while minimum intra-cluster distance was observed in cluster II (2.521). Clusters III and IV showed maximum inter-cluster distance (5.446) followed by clusters III and V (5.141) and minimum (3.379) between cluster I and IV. Based on per se performance and clustering pattern under stress conditions, genotypes HD 2967, HD 2997, HD 2733, CL 3125, WRY 544 and DEW 14 emerged as potential genotypes for future use in recombination breeding. Utilization of these parents in a hybridization programme could be useful in getting a wide spectrum of variability to isolate the desirable segregants for high yield under terminal heat stress. Grain yield had a significant genotypic positive correlation with biological yield (0.754**) and harvest index (0.507**). Biological yield had a positive significant correlation with number of spike per plant (0.415**). Harvest index showed a significant positive correlation with 1000 grain weight (0.334*) and biological yield (0.343*).The path analysis based on genotypic correlations revealed that total chlorophyll content had maximum direct effect on grain yield followed by carotenoid content, harvest index, 1000 grain weight, and days to flowering, canopy temperature depression and grains per spike. Therefore, for improving the grain yield under terminal heat stress conditions, the breeder should aim for selecting genotypes with a higher harvest index, more spikes per plant, bold grains and a stay-green characteristic (Kumar et al. 2015a).

11.18.7 Reciprocal Recurrent Selection

Reciprocal recurrent selection might be the most efficient for utilization of all three types of gene effects, viz additive, dominance and epitasis. These three gene effects were found to be responsible in varying proportions for all the crosses for seven characters in safflower. Among the epistatic interactions, dominance × dominance interaction was found to be more important than additive × additive interaction for all the crosses. Thus, for improving the characters, the reciprocal recurrent selection might be the most efficient one for utilization in breeding programmes (Shivani and Sreelakshmi 2015).

Genetic diversity of 10 soybean cultivars of *Glycine max* L. was carried out using 50 RAPD primers of the OPA, OPAC, OPC, OPB, OPG, OPL, OPH, OPR, OPX series. Twelve selected polymorphic RAPD markers produced 90 fragments of which 64 fragments were found to be polymorphic and resulted in 72.74 % polymorphism. All cultivars could be distinguished based on the RAPD profiles unweighted pair group method with the arithmetic average (UPGMA) used to construct a dendrogram. The value of similarity coefficient of the dendrogram calculated by the RAPD marker ranged from 0.313 to 0.618. In addition, cluster analysis was used to determine genetic variation among soybean cultivars. The narrow genetic base that exists among the soybean cultivars can be used for the selection of divergent parents for breeding and mapping purposes (Kalpande et al. 2015c).

11.19 Climate Change—Impact, Adaptation and Mitigation

11.19.1 Genetic Diversity in Rice Genotypes Resistant to Leaf Folder

Rice genotypes exhibit variability in terms of resistance or susceptibility to pests or diseases. The variability is evaluated by molecular markers. In rice, 20 genotypes exhibiting resistant and susceptible reactions against rice leaf folder were characterized by using 11 primers of SSR markers. Out of these 11 primers of SSR markers, 7 primers were found to be polymorphic exhibiting 78 % polymorphism with primers RM 3691, RM 600, RM 14 and RM 228. These 11 primers of SSR markers resulted in a total of 123 amplified bands. The band size was in the range of 100–1000 bp in polymorphic primers. One primer, RM 3691, resulted in a clear variation in the banding pattern between the resistant and susceptible rice genotypes for leaf folder. The Jaccard's similarity values indicate the presence of wide-range genetic diversity. In rice genotypes these values ranged from 0 to 89 %, indicating that there is a wide range of variability for resistance or susceptible to leaf folder. Further, the dendrogram exhibited a variable degree of relationship among the genotypes. There were ten genotypes that showed resistance and ten that exhibited susceptibility reactions to the leaf folder. These were grouped into two different clusters indicating that these two groups are significantly different even at the molecular level. Among these genotypes, in cluster one, two genotypes (INRC- 3021 and W 1263) were found to be closely related while in cluster II, NLR 145 and Sri Satya were closely related with an 86–89 % similarity index (Kamakshi et al. 2015).

11.19.2 Expression of Bt Cry Toxins

Extent of Cry proteins in leaves is crucial to the protection expected against insects feeding on cotton leaves. Bt toxin levels in cotton genotypes at two growth stages, viz the vegetative and reproductive stages, and in different plant parts (apical leaves, bottom leaves and squares) were quantified. The quantitative levels of Bt toxin in cotton genotypes were found to be higher during the vegetative stage (90 DAS) and decreased as the reproductive stage (120 DAS) approached. Further, it was observed that Cry 1Ac levels were higher than those of Cry 2A throughout the growth phases and the Cry toxin level decreased as the plant attained maturity (Tokas et al. 2015).

11.19.3 Molecular Characterization of Cultured Bacteria for Antivibrio Property

Microorganisms, especially bacteria, utilize carbon as a source of energy for various metabolic activities. One of the locally available carbon sources of energy is cassava waste. It is cheaply available in local areas. Several studies have confirmed that it is more effective in reducing metabolites and increasing shrimp growth. However, although it is a cheap substrate, a toxic chemical (viz hydrocyanic acid) is present in it, affecting the metabolic activities of the organisms. The mechanisms behind the toxicity of hydrocyanic acid are not well understood. A study that was conducted to study the leaching toxic effect of hydrocyanic acid from cassava waste indicated that during the soaking of cassava waste in brackish water, it results in the formation a bio-film over the waste material. This bio-film was formed due to the microbial activity. The hydrocyanic acid did not leach in to the water due to the formation of this bio-film; as a result, its toxicity for the shrimp culture could not be detected.

Bacteria from the bio-films were cultured and characterized using 16sRNA universal primers and NCBI BLAST of isolates. The results confirmed the presence of five *Shewanella* species (1386–1413 bp) having the potential for antivibrio property. Certain in vitro studies of the antivibrio property of *Shewanella* species isolates revealed that not all the Shewanella species isolates exhibited the antivibrio property against *Vibrio harveyi.* An isolate of *Shewanella* species isolate at a concentration of 105–6 cfu ml^{-1} could inhibit the growth of pathogenic *Vibrio parahaemolyticus*. The study showed that cassava waste was non-toxic and resulted in improvement of shrimp growth due to the antivibrio property of bacteria that developed on the bio-film of the cassava waste (Lalitha et al. 2015).

11.19.4 Genetic Engineering for Disease Resistance

Plants show resistance to certain pathogens and pests, and this resistance has no doubt played a key role in crop protection since the dawn of agriculture. Use of resistant varieties of host plants is considered the cheapest, easiest, safest and most effective device to control plant diseases. Moreover, for a variety of diseases, such as those caused by vascular pathogens and viruses, which cannot be adequately controlled by any available means, the use of resistant varieties provides the only means to obtain acceptable yields. There are several conventional methods of breeding for disease resistance, but it is a crucial development. Biotechnological advances in the field of tissue culture technology and recombinant DNA technology (genetic engineering) have opened a completely new array of possibilities to raise resistance in hosts to diseases.

Genetic engineering, also called genetic modification, is human manipulation of an organism's genetic material in a way that does not occur under natural conditions. Genes expected to confer resistance are isolated, cloned and transferred into the crop in question. There are numerous examples of plants engineered (transgenic plants) for improved resistance to pathogens. In the case of fungal and bacterial pathogens, transgenes coding for toxin inactivation, antibacterial peptides, lysozyme, phytoalexins, PR-proteins, RIP (ribosome inactivating proteins) and chitinases have been employed for the expression of resistance. In the case of viral pathogens, several transgenes have been evaluated—virus coat protein gene, DNA copy of viral satellite RNA, defective viral genome, antisense construct of critical viral genes and ribozymes. The viral coat protein gene approach is the most successful. However, the large majority of transgenes are still at the experimental stage and are under trials in model crops such as tobacco and potato (Deepthi et al. 2015).

11.19.5 Resistance Gene Candidates

Degenerate primers based on the conserved motifs of plant disease resistance (R) genes were used to isolate analogous sequences called resistance gene candidates (RGCs). The NBS, LRR and kinase domain are characteristic features of many plant resistance genes. The NBS domain has many conserved amino acid sequences among different R genes. Chilli (*Capsicum annuum* L.) production is seriously affected by root knot nematode disease, for which no resistant source is available in the cultivated germplasm. The NBS domain was analysed in one of the studies on chilli by the use of degenerate primers designed to the conserved region after PCR amplification.

The RGCs that were isolated from chilli showed sequence similarity with the NBS region of other known R genes, viz RPM1, CaMi, RPS2, N, L6, Lettuce R gene and RPP5. Multiple alignment of isolated chilli RGCs using CLUSTALW grouped these into two classes. An amino acid signature in the conserved motifs comprising the NBS domain clearly distinguished the two classes. All chilli RGCs belonged to the class of the R gene with the LZ (leucine zipper) domain in the NBS region and TIR domain and three RGCs (CaRGC4, CaRGC8, CaRGC10) of chilli showed separate grouping.

Sixty RAPD markers were analysed by bulk segregant analysis to identify the resistant and susceptible germplasm of chilli wherein none have resulted in any polymorphism between the bulks (Khagandramani et al. 2015).

11.19.6 Molecular Variability of Fusarium udum Isolates

Wilt is one of the most widespread red gram diseases in the country and has been a constraint on red gram productivity. Morphological as well as molecular studies can be utilized for assessing the pathogen diversity. These studies were used to characterize the 14 isolates of *Fusarium udum* collected from red gram cultivars of different red gram-growing regions in the Southern Telangana Zone (STZ) of erstwhile Andhra Pradesh, India (presently in Telangana state) to assess this pathogen diversity. All the isolates exhibited considerable morphological and cultural variability in terms of colony growth, pigmentation and external texture. RAPD primers (98) used to test the molecular variability among the isolates indicated that though all the markers could detect high levels of polymorphism among the fungal pathogen isolates, the degree of polymorphism varied depending on the marker that was selected. OPK and OPI series markers detected high levels of polymorphism among the *Fusarium udum* isolates. NTSYSpc (version 2.0) cluster analysis based on Jaccard's similarity coefficient grouped all the isolates into two major distinct groups with a minimum of 5 % similarity index. This suggested that there was the existence of a minimum of two specific races of the pathogen that were prevailing in the pigeon pea growing areas of STZ. The maximum variation was seen among the pathogen isolates collected from Rangareddy compared to others. Fu11 and Fu12 isolates of *Fusarium udum* showed 50 % similarity with most of the RAPD markers. M3 and M4 isolates of the Mahaboobnagar district were grouped separately from the rest of the isolates of Rangareddy district (at a 20 % similarity index) with around 60 % similarity among the two isolates, suggesting high levels of variability among the isolates that were collected from the same agroclimatic region of Andhra Pradesh (Suresh et al. 2015).

11.19.7 Molecular Characterization for Karnal Bunt Resistance

The increasing wheat yields in the country of India have opened the option of exports. For Indian wheat to be exported readily, there is a need to satisfy the quarantine requirements of international standards.

Karnal bunt (KB) disease of wheat has been a hindrance to wheat export. KB caused by *Neovossia indica* (Mitra) Mundkur (syn. *Tilletia indica*) is an important wheat disease with implications for wheat quality and trade. The disease reduces grain quality and inflicts changes on the chemical composition of infected grains. The infected portions of the kernel are replaced with masses of dark, fishy-smelling fungal teliospores, rendering the grains unfit for human consumption. Although yield losses are usually low, international quarantine policies against the disease may restrict the free flow of the global wheat trade.

Currently, resistance breeding is being pursued as the main strategy for its management. The development of resistant varieties through resistance breeding is difficult because of the considerable influence of environment on disease expression and limited variability for KB resistance in hexaploid wheat.

Near-isogenic lines (NILs) were developed as a breeding strategy on the background of the famous wheat cultivar PBW 343. A backcross breeding approach was utilized to incorporate the KB resistance from an already registered genetic stock KBRL 22 in the variety PBW 343. The NILs so developed were evaluated for resistance at every generation of selfing after six backcrosses. One of these resistant lines, KB 2012–03, was evaluated for yield across five different locations in the northwestern plains zone of India. Simultaneously, 194 SSR markers were used to characterize the parents, PBW 343 and KBRL 22 and the NIL KB 2012–03. Out of these, 47 markers were found to be polymorphic between the parents. Eleven SSR markers were able to characterize the regions specific to KBRL 22. The multi-location yield testing also revealed that on an average basis the NIL KB 2012–03 (42.3 q ha^{-1}) yield was significantly superior to the parental check PBW 343 (35.6 q ha^{-1}). Thus, the lines identified in this study can be utilized as a source of KB and high yield (Kumar et al. 2015c).

11.19.8 Suppression Subtractive Hybridization

Stress-inducible genes in response to gradual water-deficit, high temperature and salt stresses can be identified through suppression subtractive hybridization. Signalling among the different pathways may induce various types of genes under the different types of abiotic or biotic stresses. Pearl millet is an important cereal crop that withstands drought, high temperature and salt stresses and thus has an advantage as a source of desirable genes for genetic engineering. Despite this, it is considered as an orphan crop because of its vast untapped potential. Hence, any progress for the molecular perception of different abiotic stress tolerance mechanisms in this crop may provide a basis for developing multiple stress-tolerant genotypes. Three cDNA libraries were constructed using a suppressive subtractive hybridization (SSH) technique to identify novel genes expressed under gradual water-deficit (dehydration in dessicator), high temperature (42 °C) and salt (250 mM NaCl) stresses in pearl millet (ICMR 356) seedlings. Of the 202, 158 and 141 ESTs analysed from water-deficit, high temperature and salt stressed libraries, 77 %, 83 % and 78 % of the clones were identified to be stress responsive in nature, respectively, while the uncharacterized genes contributed to 13–15 % of the transcriptome that provided new candidate genes for investigation to elucidate their role in stress tolerance. Classification of commonly and differentially expressed ESTs as represented in the Venn diagram revealed the induction of 104, 102 and 110 uniESTs specifically in water-deficit, high temperature and salt stresses, respectively, while 8 genes [ERD1 (early responsive to dehydration 1), zinc finger domain-containing protein, calcium-dependent protein kinase, glutathione-S-transferase, phosphoinositide phosphatase protein, trypsin proteinase inhibitor, phospholipase D delta and ubiquitin protein ligase) were commonly induced under all the three stresses. This indicated the crosstalk between the signalling pathways (Vijayalakshmi et al. 2015).

11.19.9 SSR Markers in Stigma Exertion in Maintainer Lines of Hybrid Rice

In order to study the stigma exertion in maintainer lines of hybrid rice, 454 SSR markers were used; out of these, 118 markers exhibited polymorphism. Among these, the three best polymorphic markers on each chromosome covering all 12 chromosomes were employed for selective genotyping of 188 F_2 mapping population derived from APMS-6B, a low stigma exertion parent, and BF-16B, a high stigma exertion parent, and one marker on each chromosome and two reported markers were used for single marker analysis. For phenotypic study, parents, F_1 and F_2 population were evaluated for stigma exertion trait in order to understand the inheritance as well as linkage with SSR markers. A co-segregation analysis with respect to marker (based on the amplification pattern of the marker) and trait phenotype revealed that three markers RM20060 on chromosome 6, RM5647 on chromosome 8 and RM4771 on chromosome 10 showed least recombination frequency and these three were markers found to be linked to the stigma exertion trait (Sruthi et al. 2015).

References

Abubakker S, Jabbar AB, Patil VS et al (2015) Genetic variability study for yield contributing and drought resistance characters in groundnut (*Arachis hypogaea* L.). In: Compendium of abstracts of the 2nd international conference on bio-resource and stress management, ANGRAU & PJTSAU, Hyderabad, India, 7–10 January 2015

Bhadru D, Krishna L, Pasha ML (2015) Effect of environment on genetic parameters of rice (*Oryza sativa* L.). In: Compendium of abstracts of the 2nd international conference on bio-resource and stress management, ANGRAU & PJTSAU, Hyderabad, India, 7–10 January 2015

Bhalerao SP, Boratkar MV, Jawale LN (2015) Genetic variability, correlation and path analysis for yield, yield contributing and quality traits in wheat (*T. aestivum*) genotypes. In: Compendium of abstracts of the 2nd international conference on bio-resource and stress management, ANGRAU & PJTSAU, Hyderabad, India, 7–10 January 2015

Bodkhe RB, Pole SP, Mahajan RC et al (2015) Genetic variation and trait relationships in soybean (*Glycine max* L. Merill.). In: Compendium of abstracts of the 2nd international conference on bio-resource and stress management, ANGRAU & PJTSAU, Hyderabad, India, 7–10 January 2015

Burse N, Gawande PP, Solanke A (2015) Correlation and path coefficient analtsis studies of yield components in okra (*Abelmoschus esculentus* (L.) Moench). In: Compendium of abstracts of the 2nd international conference on bio-resource and stress management, ANGRAU & PJTSAU, Hyderabad, India, 7–10 January 2015

Das A, Sarkar A (2015) Genotypic variability and path analysis for nodulation and seed yield in cowpea (*Vigna unguiculata* L. Walp). In: Compendium of abstracts of the 2nd international conference on bio-resource and stress management, ANGRAU & PJTSAU, Hyderabad, India, 7–10 January 2015

Das AS, Sarkar N, Das B et al (2015a) Response to slection for basal branching in bilocular segregants in *Brassica rapa* L. var. Yellow Sarson. In: Compendium of abstracts of the 2nd international conference on bio-resource and stress management, ANGRAU & PJTSAU, Hyderabad, India, 7–10 January 2015

Das R, Pandravada SR, Sujatha M (2015b) Screening for nutrient rich finger millet (*Eleusine coracana* (L.)

Gaertn) from diverse source of germplasm lines. In: Compendium of abstracts of the 2nd international conference on bio-resource and stress management, ANGRAU & PJTSAU, Hyderabad, India, 7–10 January 2015
Deepthi P, Reddy ESS, Shukla CS et al (2015) Genetic engineering for disease resistance. In: Compendium of abstracts of the 2nd international conference on bio-resource and stress management, ANGRAU & PJTSAU, Hyderabad, India, 7–10 January 2015
Deosarkar DB, Patil HV, Kalpande HV (2015) Identification of potential parents and hybrids in intraspecific and interspecific crosses of upland cotton. In: Compendium of abstracts of the 2nd international conference on bio-resource and stress management, ANGRAU & PJTSAU, Hyderabad, India, 7–10 January 2015
Deshmukh DT, Zarea BA (2015) Stability analysis in forage sorghum [Sorghum bicolor (L.) Moench]. In: Compendium of abstracts of the 2nd international conference on bio-resource and stress management, ANGRAU & PJTSAU, Hyderabad, India, 7–10 January 2015
Deshmukh JD, Deosarkar DB, Kalyankar SV et al (2015) Stability analysis in cotton (*Gossypium hirsutum* L.). In: Compendium of abstracts of the 2nd international conference on bio-resource and stress management, ANGRAU & PJTSAU, Hyderabad, India, 7–10 January 2015
Dhuppe MV, Khanpate VB, Umate, UM et al (2015) Combining ability studies in sesame (*Sesamum indicum* L.). In: Compendium of abstracts of the 2nd international conference on bio-resource and stress management, ANGRAU & PJTSAU, Hyderabad, India, 7–10 January 2015
Diyali S, Bhanu Priya, Mukherjee S et al (2015) Character association and path coefficient analysis among yield attributing traits in bread wheat. In: Compendium of abstracts of the 2nd international conference on bio-resource and stress management, ANGRAU & PJTSAU, Hyderabad, India, 7–10 January 2015
Gaikwad AJ, Deshmukh DT, Sakhare SB (2015) Genetic divergence studies for fodder and related traits in forage sorghum (*Sorghum bicolor* (L.) Moench). In: Compendium of abstracts of the 2nd international conference on bio-resource and stress management, ANGRAU & PJTSAU, Hyderabad, India, 7–10 January 2015
Gajghate R, Sharma RK, Jain N et al (2015) Validation of known molecular markers associated with QTLs for terminal heat tolerance traits in bread wheat. In: Compendium of abstracts of the 2nd international conference on bio-resource and stress management, ANGRAU & PJTSAU, Hyderabad, India, 7–10 January 2015
Gangurde SS, Moharil MP, Ingle AU et al (2015) C_4 rice: a step towards evergreen revolution. In: Compendium of abstracts of the 2nd international conference on bio-resource and stress management, ANGRAU & PJTSAU, Hyderabad, India, 7–10 January 2015
Gedam SR, Solanke A, Dudhate S et al (2015) Genetic analysis of CMS based hybrids in upland cotton (*Gossypium hirsutum* L.). In: Compendium of abstracts of the 2nd international conference on bio-resource and stress management, ANGRAU & PJTSAU, Hyderabad, India, 7–10 January 2015
Ghorade RB, Kalpande VV, Bhongle SA et al (2015a) Path coefficient analysis of grain yield and yield contributing characters in *kharif* sorghum. In: Compendium of abstracts of the 2nd international conference on bio-resource and stress management, ANGRAU & PJTSAU, Hyderabad, India, 7–10 January 2015
Ghorade RB, Kalpande VV, Bhongle SA et al (2015b) Some promising lines and testers for grain yield and yield components in *rabi* sorghum under drought condition. In: Compendium of abstracts of the 2nd international conference on bio-resource and stress management, ANGRAU & PJTSAU, Hyderabad, India, 7–10 January 2015
Ghorade RB, Kalpande VV, Bhongle SA et al (2015c) Development of some promising high yielding and early maturing cross combinations in *kharif* sorghum. In: Compendium of abstracts of the 2nd international conference on bio-resource and stress management, ANGRAU & PJTSAU, Hyderabad, India, 7–10 January 2015
Ghorade RB, Kalpande VV, Kale P et al (2015d) Studies on some of the good general combiners for development of dual purpose *kharif* sorghum hybrids. In: Compendium of abstracts of the 2nd international conference on bio-resource and stress management, ANGRAU & PJTSAU, Hyderabad, India, 7–10 January 2015
Ghorade RB, Kalpande VV, Ranjith P et al (2015e) Path coefficient analysis for shoot fly resistance traits in *kharif* sorghum. In: Compendium of abstracts of the 2nd international conference on bio-resource and stress management, ANGRAU & PJTSAU, Hyderabad, India, 7–10 January 2015
Gomashe S, Ganapathy KN, Jain SK et al (2015) Photoperiod response, stability and GGE biplot analysis in sorghum (*Sorghum bicolor* (L.) Moench). In: Compendium of abstracts of the 2nd international conference on bio-resource and stress management, ANGRAU & PJTSAU, Hyderabad, India, 7–10 January 2015
Gopal GR, Gawande NG, Patil AE et al (2015) Genetic variability, correlation and path analysis studies in safflower (*Carthamus tinctorius* L.). In: Compendium of abstracts of the 2nd international conference on bio-resource and stress management, ANGRAU & PJTSAU, Hyderabad, India, 7–10 January 2015
Hariprasanna K, Shashikanth D, Suresh P et al (2015a) Genetic variability for morpho-agronomic descriptors in foxtail millet (*Setaria italica* (L.) P. Beauv.) accessions. In: Compendium of abstracts of the 2nd international conference on bio-resource and stress management, ANGRAU & PJTSAU, Hyderabad, India, 7–10 January 2015

Hariprasanna K, Suresh P, Aruna C et al (2015b) Stability in the expression of morpho-agronomic descriptors over seasons in sorghum [*Sorghum bicolor* (L.) Moench]. In: Compendium of abstracts of the 2nd international conference on bio-resource and stress management, ANGRAU & PJTSAU, Hyderabad, India, 7–10 January 2015

Hazarika BN (2015) Genetic diversity of citrus in North East India. In: Compendium of abstracts of the 2nd international conference on bio-resource and stress management, ANGRAU & PJTSAU, Hyderabad, India, 7–10 January 2015

Ingole DG, Lande SS, Gawande NG et al (2015) Path analysis studies in Safflower (*Carthamus tinctorius* L). In: Compendium of abstracts of the 2nd international conference on bio-resource and stress management, ANGRAU & PJTSAU, Hyderabad, India, 7–10 January 2015

Jadhav PA, Kathale MN, Gulave MS et al (2015a) Induced quantitative mutagenesis in okra (*Abelmoschus esculentus* (L.) Moench). In: Compendium of abstracts of the 2nd international conference on bio-resource and stress management, ANGRAU & PJTSAU, Hyderabad, India, 7–10 January 2015

Jadhav PA, Kathale MN, Gulave MS et al (2015b) Mutagenic effectiveness and efficiency of gamma rays and ethyl methane sulphonate in okra (*Abelmoschus esculentus* (L.) Moench). In: Compendium of abstracts of the 2nd international conference on bio-resource and stress management, ANGRAU & PJTSAU, Hyderabad, India, 7–10 January 2015

Janaki M, Ramana CV, Naidu LN et al (2015) Genetic divergence in Chilli (*Capsicum annuum* L.). In: Compendium of abstracts of the 2nd international conference on bio-resource and stress management, ANGRAU & PJTSAU, Hyderabad, India, 7–10 January 2015

Joshi T, Singh H, Singh L et al (2015) Emerging trends in biofortification- one step solution to combat hidden hunger. In: Compendium of abstracts of the 2nd international conference on bio-resource and stress management, ANGRAU & PJTSAU, Hyderabad, India, 7–10 January 2015

Kahate NS, Toprope VN, Kathale MN et al (2015) Variability and heritability studies for yield and late leaf spot resistance in groundnut (*Arachis hypogaea* L.). In: Compendium of abstracts of the 2nd international conference on bio-resource and stress management, ANGRAU & PJTSAU, Hyderabad, India, 7–10 January 2015

Kalpande HV, Chinchane VN, Chavan, SK et al (2015a) Heterosis and combining ability for yield and fibre quality in *desi* cotton (*Gossypium arboreum* L.). In: Compendium of abstracts of the 2nd international conference on bio-resource and stress management, ANGRAU & PJTSAU, Hyderabad, India, 7–10 January 2015

Kalpande HV, Chavan SK, Deshmukh JD et al (2015b) Molecular evaluation of elite cotton germplasm using DNA marker. In: Compendium of abstracts of the 2nd international conference on bio-resource and stress management, ANGRAU & PJTSAU, Hyderabad, India, 7–10 January 2015

Kalpande HV, More AW, Chavan SK et al (2015c) Genetic diversity analysis of soybean (*glycine max* l.) cultivars revealed by pcr based molecular markers. In: Compendium of abstracts of the 2nd international conference on bio-resource and stress management, ANGRAU & PJTSAU, Hyderabad, India, 7–10 January 2015

Kalpande VV, Ghorade RB, Bhongle SA et al (2015d) Sorghum genotype SPV 1786 with high yield and excellent. In: Compendium of abstracts of the 2nd international conference on bio-resource and stress management, ANGRAU & PJTSAU, Hyderabad, India, 7–10 January 2015

Kamakshi N, Rajasekhar P, Ramachandra Rao G et al (2015) Characterisation of genetic diversity of rice genotypes showing resistance to rice leaffolder using SSR markers. In: Compendium of abstracts of the 2nd international conference on bio-resource and stress management, ANGRAU & PJTSAU, Hyderabad, India, 7–10 January 2015

Kanatti A, Radhika K, Padma V et al (2015) Intra-population variability for grain iron and zinc densities in commercial open-pollinated varieties of pearl millet. In: Compendium of abstracts of the 2nd international conference on bio-resource and stress management, ANGRAU & PJTSAU, Hyderabad, India, 7–10 January 2015

Karadi A, Kajjidoni ST (2015) Genetic assessment of minicore collections for grain quality and productivity related traits in sorghum during postrainy season. In: Compendium of abstracts of the 2nd international conference on bio-resource and stress management, ANGRAU & PJTSAU, Hyderabad, India, 7–10 January 2015

Kathale MN, Gulave MS, Jadhav PA et al (2015a) Heterosis and yield contributing traits in maize. In: Compendium of abstracts of the 2nd international conference on bio-resource and stress management, ANGRAU & PJTSAU, Hyderabad, India, 7–10 January 2015

Kathale MN, Chavan AA, Kahate NM et al (2015b) Line × Tester Analysis in Pearl millet (*Pennisetum glaucum* L.). In: Compendium of abstracts of the 2nd international conference on bio-resource and stress management, ANGRAU & PJTSAU, Hyderabad, India, 7–10 January 2015

Kathale MN, Chavan AA, Kahate NM et al (2015c) Heterosis in pearl millet (*Pennisetum glaucum* L.). In: Compendium of abstracts of the 2nd international conference on bio-resource and stress management, ANGRAU & PJTSAU, Hyderabad, India, 7–10 January 2015

Kathale MN, Gulave MS, Jadhav PA et al (2015d) Combining ability and yield contributing traits in maize). In: Compendium of abstracts of the 2nd international conference on bio-resource and stress management,

ANGRAU & PJTSAU, Hyderabad, India, 7–10 January 2015

Katiyar M, Prakash HG, Pandey RK et al (2015) Genetic enhancement of Lentil (*Lens culinaris* Medik) through introgression of genes from wild lens species. In: Compendium of abstracts of the 2nd international conference on bio-resource and stress management, ANGRAU & PJTSAU, Hyderabad, India, 7–10 January 2015

Khagandramani M, Sivaramakrishnan S, Raghuprakash K (2015) Cluster analysis of the RAPD marker and rgcs sequence data in chillis. In: Compendium of abstracts of the 2nd international conference on bio-resource and stress management, ANGRAU & PJTSAU, Hyderabad, India, 7–10 January 2015

Khan H, Latha HC, Sowmya HC (2015) Study of genetic variability parameters in groundnut (*Arachis hypogaea* L.) genotypes. In: Compendium of abstracts of the 2nd international conference on bio-resource and stress management, ANGRAU & PJTSAU, Hyderabad, India, 7–10 January 2015

Khichi P, Chandan PM, Chauhan J et al (2015) Performance of different varieties of garden pea. In: Compendium of abstracts of the 2nd international conference on bio-resource and stress management, ANGRAU & PJTSAU, Hyderabad, India, 7–10 January 2015

Khillari AV, Patil AE, Gawande NG et al (2015) Genetic divergence studies in lathyrus (*Lathyrus sativus* L). In: Compendium of abstracts of the 2nd international conference on bio-resource and stress management, ANGRAU & PJTSAU, Hyderabad, India, 7–10 January 2015

Krishna L, Surender RC, Sudheer Kumar S et al (2015) Correlation studies in aromatic rice (*Oryza sativa* L.). In: Compendium of abstracts of the 2nd international conference on bio-resource and stress management, ANGRAU & PJTSAU, Hyderabad, India, 7–10 January 2015

Kumar M, Sharma RK, Kumar P et al (2015a) Genetic diversity and association studies with reference to terminal heat tolerance in bread wheat (*Triticum aestivum*). In: Compendium of abstracts of the 2nd international conference on bio-resource and stress management, ANGRAU & PJTSAU, Hyderabad, India, 7–10 January 2015

Kumar MP, Girwani A, Dilip Babu J et al (2015b) Standardization of nectar recipe with red pulped. In: Compendium of abstracts of the 2nd international conference on bio-resource and stress management, ANGRAU & PJTSAU, Hyderabad, India, 7–10 January 2015

Kumar S, Mishra CN, Gupta V et al (2015c) Molecular characterization and yield evaluation of Near Isogenic Line (NIL) of wheat cultivar PBW 343 developed for Karnal Bunt resistance. In: Compendium of abstracts of the 2nd international conference on bio-resource and stress management, ANGRAU & PJTSAU, Hyderabad, India, 7–10 January 2015

Ladole MY, Deshmukh SN, Mane PN et al (2015) Effect of diverse environments on performance and stability of groundnut (*Arachis hypogaea* L.) genotypes for pod yield and its component traits. In: Compendium of abstracts of the 2nd international conference on bio-resource and stress management, ANGRAU & PJTSAU, Hyderabad, India, 7–10 January 2015

Lalitha N, Muralidhar M, Saraswathy R et al (2015) *In vitro* studies on Hydrocyanic acid toxicity and Antivibrio effect of bio film of cassava waste. In: Compendium of abstracts of the 2nd international conference on bio-resource and stress management, ANGRAU & PJTSAU, Hyderabad, India, 7–10 January 2015

Madhuri A, Singh HC, Sharma P et al (2015a) Genetic analysis of heterotic crosses in Rice (*Oryza sativa*) under rainfed ecosystem. In: Compendium of abstracts of the 2nd international conference on bio-resource and stress management, ANGRAU & PJTSAU, Hyderabad, India, 7–10 January 2015

Madhuri S, Deshmukh SB, Solanke A et al (2015b) Combining ability and heterosis for seed yield, Its Components and Quality Traits in *Gossypium hirsutum* L. In: Compendium of abstracts of the 2nd international conference on bio-resource and stress management, ANGRAU & PJTSAU, Hyderabad, India, 7–10 January 2015

Majumder J, Singh SK, Singh KP et al (2015) *In vivo* and *in vitro* mutagenesis in marigold (*Tagetes erecta* L.) using ^{60}Co gamma rays. In: Compendium of abstracts of the 2nd international conference on bio-resource and stress management, ANGRAU & PJTSAU, Hyderabad, India, 7–10 January 2015

Mathapati BR, Yamuna Rani BR, Iranacolaivalu M et al (2015) Genetic strategies for biofortification of food crops with zinc. In: Compendium of abstracts of the 2nd international conference on bio-resource and stress management, ANGRAU & PJTSAU, Hyderabad, India, 7–10 January 2015

Mohanty P, Ashok P, Sasikala K (2015a) Morphological characterization, character association and path coefficient analysis in sweet potato. In: Compendium of abstracts of the 2nd international conference on bio-resource and stress management, ANGRAU & PJTSAU, Hyderabad, India, 7–10 January 2015

Mohanty P, Ashok P, Sasikala K et al (2015b) Studies on genetic variability, correlation and divergence study in sweet potato (*Ipomoea batatas* (L.) Lam.) genotypes. In: Compendium of abstracts of the 2nd international conference on bio-resource and stress management, ANGRAU & PJTSAU, Hyderabad, India, 7–10 January 2015

Mohanty P, Ranjit D, Ashok P et al (2015c) Pheromone -trap monitoring system for sweet potato weevil *(Cylas formicarius):* significance and management. In: Compendium of abstracts of the 2nd international conference on bio-resource and stress management, ANGRAU & PJTSAU, Hyderabad, India, 7–10 January 2015

More AW, Kalpande HV, Duutamal RR et al (2015a) Heterosis for growth, yield and physiological parameters in *Kharif* sorghum hybrids. In: Compendium of abstracts of the 2nd international conference on bio-resource and stress management, ANGRAU & PJTSAU, Hyderabad, India, 7–10 January 2015

More AW, Kalpande HV, Sakhare GR et al (2015b) Breeding of heterotic *Kharif* sorghum hybrids for grain yield and grain mould tolerance. In: Compendium of abstracts of the 2nd international conference on bio-resource and stress management, ANGRAU & PJTSAU, Hyderabad, India, 7–10 January 2015

Muundiyara R, Kerkhi SA (2015) Direct and indirect selection parameters in bread wheat (*Triticum aestivum* L.). In: Compendium of abstracts of the 2nd international conference on bio-resource and stress management, ANGRAU & PJTSAU, Hyderabad, India, 7–10 January 2015

Nichal SS, Chawhan RG, Tayade SD et al (2015) Correlation of seed and seedling characters with yield of sunflower hybrids (*Helianthus annuus* L.). In: Compendium of abstracts of the 2nd international conference on bio-resource and stress management, ANGRAU & PJTSAU, Hyderabad, India, 7–10 January 2015

Nikam DP, Bhave SG, Bahurupe JV et al (2015) DNA fingerprinting of released rice varieties of Dr. B.S.K.K.V., dapoli through ssr markers. In: Compendium of abstracts of the 2nd international conference on bio-resource and stress management, ANGRAU & PJTSAU, Hyderabad, India, 7–10 January 2015

Ninghot CJ, Patil AE, Gawande NG et al (2015) Genetic diversity studies in wheat using RAPD marker. In: Compendium of abstracts of the 2nd international conference on bio-resource and stress management, ANGRAU & PJTSAU, Hyderabad, India, 7–10 January 2015

Pancholee V, Devkumar D, Meena AK et al (2015) Comparison of indigenous and exotic genotypes of sorghum (*Sorghum bicolor* (L.) Moench) for biomass production and yield parameters and effect of key growth stages and stay green character on biomass production. In: Compendium of abstracts of the 2nd international conference on bio-resource and stress management, ANGRAU & PJTSAU, Hyderabad, India, 7–10 January 2015

Patil HE, Jadeja GC (2015a) Character association of yield and yield traits under irrigated and terminal drought stress conditions in Pearl Millet [*Pennisetum glaucum* (L.). In: Compendium of abstracts of the 2nd international conference on bio-resource and stress management, ANGRAU & PJTSAU, Hyderabad, India, 7–10 January 2015

Patil HE, Jadeja GC (2015b) Correlation analysis terminal drought stress conditions in Pearl millet (*Pennisetum glaucum*). In: Compendium of abstracts of the 2nd international conference on bio-resource and stress management, ANGRAU & PJTSAU, Hyderabad, India, 7–10 January 2015

Patil AE, Gawande NG, Nand KK et al (2015a) Studies on heterosis and combining ability analysis in maize (*Zea mays* L.). In: Compendium of abstracts of the 2nd international conference on bio-resource and stress management, ANGRAU & PJTSAU, Hyderabad, India, 7–10 January 2015

Patil HV, Chavan SK, Patil VS et al (2015b) Studies on combining ability and heterosis in maize (*Zea mays* L.) for yield and yield contributing traits. In: Compendium of abstracts of the 2nd international conference on bio-resource and stress management, ANGRAU & PJTSAU, Hyderabad, India, 7–10 January 2015

Patil HV, Deosarkar DB, Arbad SK (2015c) Correlation and path coefficient analysis for yield and yield contributing and fiber characters in rainfed cotton. In: Compendium of abstracts of the 2nd international conference on bio-resource and stress management, ANGRAU & PJTSAU, Hyderabad, India, 7–10 January 2015

Patil VA, Patil AE, Gawande NG et al (2015d) Genetic evaluation for yield and yield contributing charactersin rice (*Oryza sativa* L.). In: Compendium of abstracts of the 2nd international conference on bio-resource and stress management, ANGRAU & PJTSAU, Hyderabad, India, 7–10 January 2015

Pawar VS, Marawar MW, Mhasal GS et al (2015) Exploitation of CMS based pigeonpea (*Cajanus cajan* L.) hybrids for yield and its contributing traits. In: Compendium of abstracts of the 2nd international conference on bio-resource and stress management, ANGRAU & PJTSAU, Hyderabad, India, 7–10 January 2015

Pole SP, Mathur H, Dhuppe MV et al (2015) Combining ability studies for yield and yield components in maize (*Zea mays* L.). In: Compendium of abstracts of the 2nd international conference on bio-resource and stress management, ANGRAU & PJTSAU, Hyderabad, India, 7–10 January 2015

Puttawar MR, Patil AE, Nagargoje VT et al (2015) Combining ability analysis in mustard using drought tolerant testers. In: Compendium of abstracts of the 2nd international conference on bio-resource and stress management, ANGRAU & PJTSAU, Hyderabad, India, 7–10 January 2015

Raju, CD, Raju, CS, Chandra Mohan Y et al (2015) JGL 13595 A high yielding, short duration, multiple gall midge biotype resistant rice culture suitable for Telangana State in India. In: Compendium of abstracts of the 2nd international conference on bio-resource and stress management, ANGRAU & PJTSAU, Hyderabad, India, 7–10 January 2015

Ramchander L, Shunmugavalli N, Muthuswamy (2015) Spectrum of viable mutants in M_2 generation of blackgram [*Vigna mungo*(L.) Hepper] through induced mutation. In: Compendium of abstracts of the 2nd international conference on bio-resource and stress management, ANGRAU & PJTSAU, Hyderabad, India, 7–10 January 2015

Ramulamma A, Sridevi D, Jhansilakshmi V et al (2015) Evaluation of different rice germplasm accessions for resistance to Brown Plant hopper (*Nilaparvata lugens* Stal.). In: Compendium of abstracts of the 2nd international conference on bio-resource and stress management, ANGRAU & PJTSAU, Hyderabad, India, 7–10 January 2015

Rani CS, Sudhakar C, Satish P (2015a) Response of Pigeonpea (Cajanus calan. L) varieties to different square plantings. In: Compendium of abstracts of the 2nd international conference on bio-resource and stress management, ANGRAU & PJTSAU, Hyderabad, India, 7–10 January 2015

Rani KR, Reddy KR, Raju CS (2015b) Evaluation of bitter gourd F_1 hybrids during summer season for yield and earliness. In: Compendium of abstracts of the 2nd international conference on bio-resource and stress management, ANGRAU & PJTSAU, Hyderabad, India, 7–10 January 2015

Ranjith P, Ghorade RB (2015) Assessment of genetic variability and correlation studies for yield and shoot fly resistance characters in newly derived lines of sorghum [*Sorghum bicolor* (L.) Moench]. In: Compendium of abstracts of the 2nd international conference on bio-resource and stress management, ANGRAU & PJTSAU, Hyderabad, India, 7–10 January 2015

Rao BB, Ashok P, Ramanandam G et al (2015) Assessment of genetic diversity in cassava (*Manihot esculenta* Crantz.) genotypes. In: Compendium of abstracts of the 2nd international conference on bio-resource and stress management, ANGRAU & PJTSAU, Hyderabad, India, 7–10 January 2015

Reddy PS, Patil JV (2015) Breeding *Rabi* Sorghum (*Sorghum bicolor* L. Moench) for abiotic stress tolerance. In: Compendium of abstracts of the 2nd international conference on bio-resource and stress management, ANGRAU & PJTSAU, Hyderabad, India, 7–10 January 2015

Sakhare BA, Ghorade RB, Kalpande VV et al (2015a) Combining ability effects for grain yield and its components in promising post rainy sorghum hybrids. In: Compendium of abstracts of the 2nd international conference on bio-resource and stress management, ANGRAU & PJTSAU, Hyderabad, India, 7–10 January 2015

Sakhare BA, Ghorade RB, Kalpande VV et al (2015b) High grain and fodder yielding cross combinations in post rainy sorghum. In: Compendium of abstracts of the 2nd international conference on bio-resource and stress management, ANGRAU & PJTSAU, Hyderabad, India, 7–10 January 2015

Sakhare BA, Ghorade RB, Kalpande VV et al (2015c) Some promising lines and testers for developing high yielding and early maturing post rainy sorghum hybrids. In: Compendium of abstracts of the 2nd international conference on bio-resource and stress management, ANGRAU & PJTSAU, Hyderabad, India, 7–10 January 2015

Sandeep S, Bharathi M, Reddy VN et al (2015) Genetic variability, heritability and genetic advance studies in inbreds of maize (*Zea mays*L.). In: Compendium of abstracts of the 2nd international conference on bio-resource and stress management, ANGRAU & PJTSAU, Hyderabad, India, 7–10 January 2015

Satyanarayana NH, Visalakshmi V, Mukherjee S et al (2015) Variability and character association in some Roselle (*Hibiscus sabdariffa* L.) genotypes for fibre yield and its attributes. In: Compendium of abstracts of the 2nd international conference on bio-resource and stress management, ANGRAU & PJTSAU, Hyderabad, India, 7–10 January 2015

Sayyad PR, Dhuppe MV, Pole SP et al (2015) Genetic variability studies in mungbean (*Vigna radiata* L.Wilczek). In: Compendium of abstracts of the 2nd international conference on bio-resource and stress management, ANGRAU & PJTSAU, Hyderabad, India, 7–10 January 2015

Seth S, Panigrahi J (2015) In vitro mass multiplication and assessment of genetic stability of in vitro raised *Abutilon indicum* (L.) Sweet plants using SCoT markers. In: Compendium of abstracts of the 2nd international conference on bio-resource and stress management, ANGRAU & PJTSAU, Hyderabad, India, 7–10 January 2015

Sethi K, Lenka PC, Tripathy P et al (2015) Studies of heterosis in Cashew (*Anacardium occidentale* L.). In: Compendium of abstracts of the 2nd international conference on bio-resource and stress management, ANGRAU & PJTSAU, Hyderabad, India, 7–10 January 2015

Shelar VR, Karjule AP, Gagre KC et al (2015) Screening of soybean (*Glycine Max* L.) germplasm for resistance to mechanical damage. In: Compendium of abstracts of the 2nd international conference on bio-resource and stress management, ANGRAU & PJTSAU, Hyderabad, India, 7–10 January 2015

Shivani D, Sreelakshmi Ch (2015) Detection and estimation of components of genetic variation for yield and contributing traits in safflower, *Carthamus tinctorius*. In: Compendium of abstracts of the 2nd international conference on bio-resource and stress management, ANGRAU & PJTSAU, Hyderabad, India, 7–10 January 2015

Singh KP, Dabbas MR, Singh DP et al (2015) Association of character and selection parameter for quality traits in table pea (*Pisum sativum* L.). In: Compendium of abstracts of the 2nd international conference on bio-resource and stress management, ANGRAU & PJTSAU, Hyderabad, India, 7–10 January 2015

Solanke PD, Patil AE, Jadhao SK et al (2015) Genetic divergence studies in maize (*Zea mays* L). In: Compendium of abstracts of the 2nd international conference on bio-resource and stress management, ANGRAU & PJTSAU, Hyderabad, India, 7–10 January 2015

Sowmya HC, Savithramma DL, Khan H et al (2015a) Genetic variability and association study, identification of high yielding and water use efficient genotypes in the minicore collection of peanut (*Arachis hypogaea* L.). In: Compendium of abstracts of the 2nd international conference on bio-resource and stress management, ANGRAU & PJTSAU, Hyderabad, India, 7–10 January 2015

Sowmya T, Durga KK, Venkateswaran K et al (2015b) Characterization of green gram accessions on the basis of morphological characters. In: Compendium of abstracts of the 2nd international conference on bio-resource and stress management, ANGRAU & PJTSAU, Hyderabad, India, 7–10 January 2015

Sreelatha D, Sivalakshmi Y, Anuradha M et al (2015) Evaluation of maize hybrids under different dates of sowing based on agro-climatic indices. In: Compendium of abstracts of the 2nd international conference on bio-resource and stress management, ANGRAU & PJTSAU, Hyderabad, India, 7–10 January 2015

Sreenivas G, Leela Rani P, Raji Reddy G et al (2015a) Risk analysis and management strategies to enhance grain yield of rainfed maize using DSSAT V4.6 in Southern Telangana Agro Climatic Zone of Telangana State. In: Compendium of abstracts of the 2nd international conference on bio-resource and stress management, ANGRAU & PJTSAU, Hyderabad, India, 7–10 January 2015

Sreenivas M, Reddy ML, Dorajeerao AVD et al (2015b) Sex-ratio, Apple and Nut parameters in certain F_1 hybrids of cashewnut. In: Compendium of abstracts of the 2nd international conference on bio-resource and stress management, ANGRAU & PJTSAU, Hyderabad, India, 7–10 January 2015

Sridhar M, Murali Krishna K, Sultana R et al (2015) Character association and path analysis for grain and yield component traits in popcorn (*Zea mays* var. *everta* Sturt.). In: Compendium of abstracts of the 2nd international conference on bio-resource and stress management, ANGRAU & PJTSAU, Hyderabad, India, 7–10 January 2015

Srinivas D, Anuradha Ch, Sujatha M et al (2015) Asessment of genetic diversity in rice. In: Compendium of abstracts of the 2nd international conference on bio-resource and stress management, ANGRAU & PJTSAU, Hyderabad, India, 7–10 January 2015

Sruthi E, Jayaramulu S, Madhav KM et al (2015) Identification of SSR marker associated with stigma exsertion in maintainer lines of hybrid rice. In: Compendium of abstracts of the 2nd international conference on bio-resource and stress management, ANGRAU & PJTSAU, Hyderabad, India, 7–10 January 2015

Supriya M, Vani VS, Pattanayak SK (2015) Impact of INM on vegetative and fruit characters of papaya (*Carica papaya* L.) variety Ranchi Dwarf. In: Compendium of abstracts of the 2nd international conference on bio-resource and stress management, ANGRAU & PJTSAU, Hyderabad, India, 7–10 January 2015

Suresh M, Rajeswari R, Sameer Kumar CV et al (2015) Molecular variability of *Fusarium Udum* isolates causing wilt of Pigeon pea in Southern Telangana Region, India. In: Compendium of abstracts of the 2nd international conference on bio-resource and stress management, ANGRAU & PJTSAU, Hyderabad, India, 7–10 January 2015

Tayade SD, Ratnaparkhi RD, Nichal SS (2015) Evaluation of first cycle families from recurrent selection using genetic male sterility in safflower. In: Compendium of abstracts of the 2nd international conference on bio-resource and stress management, ANGRAU & PJTSAU, Hyderabad, India, 7–10 January 2015

Thapa B, Pandey AK, Kumar N et al (2015a) Trait association studies for yield components in tomato (*Solanum lycopersicum* L.). In: Compendium of abstracts of the 2nd international conference on bio-resource and stress management, ANGRAU & PJTSAU, Hyderabad, India, 7–10 January 2015

Thapa B, Joshi T, Singh H et al (2015b) Assessment studies on genetic variability traits in different lines of tomato (*Solanum lycopersicum* L.). In: Compendium of abstracts of the 2nd international conference on bio-resource and stress management, ANGRAU & PJTSAU, Hyderabad, India, 7–10 January 2015

Tokas J, Sikka VK, Shikha Y, Singal, HR (2015) Expression of Bt Cry toxins in cotton genotypes depend on developmental stage but not on age of leaves. In: Compendium of abstracts of the 2nd international conference on bio-resource and stress management, ANGRAU & PJTSAU, Hyderabad, India, 7–10 January 2015

Tripathy P, Sethi K, Mukherjee SK (2015) Evaluation of cashew varieties under Odisha condition. In: Compendium of abstracts of the 2nd international conference on bio-resource and stress management, ANGRAU & PJTSAU, Hyderabad, India, 7–10 January 2015

Umarani E, Radhika K, Subbarao LV et al (2015) Evaluation of land races of rice (*Oryza sativa* L). In: Compendium of abstracts of the 2nd international conference on bio-resource and stress management, ANGRAU & PJTSAU, Hyderabad, India, 7–10 January 2015

Vijayalakshmi T, Varalaxmi Y, Maheswari M et al (2015) Identification of stress-inducible genes through suppression subtractive hybridization in pearl millet expressed in response to gradual water-deficit, high temperature and salt stresses. In: Compendium of abstracts of the 2nd international conference on bio-resource and stress management, ANGRAU & PJTSAU, Hyderabad, India, 7–10 January 2015

Yerra A, Vanaja M, Kumar GV et al (2015) Variability in performance of yield and yield attributes of black gram (*Vigna mungo* L. Hepper) genotypes under high temperatures. In: Compendium of abstracts of the 2nd international conference on bio-resource and stress management, ANGRAU & PJTSAU, Hyderabad, India, 7–10 January 2015

Post-harvest Technology for Reducing Stress on Bioresource: Recent Advances and Future Needs

12

Somesh Sharma, Ashok Kumar Thakur, and Ratikanta Maiti

Abstract

Diverse agroclimatic conditions afford ample opportunities to grow a large number of agricultural crops. The present day agricultural shifts its focus towards quality, quantity and processing and value addition. In order to deliver a quality product to the market and ultimately to the consumer, to command buyer attention and to give the grower a competitive edge, proper post-harvest management is required. Nearly, 20–25 % of fruits and 30–40 % of vegetables are wasted due to inaccurate post-harvest practices during harvesting packaging, storage, grading, etc. This wastage can be reduced to some extent through the adoption of proper scientific methods such as by cleaning, washing, sorting and grading, waxing, packing, pre-cooling, curing, chemical treatment, irradiation, harvesting at right maturity stage, free from injuries, properly storing and grading, right packing, temperature and humidity of storage, cleanliness of storage, chemical composition of produce and variety. In this chapter, the recent trends in research and developmental work done in the field of post-harvest management of agricultural commodities especially perishable ones like fruits, vegetables, and ornamental plants have been summarized. The literature has been presented and discussed under different sections include the extent of post-harvest losses, post-harvest physiology, handling during harvest and post-harvest, packaging, processing, value addition, quality assurance, marketing, etc. The text described here provides a gist of research trends and new avenues in the field of post-harvest technology.

S. Sharma (✉)
School of Bioengineering and Food Technology, Shoolini University of Biotechnology and Management Sciences, Solan, HP, India
e-mail: someshsharma@shooliniuniversity.com

A.K. Thakur
Department of Seed Science and Technology, Dr Y S Parmar University of Horticulture and Forestry, Solan, HP, India
e-mail: ashok.horticulture@gmail.com

R.K. Maiti
Facultad de Ciencias Forestales, Universidad Autónoma de Nuevo León, Carr. No 85, km 145, NL 67700 Linares, Mexico
e-mail: ratikanta.maiti@gmail.com

R.K. Maiti et al. (eds.), *Bioresource and Stress Management*, DOI 10.1007/978-981-10-0995-2_12

12.1 Introduction

The term post-harvest means the use of science and technology for the management of food commodities and ornamental plants at post-harvest level and include packaging, transportation, storage, processing, development of improved product, preservation and marketing. It also means the handling of an agricultural product after harvest to prolong storage life, freshness and an attractive appearance. India is the world's second largest producer of fruits and vegetable. In order to deliver a quality product to the market and ultimately to the consumer, to command buyer attention and to give the grower a competitive edge, proper post-harvest management is required. The vegetables and fruits gets various physiological changes after harvesting; the changes include utilization of energy reserves through respiration, changes in biochemical composition, changes in texture, physiological loss in weight, shriveling and shrinkage, rotting or losses through spoilage and increased ethylene production.

In addition to these metabolic changes, there may be losses in quality through mechanical damage, pest and diseases, through physiological disorders, high or low temperatures and faulty storage atmosphere. Nearly, 20–25 % of fruits and 30–40 % of vegetables are wasted due to inaccurate post-harvest practices during harvesting packaging, storage, grading, etc. This wastage can be reduced to some extent through implementation of proper and scientific methods by cleaning, washing, sorting, grading, waxing, packing, pre-cooling, curing, chemical treatment, irradiation, harvesting at right maturity stage, free from injuries, properly storing and grading, right packing, temperature and humidity of storage, cleanliness of storage, chemical composition of produce, kinds of crop and variety. Proper storage practices include duration of storage, temperature, relative humidity, air circulation and maintenance of space between containers for adequate ventilation and avoiding incompatible product mixes. Pre-storage treatments given to a commodity (fruits and vegetables) generally after harvesting to reduce post-harvest losses and enhance storage life and retain quality protect vegetables and fruits from fluctuating weather conditions and help in precision planning. The stored vegetables and fruits can be used economically for making various products. These methods ensure confirmed supply of fresh vegetables and fruits in off-season.

Post-harvest management can be considered as second production operation to add values to the products and the basic means for effective marketing (Nag et al. 2015). Food processing is a sun rising sector in the Indian economy that has gained importance in recent years. It constitutes around 9 % and 11 % of the GDP in manufacturing and agricultural sector, respectively, as per the annual report of ministry of food processing industry, 2013–2014. Jainuddin et al. (2015) reported the performance of Indian food processing sector by analyzing important elements like share of FDI in FPI, bank credit deployment to FPI, export share, employment status, registered and unregistered FPI units, etc. The percentage share of gaur gum and groundnut to total export of processed products was found to be higher over the 3 years but annual growth in export was moderately decreasing in both (21.76 % for groundnut and 7.75 % for gaur gum) commodities but in case of milled and pulses products, it has increasing at the rate of 56.35 % and 40.57 %, respectively. The annual growth in export value of processed products was found to be highest in case of cocoa (80.10 %) followed by milled (67.41 %) products. The annual growth in credit deployment to food processing sector was moderately higher (around 30 %) as compared to total credit disbursement to industry sector (23.13 %). The share of foreign direct investment in FPI has shown ups and downs over the year but it has increased drastically in the year 2013–2014. The per annum growth in FDI for food processing sector (22.91 %) was found to be marginally same pattern as that of FDI for total industrial sector (25.64 %) of the country. The study also found that among the states, Andhra Pradesh (14.52 %) has generated highest employment opportunity followed by Maharashtra (12.87 %) state in the food processing sector whereas Uttar Pradesh has the highest number of food processing units in India. The study also indicated that highest number of food processing units were found in grain mill products with 75,704 (49.27 %) followed by food products with 22,821 (14.85 %) and fruits and vegetables with 14,066 (9.16 %).

12.2 Post-harvest Losses

All those losses which occur after the healthy commodity has been separated from the stalk or stem or other medium from which it is originated and nourished to growth. Further, these are the losses that occur between the periods of separation of the healthy commodity from this parent plant until they are consumed. Perishable horticultural commodities are liable to post-harvest losses arising due to inappropriate management of physical, physiological and microbiological deterioration during their post-harvest life. These factors determine the quality of the product. Temperature, humidity and composition of the atmosphere and the physical stress derived from physical damage in the plant tissues are the most important factors that plays the pivotal role in post-harvest quality.

The thorough knowledge of the post-harvest respiratory metabolism, ripening, ethylene synthesis, post-harvest handling, crop quality, food safety and post-harvest pathologies is essentially required to manage the post-harvest quality. Raisin is predominantly infected by black *Aspergilli* throughout the globe. Sefloo et al. (2012) reported that *Aspergillus* section Nigri was the most common species on raisins in Khorasan-e-Razavi stock storage and markets. However, *Aspergillus niger*, *A. awamori*, *A. carbonarius*, *A. tubingensis*, *A. foetidus* and *A. aculeatus* were identified. Green mould rot caused by *Penicillium digitatum* is one of the most prominent post-harvest rots of kinnow (*Citrus deliciosa*) fruits during storage and marketing (Sharma et al. 2012). Yasir et al. (2011) reported that among the factors that cause reduction in its production and productivity area, distance and number of labour significantly affected at producer level. However, all variables except area were found non-significant at the pre-harvest contractor level.

12.3 Post-harvest Physiology

The basic fact in the post-harvest handling of produce is that harvested crops are the "living entities". It is known fact that produce is living, biological entity when it is attached to the growth of parent plant in its agriculture environment. However, even after its harvest, the harvest produce is still living as it continues to perform the metabolic reactions and maintain the physiology systems that were present when it was attached to plant. An important aspect of plants, especially the horticulture commodities, is that they respire even after harvested from the parent plant. They respire by taking up oxygen and giving of carbon dioxide, heat and even lose water. However, when attached to plant the losses are due to respiration and transpiration is replaced from the flow of sap, which contains water, photosynthates and minerals. Respiration and transpiration even continue after harvest even if the plant has been removed from its parent plant. Therefore, losses of respirable substrates and moisture are not made up and deterioration has commenced.

The post-harvest life of most horticultural perishable commodities like cut flowers, fruits, vegetables and potted flowering plants is relatively shorter. Arrays of post-harvest technologies are employed to extend the shelf live by managing or controlling various physiological processes. The vase life is inversely related to respiration rate, therefore quick cooling to the lowest safe temperature has the pivotal role in extension of post-harvest life of perishable crops. Additionally, low temperatures also manage senescence, water loss, ethylene, leaf yellowing and diseases. However, the risk of chilling injury is associated with low temperature treatments. Chemical strategies include the application of abscisic acid to reduce water loss, pre-treatment with the volatile ethylene inhibitor 1-methyl cyclopropene (1-MCP), treatment with gibberellins or cytokinins (CKs), hypochlorite, etc. Floral senescence is an active process with many of the hallmarks of programmed cell death. Molecular analysis has revealed a large number of candidate genes with possible roles in senescence and remobilization. Virus-induced gene silencing has been used to evaluate the potential role of some of these genes, particularly regulatory genes such as transcription factors and kinases, although none has yet been identified as a key controller. Ornaments are particularly suited to testing transgenic strategies for extending shelf life. The results of experiments using constructs where inducible promoters are used to drive genes that extend flower life is reported. Of particular interest is the dramatic extension of longevity resulting from silencing a component of the 26S proteasome, which indicates the importance of targeted protein degradation in control of floral senescence and could serve as a strategy for extending the life of ethylene-insensitive ephemeral flowers. Future research will undoubtedly focus on providing better germplasm by using traditional, genomic assisted and/or molecular breeding approaches for improving the post-harvest performance of ornamentals. Lee et al. (2012) observed that 1-methyl cyclopropene (1-MCP) treatment to Asian pear (*Pyrus pyrifolia* Nakai) harvested at 130 and 140 DAFB effectively delayed firmness loss during storage at 25 °C. 1-MCP treatment also reduced respiration rate to the tune of 50 % at 14 days of shelf life. It completely checked the incidence of internal browning and remarkably reduced pithiness and core browning, however promoted the flesh spot decay to some extent.

The tomatoes harvested at matured green stage were better during storage and storing in the ECS prolongs the shelf life (Muhammad et al. 2011). The heptose pool is an important source of carbon and/or energy in avocado fruit (Blakey et al. 2012). The low concentrations of two heptoses, D-mannoheptulose and perseitol and the quick reduction in their levels during post-harvest indicate that optimal post-harvest management in terms of cold storage and ripening temperature regimes is necessary to reduce the post-harvest disorders and maintain high-quality avocado fruit. The knowledge of ethylene receptors and its signal transduction pathway is vital for manipulating and regulating ethylene sensitivity by manipulating the genetic makeup of the crop. Agarwal et al. (2012) recommended that the past 2 decades have been rewarding in terms of deciphering the ethylene signal transduction and functional validation of the ethylene receptor and downstream genes involved in the cascade. The ethylene perception, receptor-mediated regulation of ethylene biosynthesis, role of ethylene receptors in flower senescence, fruit ripening and other effects induced by ethylene are emphasized upon. The expression behaviour of the receptor and downstream molecules in climacteric and non-climacteric crops is also elaborated upon. Possible strategies and recent advances in altering the ethylene sensitivity of plants using ethylene receptor genes in an attempt to modulate the regulation and sensitivity to ethylene have also been discussed. Not only will these transgenic plants be a boon to post-harvest physiology and crop improvement but it will also help us in discovering the mechanism of regulation of ethylene sensitivity (Agarwal et al. 2012).

Yang et al. (2011) studied the ox horn pepper fruit to explore the effect of UV-C treatment on the sensory quality, nutritional quality and physiological index. The objective was to provide reference for short-term storage and to prolong shelf-life of ox horn pepper fruit. The dose of 0.9 kJ m^{-2} was the best treatment for protecting ox horn pepper fruit from rotting and weight loss, and there was a significant difference compared with the control. The soluble solid and Vc content of ox horn pepper fruits treated by 0.9 kJ m^{-2} UV-C were significantly higher than the control on the 15 day. Treated group could maintain the content of chlorophyll, restrain the respiration rate and relative permeability increasing and reduce the accumulation of MDA; the activities of SOD, POD and CAT on ox horn pepper fruits were higher than that of the control. Hence, Post-harvest senescence of ox horn pepper fruit was held back and fruit quality was maintained by UV-C treatment. Hu et al. (2011) reported that 90 % ripening attained about 42 days after female flowering is the proper harvest time for Hami melon cultivar Huangpi 9818 based on the post-harvest quality parameters like SSC after 14 days in storage (16.99 %), organic acids (0.83 %) and vitamin C (28.86 10^{-2} $mg.g^{-1}$).

Kuhn et al. (2011) focused on how to reduce the length of the post-harvest sunning period to maximize both colour development and fruit quality. It was demonstrated that 14 mg N l^{-1} resulted in a low yield, small fruit, low vegetative growth and low N content in leaf dry matter, but good fruit colour development. Excessive N fertigation (224 mg N l^{-2}) did not produce the significantly highest yield or the largest fruit, but it did enhance vegetative growth and resulted in less fruit colouration. N fertigation influenced optimal picking date; higher N fertigation delayed the time of picking and increased the incidence of green fruit. Fruit from the high-N treatments required more days of post-harvest sunning to obtain the comparable red skin colour than fruit from the lower N-treatments. However, when all treatments were picked at ca. 20 % fruit coloration it was not possible to reduce the length of post-harvest sunning period by N supply. Most of the post-harvest colour development occurred within the first 6–9 days following harvest; colour developed on both sides but was enhanced on the shade side of the fruit. The shade side of the fruit had a greater potential for better red colour development than the blush side of the fruit. The content of soluble solids, fruit firmness and contents of minerals were not influenced by N supply. The sunning period significantly reduced fruit firmness. Grass cover competed with the tree for N uptake with the narrow herbicide strip producing the smallest yield, the smallest vegetative growth and the reddest fruit. The wider the herbicide strip the lower the N fertigation necessary for optimal fruit production.

Sharma et al. (2012) reported following antagonists (Table 12.1) for control of green mould rot (*Penicillium digitatum*).

Based on visual evaluation for the incidence of biotic (diseases and pests) and abiotic (physiological disorders and mechanical damage) injuries after a storage at 25 °C and 85 % of UR for 14 days, Fischer et al. (2012) suggested the adoption of phytosanitary management measures and adjustments in post-harvest processing aiming fruits with better quality in both organic and conventional system of production. In general, the changes of ripening and cell wall metabolism parameters in the tomato fruits treated with hot water for 10 min at 55 °C after low-temperature storage exhibited a comparable pattern to that of non-cold-stored fruit (Zhang et al. 2012).

Table 12.1 Antagonists and their efficacy against green mould rot (*Penicillium digitatum*)

Sr. No.	Antagonists	% Growth inhibition
1	*Debaryomyces hansenii*	89.63
2	*S. pararoseus* (KFY-1)	83.15
3	*Pseudomonas fluorescens* (Pf-1)	82.41
4	*B. subtilis*	74.44
5	*P. fluorescens* (CHAO)	69.82

Regulation of respiration rate is the basis of modified atmosphere or control atmosphere (MA or CA) storage system. Accurate mapping of respiration process is the key to success of these. Models developed by Pandey and Goswami (2011) based on the principle of enzyme kinetics with Arrhenius type and four-parameter second-order polynomial model for capsicum showed better agreement with the experimental data than the non-linear model developed.

12.4 Harvesting

Fresh commodities especially fruits and vegetables are highly perishable in nature may be unacceptable for consumption if not handled properly during harvesting. These commodities pass through a long channel before their use, which may lead to number of undesirable physico-chemical changes in their composition. Commodities that are processed into valuable products also require essentially a right stage of picking. So for the effective post-harvest management of commodities to provide their acceptable quality for the consumption to a maximum extent, it is very much essential to understand about their optimum maturity and stage of harvesting, post-harvest handling operations and the physiological changes in their composition which take place during ripening, transportation and marketing. Harvesting is the separation of commodity from its parent plant. Removal of fish from water is also harvesting. Harvesting with improper method results in the damage of crop by bruising which can be caused by compression, impact or vibration. Therefore, during harvesting factors such as delicacy of crop, maturity, time and method of harvesting mode of packaging and transportation, economy of the operations should be taken into consideration. Post-harvest losses of fruits ranges between 22 and 25 % and majority of which occurs at the time of harvesting.

Dehulling and deseeding of tamarind from dehuller-cum-deseeder was 4–5 times faster at dehulling operation and 11–13 times faster at deseeding operations than manual processing when operated at 4 cm rings clearance and 5 kg h^{-1} feed rate (Aruna et al. 2015). Elkins et al. (2011) reported that platform-harvested fruit had 57 % fewer stem punctures. The other advantages are:

- Night-time lights
- Overhead shading
- Increased vertical reach
- Re-configured conveyors to target a productivity goal of 6 bins $worker^{-1}$ day^{-1}

However, major barriers to widespread adoption are:

- Capital and maintenance costs
- Orchard renovation expense
- Fruit sorting challenges
- Lack of imminent drastic labour shortage
- Post-harvest Handling

The agricultural commodities being living organs continue to respire even after harvesting. In addition to degradation of respiratory substrates, a number of changes in taste, colour, flavour, texture and appearance take place in harvested commodities that makes them unacceptable for consumption by the consumers if these are not handled properly. The improper temperature of storage may lead to disturbance in the normal metabolism of the harvested organs. It is well established that the quality of the harvested commodities cannot be improved further but it can be retained until their consumption if the rate of metabolic activities are reduced by adopting the appropriate post-harvest handling operations. There are a number of storage techniques that are being used for commodities that are highly perishable in nature such as fruits and vegetables. The quality of stored potato (*Solanum tuberosum*) tubers is largely determined by the initial quality of tubers. Bethke and Busse (2010) reported that vine-kill treatment and tuber maturity at harvest have long-term effects on tuber quality during the storage. Puerta-Gomez et al. (2013) developed the model for the estimation of growth rate with an experimental linear cooling profile (slow cooling) in baby spinach leaves reported that the theoretical minimum temperature for growth was 5.88 °C and 4.76 °C for *Salmonella* and *E. coli*, respectively. Gustafson et al. (2012) quantified anthocyanin (ANC), proanthocyanidin (PAC) and chlorogenic acid (CA) concentrations in wild blueberry fruit (WBB) exposed to a variety of post-harvest handling practices relevant to consumers and to industry. Additionally, we analyzed the bioactive potential of WBB subjected to common culinary preparations such as baking, boiling and microwaving. Levels of ANC, PAC and CA in individually quick frozen (IQF) wild blue berry (WBB) that had been subjected to temperature fluctuations, which are often encountered during distribution and handling for retail sales, dropped by approximately 8, 43, and 60 %, respectively, compared to an IQF WBB composite that was stored continuously from harvest at −80 °C. Baking IQF wild blue berry (WBB) reduced anthocyanin (ANC), proanthocyanidin (PAC) and chlorogenic acid (CA) concentrations by 11.2, 14.6 and 10.6 %, respectively, while boiling decreased ANC, PAC and CA concentrations by a minimum of 7.4, 14.4 and 36.8 %, respectively. However, microwaving IQF WBB for 1 min increased ANC concentrations by 12.9 % but exposure to 3 and 5 min resulted in significant decreases (29.8 and 81.6 %, respectively) (Gustafson et al. 2012).

Duvenhage et al. (2012) studied the packed cartons of plum fruit with three doses of irradiation (400, 900 and 1400 Gy) where after the fruit was cold stored for 42 days with a dual temperature regime and then 7 days to simulate shelf-life conditions and found that the respiration rate of the fruit generally increased as the irradiation dose increased, except after shelf-life where it decreased after reaching a maximum at 400 Gy. Even at the highest dose most of the quality parameters measured at the end of cold storage was above the minimum arrival standards for overseas markets. However, shrivel was higher than the maximum allowed. It is suggested that refinement of the system using insect-proof bags and irradiation would be an effective non-toxic alternative to currently used mitigation treatments for plums (Duvenhage et al. 2012). King chilli (*Capsicum chinense* Jacquin) fruits treated with calcium chloride (0.5 %) packed in non-perforated bags and stored under refrigeration at 5 °C recorded lowest physiological loss in weight (2.55 %) and change in fruit colour (10 %) and highest value of ascorbic acid content (120.5 mg 100 g^{-1} of fruit) at 9th day after storage and longest shelf life (9 days) (Rongsennungla et al. 2012).

In order to determine the effectiveness of ethanol, Bal et al. (2011) examined quality variations due to menthol and thymol treatments, during the cold storage of cv. Kozak Siyah. For this aim, grape clusters were dipped into solution containing 40 % ethanol for ethanol treatment, on the other hand; 0.1 ml menthol and thymol absorbed gauzes were put near the clusters for treatments of menthol and thymol. Besides, pads containing sulphur and control grape (without treatment) were also taken. Later, all clusters were packed with polyethylene bags and stored at 0–1 °C and conditions of 85–90 % relative humidity; some quality characteristics were examined at 30 day interval. As a result of findings, better results from treatments of ethanol, menthol and thymol than control were found out during the storage of modified atmosphere. In the research, decreases in total phenolic compound content, tannin content and anthocyanin content were observed at the end of the storage period. Among the treatments, it was determined that menthol treatment showed as same effect as sulphur treatment and would be used as an alternative method for sulphur treatment.

Reduced relative water content of *Oncidium* flowers treated with pulsing STS (2 mM) was observed by Mattiuz et al. (2012), in study on treatment with pulsing solutions containing silver salts, 8-HQC and sucrose. The soluble carbohydrates and reducing sugars content decreased, and higher contents were measured in flowers treated with STS (2 mM). All parameters related to flower coloration were reduced in all treatments. Similar results were obtained for carotenoid content, except for the STS treatment (2 mM), which differed significantly from the other treatments on the 8th and 12th days of vase life. The inflorescences longevity varied from 12 days (distilled water) to 15 days (STS 2 mM).

Post-harvest treatment with chlorine solution (200 ppm sodium hypochlorite) for 15 min has the potential to control decaying incidence, prolong shelf life (up to 22 days) and preserve valuable attributes of post-harvest tomato fruits (Sood et al. 2011). Oligochitosan treatment (pre-harvest 0.05 % oligochitosan) was most effective in maintaining the fruit quality of apricot, increasing storage period of fruits by modulating the ripening and senescence of fruit cells (Liu et al. 2012). Prompt pre-cooling and cold environment (0 °C) during storage reduces decaying, respiration, ethylene production, total acid content falling, SSC content falling and pubescence rate of *Zizyphus jujuba* Mill., resulting in maintaining the membrane permeability of fruit cell, hardness and edible rate (Pan et al. 2010).

Semper fresh treatment could maintain the storage quality such as soluble solids content (SSC), vitamin C content and titratable acid in fruits (Li and Chen, 2010). The loss of water and decay rate of fruits were also been inhibited. The activity of some antioxidant enzymes (SOD and CAT) also indicated that Semper fresh treatment can prevent the quality degradation of Newhall navel orange. The optimum concentration was 15 g l^{-1} treatment.

The prevention of litchi pericarp from browning and losing water is most important and difficult problems for post-harvest storage of litchi. The production of litchi and its current post-harvest storage techniques in Guangxi have been studied. The post-harvest biology of local varieties in Guangxi was supposed to be further studied by learning experience from China and other countries. Keeping view of maturity and senescence, low temperature and atmosphere that affects the post-harvest of litchi fruit, practical storage technologies meeting the needs of litchi production in Guangxi have been put forward to make the post-harvest storage characteristic and diversified (Yao et al. 2010).

Oligochitosan treatment can also prolong storage period of fruits by modulating the ripening and senescence of fruit cells. The qualities and indexes of Saimaiti apricot fruits treated by pre-harvest 0.05 % oligochitosan were superior to other testing group. The effect of the storage system on the chemical composition and thermal properties of pineapple fruit (*Ananas comosus* L.) of cv. Josapine was investigated. The fruits were stored in a refrigerator, 10 ± 1 °C, 35 % H_2O and a humidity chamber, 10 ± 1 °C, 75 % H_2O. The properties of the fruit juice were examined every 4 days for changes in proximate analysis, total soluble solids, colour, freezing point, water activity, pH, sugar contents, thermal conductivity and density (Liu et al. 2012).

The statistical analysis indicates that total soluble solids are indicative of the ripening of the pineapple fruit Shamsudin et al. (2011). 1-Methylcyclopropene (1-MCP) significantly decreases synthesis of ethylene and maintains firmness of apples for a long period of time, whereas the synthesis of ethylene in pears is only slightly decreased or remains unchanged. The application of 1-MCP effectively prevents negative physiological changes during storage of most fruit kinds but some negative consequences of its usage are also possible.

For sugar snap peas, however, low temperature in normal air is the best storage condition and tolerance to modified atmosphere is limited. Modified atmospheres are routinely used to control discoloration in cut lettuce products, but other aspects of quality may be negatively affected (off-doors, texture, ascorbic acid concentrations). Spinach is a leafy green which is little benefited by modified atmospheres and will only tolerate moderate O_2 and CO_2 levels (about 5 %). Monitoring changes in ammonia concentrations in addition to measurement of ethanol and acetaldehyde helps to determine stressful atmospheres in green and leafy vegetables (Cantwell and Kasmire 2002a, b).

The harvested fruits of mango (Langra and Samar Bahisht Chaunsa) varieties showed significant variations in quality characteristics by varying harvest stage, storage conditions and their combinations. The contents of vitamin C and acidity were highest in fruits harvested at 80 (early stage) days after fruit set (DAFS) and sugar contents in fruits harvested 110 DAFS (late stage). The weight loss was highest and shelf life was longest for fruits harvested at 80 DAFS (early stage) and waste percentage was lowest for fruits harvested 95 DAFS (mid stage). The waste percentage, weight loss, pH, total soluble solids, carotenoids and total sugar increased the percentage of acidity and vitamin C was decreased with storage time or ripening process, irrespective of maturity stages. The ripening rate was increased and the shelf life was decreased with the increase in storage temperature. The skin colour, total soluble solids, sugar contents and carotenoids were well correlated (Baloch and Bibi, 2012).

12.5 Storage

Diffusion channel is a potential system to extend the shelf-life of tomato. Optimization of the length and diameter of the diffusion channel for maintain a steady-state O_2 concentration in airtight storage chambers at different temperatures viz., 10, 20 and 30 °C. Combinations of length of 60, 120, 180 and 240 mm and diameter of 3, 6, 9 and 12 mm were tested by Kandasamy et al. (2015). Both length and diameter had a significant effect on the steady-state O_2 concentration. When the diameter was 3 and 6 mm, increasing the length had no significant effect on steady-state gas concentration. Similarly, no significant difference in the gas concentration was observed by changing the diameter was observed when the length of channel was less than 120 mm and greater than 180 mm. Diffusion channel length of 180 mm and diameter of 9 mm maintained the O_2 concentration as low as 6–8 % and CO_2

concentration as high as 16–18 %. Respiration rate of tomato decreased with increase in length. Temperature also had a significant influence on respiration rate that was directly proportional to the temperature. The respiration rates of 1.95–8.36, 2.5–11.43 and 4.2–14.3 mg O_2 kg^{-1} h^{-1} at 10, 20 and 30 °C, respectively, was sustained under the diffusion channel system irrespective of length and diameter of the channel. Quality parameters such as firmness, acidity and ascorbic acid were found to be significant ($p \leq 0.05$) by temperature, channel length and diameter. Total sugars and lycopene content had a significant difference with temperature and diameter but not with length of channel. The colour change (ΔE) was directly related to lycopene formation. Temperature had a significant effect on TSS but length and diameter of channel had no significant effect. The pH was not significant with temperature, diameter and length of channel. Quality evaluation indicated that tomato stored at 10 °C for 45 days and 38 days at 20 °C obtained best results using diffusion channels of length 180 mm and diameter of 9 mm. Tomato stored under these conditions had good appearance, colour and good marketability conditions.

Bamboo battens storage structure with AC sheet roofing with bottom and side ventilation retained the maximum total soluble solids (TSS), ascorbic acid retention, total sugars (reducing and non-reducing) compared to other methods of storage (Rekha et al. 2015). The maximum percentage of marketable bulbs was recorded in bulbs stored in bamboo battens storage structure with AC sheet roofing with bottom and side ventilation followed by bulbs stored in wooden battens storage structure with galmenium sheet roofing and with bottom and side ventilation at 3 months of storage.

In stored strawberries, pH, SS, relation SS/TA, decay and mass loss have increased (Avila et al. 2012) during storage. The firmness varied irregularly, TA and H decreased significantly. The values of volatile compounds (VC) increased with the exception of methyl acetate (MA). Ethanol (Et), acetaldehyde (AA) and ethyl acetate (EA) correlated positively with decay for organic cultivated 'Camino Real'. In a general way, the MA correlated inversely with decay with the exception of conventional cultivated 'Camino Real'. Strawberries from both cultures can be stored for 5 days without losing quality regardless the production system.

The cool and wet storage of premature spikes of *Consolida ajacis* cv. Violet Blue for 72 h in distilled water followed by placing them in 0.2 M sucrose improves the cut flower vase life and can be used as an effective post-harvest storage system for this beautiful cut flower (Waseem et al. 2009).

A high respiratory rate is characteristics of 'Maxi Gala' and 'Brookfield' apples in all storage conditions. The 'Brookfield' apples presented high pulp cracks when stored under a HRH environment. The temperature of 1.0 °C decreases the ACC oxidase enzyme activity and ethylene production and maintains higher flesh firmness. The LRH maintains higher flesh firmness and total soluble solids and decreases the flesh breakdown. The fruits of the 'Brookfield' cultivar presented a lower storage potential in relation to the other cultivars (Weber et al. 2012).

Apples stored at 1.0 °C in standard CA (1.0 kPa O_2 and 2.0 kPa CO_2) showed less ethylene production, respiration, ACC oxidase activity and lower breakdown and mealy pulp incidence. Increasing in CO_2 concentration at the end of the storage period was not more efficient than standard CA in maintenance fruit quality during shelf life, but did not cause damage in the fruits. The decrease in O_2 and/or temperature during storage reduces the fruit decay incidence (Brackmann et al. 2012). Bahri et al. (2012) reported that calcium chloride was the best CMA for preserving water content and TSS of lettuce during ambient storage. In addition, water content and TSS decreased by increasing STP for ambient stored lettuce.

Respiration rate and ethylene production of fruit applied with AVG were greatly reduced more than those in control. Flesh firmness and acidity of fruit applied with AVG were higher than those in control. Flesh browning occurred in all of the control fruit regardless of harvest date. However, AVG-treated fruits were free of flesh browning except for late harvested fruits. At 185 DAFB, 'Fuji' apples were harvested and pre-storage heat treatment was done for 3 days at 38 °C and 6 h at 46 °C. Heat treatment at 38 °C and 46 °C prior to CA storage greatly reduced respiration rate and ethylene production. The incidence of flesh browning were 35 %, 14 % and 5 % in control fruit, heating at 38 °C and 46 °C, respectively. The titratable acidity was lower at heated apple than at controlled one (Kweon et al. 2012).

Storage insects are aerobic organisms requiring oxygen for their survival. Therefore, they respond to altered atmospheric gas compositions containing low O_2 or high CO_2. The lower the grain moisture content (m.c) and the corresponding intergranular humidity, the higher the mortality due to the desiccation effect on insects caused by low O_2 or elevated CO_2 concentrations. To achieve insect control, the temperature of the grain should be above 21 °C. Hermetic storage is based on the principle of generation of an oxygen-depleted, carbon dioxide-enriched interstitial atmosphere caused by the respiration of the living organisms in the ecological system of a sealed storage. A sufficiently low oxygen and elevated CO_2 atmosphere is created through a natural metabolic process based on insect respiration and in cases where the commodity has sufficiently high moisture, the respiration of the microorganisms within a sealed storage system. An O_2 ingress rate of 0.05 % day^{-1} is sufficient to arrest the theoretical weight loss at a level of 0.018 % over 1-year storage period. At this ingress rate, the possibility of a residual surviving insect population is eliminated. This low O_2 ingress level could serve as a guideline for the sealing specifications of structures appropriate to the hermetic storage method. The applications for which hermetic technology has been most widely accepted are (a) for long-term storage of cereal grains, primarily rice, corn, barley, and wheat; (b) for long-term storage of a variety of seeds to preserve germination potential and vigour and (c) for quality preservation of high-value commodities, such as dried fruits (Navarro, 2012).

12.6 Packaging

Packaging is one of the most important processes to maintain the quality of product for storage, transportation and end use. It also prevents quality deterioration and facilitates distribution and marketing. The basic functions of packaging are protection, containment, information and convenience. The first examples of packaging were natural objects – shells, gourds and leaves – used to contain food and drink so that it could be consumed in a place away from where it was obtained. There are different types of packaging systems such as individual packaging, external packaging. The importance of packaging is that it reduces the product damage and food spoilage, reduces the risk of tempering and adulteration, presents food in a hygienic and often attractive way, promotes goods in competitive market place, increases consumer choice and saves energy. There are different packaging materials used for horticulture crops some of them are paper board, corrugated fibre board cartons, plastic crates and wooden crates. The novel food packaging techniques include biodegradable coatings, modified atmosphere storage, etc.

Guava is one of the most delicious and nutritious fruits, liked by the consumers for its refreshing taste and pleasant flavour. The climacteric nature of guava fruit results in rapid ripening and very short shelf life ranging from 2 to 3 days at room temperature. The fruit ripening in guava is characterized by loss of green colour, softening, shrinkage, loss of brightness and rot development. Chitosan, an edible coating, protects perishable produce from deterioration by reducing transpiration, respiration and maintaining the textural quality. Fruit coating with 1 % chitosan effectively extended the shelf life of guava up to 7 days in terms of delayed ripening, higher firmness retention and lesser physiological weight loss (Krishna and Rao, 2015). Cantwell and Kasmire (2002a,b) summarized the following post-harvest technologies for highly perishable crop produce (Table 12.2).

Apricot cultivars show variable response to packaging films and optimal permeability may also differ (Gouble et al. 2012). Baby corn is

Table 12.2 Postharvest technologies for highly perishable crop

Sr. No.	Crop	Technology
1	Broccoli	Increased CO_2 concentration
2	Sugar snap peas	Low temperature
3	Spinach	Moderate O_2 and CO_2 concentrations (about 5 %)

a highly perishable commodity owing to higher respiration rate. The availability of a range of new food good polymeric films with different permeabilities to the atmospheric gases has revived the interest in packaging in sealed bags. Cobs packed in 0.2 % ventilated polybags irrespective of gauges were found to be promising. Cobs packed in unventilated polybags under all gauges were subjected to anaerobic respiration and tended to produce off odours and off flavours.

Santos et al. (2011)) used of cassava starch at 2 % and corn starch biofilms at 4 % and observed reduction in the mass loss, maintained the firmness, improved the visual aspects and prolonged the storage period without decreasing the fruits quality. The flowers packed in polyethylene and stored in cold store at 10 °C expressed maximum flower diameter (5.94 cm), fresh weight of flowers (16.24 g), minimum physiological loss in weight (0.69 %) and minimum abscised petals flower^{-1} (2.17). The moisture content played the critical role (Kumar et al. 2011). Incedayi and Suna (2012) treated cauliflower florets with NaCl (1 %) + Na-metabisulphite (1500 ppm) solution and chlorinated (150 ppm, pH 6.9). They were then rinsed with tap water. In this manner, one group of vegetables was treated with citric acid solution (1.5 %) and the other group with Ca-ascorbate (0.5 %) + citric acid (1 %) solution. After centrifugation, the samples were packed in 20 % atmospheric air +80 % N_2 and 20 % atmospheric air +70 % N_2 + 10 % CO_2 conditions with biaxially oriented polypropylene (BOPP) film. After 15 days of storage at 4 ± 2 °C ascorbic acid, total phenolics and total carotenoids contents have decreased; whereas aerobic mesophilic, psychrophilic, total coliform bacteria counts and polyphenol oxidase activity have increased. The losses of ascorbic acid, total phenolics, total carotenoids and antioxidant activity were determined as 11.54 %, 7.56 %, 29.08 % and 17.36 %, respectively. Generally, greater reductions occurred in blank samples which were not treated with any chemicals, but were packed under modified atmosphere conditions. A little weight loss, total dry matter loss and polyphenol oxidase activity increase were found in samples packaged in carbon dioxide enriched atmosphere and treated with Ca-ascorbate + citric acid. This condition was also found to be more effective in limiting the quality losses and preventing organoleptic degradation. On the contrary, control samples were generally rejected by the panelists at 95 % probability level because of their deleterious odour and colour.

The use of modified atmosphere through the development of fruit with cassava starch is called edible films and has been widely used for the reduction of postharvest losses by reducing the metabolic activity and loss of water, improving its commercial aspect and reflecting in the increase of the commercialization period. The purpose of this study was to evaluate the use of edible cassava biofilms in maintaining post-harvest tomatoes and bell peppers. The fruits were immersed in a suspension of cassava starch biofilms at concentrations of 0 % (control), 3 % and 4 %, and it was also carried out an assessment covering the fruit with polyethylene film as a way of comparison between the edibles. The tomatoes and the bell peppers were stored at 12 and 24 °C for 12 and 7 days, respectively. It can be concluded that tomatoes covered with cassava starch biofilm at 3 % had a better appearance than the control fruits and the ones covered with polyethylene film. Under the conditions of the experiment, the tomato fruit ripening proceeded normally. In the experiment with bell peppers, it can be concluded that the temperature of the 12 °C was effective in maintaining fruit quality, especially for treatments with the use of edible biofilms for 12 days post-harvest (Santos et al. 2011).

Chitosan coatings inhibit the activity of CAT and POD and reduced their peak activity. The squash with chitosan treatment could reduce the occurrence of chilling injury and the decline of storage quality. The group with 10 g l^{-1} chitosan gave the best inhibition of chilling injury (Yuan et al. 2012). Hameed et al. (2012) concluded that mangoes should be harvested at the advance stage of maturity, and the use of polyethylene bags for packaging of fruit in ambient conditions is not practicable to extend the shelf life of mangoes but it might be applicable after further investigation in controlled low temperature and high humidity. Sidhu et al. (2012) concluded that garlic variety PG-01 has longer storage life. Garlic cloves could be stored in LDPE film for 5 days at room temperature and more than 15 days at refrigerated conditions. Statistical analysis showed that storage period, storage condition and packaging material have a significant effect at 5 % level of significance on the moisture content and colour of the samples, whereas variety has a non-significant effect on the quality parameters. Treating the flowers in boric acid 4 %, packing in PP 60 micro and CFB packaging was found significant and extended the shelf life of flowers to 192.32 h. Visual quality and the physiological parameters were highly significant in this package when compared to other packages and control (Thamaraiselvi et al. 2010).

The kagzi lime fruits stored in 300 gauge polypropylene bags recorded significantly minimum pH up to 28 days (Jadhao et al. 2015). As the levels of ventilation increased, pH of kagzi lime also increased. The acidity of kagzi lime fruits decreased significantly with the days of storage period. Fruits stored in 300 gauge polypropylene bags registered comparatively highest acidity on 7, 14, 21 and 28 days of storage followed by fruit stored in 200 gauge polypropylene bags. The lowest brix/acid ratio of fruits was recorded in 300 gauge polypropylene bags (1.12) followed by polypropylene bags of 200 gauge (1.16) at the end of storage. The ascorbic acid of fruits decreased with the enhancement of storage period. Among different gauges of polypropylene bags, fruits stored in 300 gauge polypropylene bags recorded highest ascorbic acid content as compared to fruits stored in 200 gauge polypropylene bags. The fruits stored in 300 gauge polypropylene bags registered minimum reducing sugar (0.70, 0.91 and 1.27 %) followed by fruits stored in 200 gauge polypropylene bags (0.72, 0.95 and 1.31 %) respectively on 7, 14 and 21 days. Highest reducing sugars were registered in 4 % ventilated polypropylene bags viz., 0.88 % (7 days) and 1.10 % (14 days). The fruits stored in polypropylene bags with 0.5 % ventilation recorded lowest non-reducing sugar (0.81 %) followed by 1 % ventilated polypropylene bags (0.91 %) (Jadhao et al. 2015).

12.7 Preservation

One of the major advances in human history was the ability to preserve food. It was the prerequisite to man settling down in one place, instead of moving from place to place in the never-ending hunt for fresh food. The earliest preservation technologies developed were drying, smoking, chilling and heating. Later on the art of controlling, these technologies was developed. The work of Pasteur in the nineteenth century then made it possible to understand the real mode of operation of preservation techniques such as heating, chilling and freezing, providing the basis for more systematic monitoring and control. The use of various compounds such as salt and spices to preserve foods was also used in ancient times. Unfortunately, the gradual use of a wider range of chemicals for preservation such as boron or cumarine sometimes led to misuse (Zeuthen and Sùrensen 2003). The shelf-life of fresh sugarcane juice is limited due to enzymatic browning caused by polyphenol oxidase (PPO) enzyme. High pressure processing is one of the novel food preservation methods used as an alternative to thermal treatment. Sreedevi and Rao (2015) observed that the inactivation of PPO in sugarcane juice increased with increase in either pressure or dwell time. Maximum 21 % inactivation of PPO was obtained at 400 MPa and 25 min. During the sauerkraut production by conventional fermentation of cabbage causes loss of glucosinolates (GSs). The thermal treatment (blanching) followed by fermentation (4 % brine at 25 °C) by the probiotic strain *Lactobacillus paracasei* LMG

P22043 retained 27.2 ± 2.3 micro mol 100 g^{-1} GSs 71 h after fermentation, out of the initial 35 % of the total GSs before fermentation (Sarvan et al. 2013). Mushrooms are especially sensitive to senescence, browning, water loss and microbial attack. Gamma-irradiation was previously tested in wild *Lactarius deliciosus*, being verified that its application up to 1 kGy did not imply significant changes in chemical parameters (Fernandes et al. 2013).

Fruits and vegetables consumption has risen noticeably during recent decades, leading to a greater frequency of food borne illnesses associated with fresh produce. Novel industrial applications and improvements in ozone technology together with new regulatory actions worldwide have emerged in recent years, making its use in the food industry easier. This technology has attracted considerable commercial interest, especially because ozone does not leave any residues on the treated produce and it is accepted by many organic grower organizations. However, discrepancies regarding the efficacy of this technology are often found in the bibliography and further research is still needed. Method of ozone generation and application, concentration and exposure time to the gas as well as the way in which produce is packed was studied. In this sense, standardization in the working conditions and in the units to measure ozone concentration will be useful to better understand the mode of action and the effects of ozone on food products. Consequently, it would be possible to improve its potential as a sanitizer in the food industry (Horvitz and Cantalejo, 2014).

The *burfi* prepared with addition of orange pulp in sweetened *khoa* is popularly known as *Santra burfi* in Maharashtra and it has great commercial potential owing to its typical taste. The *santra burfi* was prepared by varying the rates of orange pulp addition and was tested for various textural properties such as hardness, cohesiveness, gumminess, chewiness, adhesiveness and springiness with TA-XT2i Texture Analyzer using two-bite compression. The hardness was found to have positive correlation with proteins, fat and ash content, while the moisture and level of orange pulp had negative correlation. Similar trends were observed for springiness, gumminess, chewiness and cohesiveness with the exception of ash. On the contrary, the mean adhesiveness showed negative correlation with protein, fat and ash content and shown positive correlation with moisture content and level of orange pulp (Nikam et al. 2015).

Scanning electron microscopy showed oval and irregular shaped potato starch granules with average diameter of 15 micro m and the granule diameter increased after storage. Peak viscosity was lower after storage at 8 °C and higher at 16 °C. Hot paste viscosity decreased while breakdown viscosity and set back viscosity increased after storage and there was no significant change in cold paste viscosity. A significant decrease in pasting time and increase in pasting temperature was observed after storage. Phosphorus content showed significant positive correlation with peak viscosity ($r = 0.452$, $p < 0.05$) and breakdown viscosity ($r = 0.685$, $p < 0.01$) and a negative correlation with amylose content ($r = -0.674$, $p < 0.01$). 'Kufri Sindhuri' starch showed significantly ($p < 0.05$) higher peak, hot paste, breakdown and cold paste viscosity. The X-ray diffraction pattern of starch showed a distinctive maximum peak at around 17 degrees, 2 theta and it was not affected by the cultivar or storage temperature.

Sharma and Sharma (2015) developed and analysed blend of 65 % *Aloe vera* juice +35 % aonla juice and found good amount of ascorbic acid (78.65 mg 100 g^{-1}) and phenolics (32.15 mg 100 g^{-1}) compared to control sample (100 % *Aloe vera*). It also possessed high antioxidant potential (70.85 %) and strong antimicrobial activity (32 mm inhibition zone). The calculated energy value of the developed product was recorded to be 53.65 Kcal 100 g^{-1}, which was significantly low compared to the control sample (165 Kcal 100 g^{-1}). Citrus fruits are well known for their refreshing fragrance, thirst quenching ability providing adequate vitamin C as per recommended dietary allowance and phytochemicals which play the role of neutraceuticals, such as carotenoids (Lycopene and β-carotene), limonoids, flavanones (Naringins and rutinoside) and vitamin-B complex (Sree, 2015; Ladaniya, 2008).

Pomegranate fruit can be processed to delicately flavoured RTS beverage. Further, pomegranate juice can be blended with other fruit drinks (Yadav et al. 2006). Ready-to-serve (RTS) prepared from sweet orange (*Citrus sinensis* Osbeck) variety Sathgudi using sugar substituted with fructose, 50 % sucrose + 50 % fructose and 50 % sucrose + 50 % sucralose were rated as best recipes (Byanna and Gowda, 2012). The RTS beverage had storage stability up to 6 months. During storage period, the TSS, acidity and reducing sugars increased, while pH, total sugars, non-reducing sugars and ascorbic acid decreased (Byanna and Gowda, 2012). Aviram et al. (2002) found that pomegranate juice flavonoids inhibited low-density lipoprotein oxidation and cardiovascular diseases in humans. Pomegranate juice is thirst quenching and is the best for patients suffering from leprosy, high cholesterol levels and heart, kidney and tuberculosis patients. Extracts of fruit has antiviral (Poliovirus) activity (Konowalchuk and Speirs, 1978).

The blending of fruit drinks could be an economic requisite to utilize the fruits profitably, as some varieties of fruits used for processing may not have otherwise favourable characteristics such as colour, aroma and mouth feel. There is continuous increase in TSS, acidity and sugars contents of tamarind jelly; however satisfactory keeping quality was observed up to 90 days in the jelly prepared using sugar (100 %) followed by a combination of sugar (40 %) + liquid glucose (60 %) (Supe and Saitwal 2015). Adarsh et al. (2015) observed the maximum retention of ascorbic acid, least browning, reducing sugars and total sugars content of dehydrated slices were observed in 1 % sodium hydrogen sulphite ($NaHSO_3$). The mean scores for organoleptic characters, such as colour and appearance, texture, taste and flavour were found maximum in 1 % $NaHSO_3$ pre-treated slices, whereas the highest score for overall acceptability was recorded in 1 % $NaHSO_3$ + 0.5 % citric acid pre-treated samples. Jiang et al. (2010) standardized explosion puffing conditions for sweet potato chips and reported that the puffing temperature is 91 °C, vacuum drying temperature is 75 °C and vacuum drying time is 47 min is optimum with respect to various quality indices. Variety of oil palm, processing method and storage history have significant effect on specific heat capacity of the samples. The specific heat capacity increased in the temperature range from −10 to 10 °C followed by a decrease to about 80 °C. The specific heat capacity of cold processed palm oil was generally lower, ranging from 1.29 to 5.26 J g^{-1} °C, while that of hot processed samples ranged from 1.80 to 6.24 J g^{-1} °C (Alakali et al. 2012). Sandhu et al. (2013) tries to explain the complex mass transfer mechanisms taking place during deep fat frying with respect to real time pressure variations inside potato discs and chicken nuggets. Frying experiments were performed at two temperatures of 175 °C and 190 °C for 200 and 240 s for potato discs and chicken nuggets, respectively. The guage pressure increased rapidly above the atmospheric pressure immediately after the samples were introduced into the hot oil. The rise in pressure was greater in potato discs with greater initial moisture content. This was expected due to sudden moisture flash-off. As frying progressed, the temperature inside the samples increased whereas the guage pressure started decreasing and became negative. The onset of negative pressure was observed during initial stages of frying for chicken nuggets, but in the middle of frying for potato discs. The negative pressure values before the product is taken outside the fryer may cause increased oil uptake during frying itself. During the post frying cooling, the pressure further decreased and reached negative values. The negative pressure is expected to have caused rapid absorption of surface oil during both frying and cooling stages.

Initial size of cauliflower, drying air temperature and velocity governs the drying time; however rehydration ratio is dependent on combined effect of temperature and airflow velocity. Vitamin C content of the dried cauliflower is affected by temperature only and non-enzymatic browning is a function of temperature, airflow velocity and interaction of temperature and airflow velocity Gupta et al. 2013. Gupta et al. (2013) standardized of the drying process parameters (temperature 60.10 °C, air flow 59.28 m min^{-1}, size 3.35 cm) and attainments in terms of time, vitamin C content, rehydration ratio and browning values were 491.22 min, 289.86 mg 100 g^{-1}, 6.91 and 0.14, respectively, with the desirability factor of 0.787.

Ledbetter (2012) has provided high quality products for nearly a century, annually accounting for approximately 20 % of available tonnage to California's dry apricot industry. The 'Patterson' cultivar currently dominates California dry apricot sales, but the cultivar is not without faults. Newer cultivars and breeding accessions are being evaluated for quality attributes when processed as dry product. Apricot cultivars 'Patterson,' 'Golden sweet' and breeding selection Y117 were evaluated for changes in CIELAB coordinates L*, a* and b* as well as colour components chroma and hue throughout a 7 month storage period at 25 °C. Fresh fruit samples of the three accessions differed significantly in initial quality characteristics (flesh firmness, soluble solid content (Brix) and juice acidity), with cultivar 'Golden sweet' being harvested at a more fully mature state. Drying ratios of the three accessions varied significantly and ranged from 3.96 (Y117) to 4.61 ('Patterson'). Pre-storage dry apricot halves from the three accessions were visually distinct and colour coordinates L*, a*, b*, chroma and hue all varied significantly among the accessions. Values of L*, a*, b* and chroma decreased throughout the storage period for all apricot accessions, but the rate of decrease was accession dependent. Hue values of dried apricot halves remained relatively constant during the 7 month storage period, regardless of accession. Colour degradation was most pronounced in cultivar 'Patterson', holding sufficient visual appeal for marketability between 4 to 5 months in storage. In contrast to 'Patterson', dried halves of 'Golden sweet' and Y117 were deemed marketable at the end of 7 months storage and therefore suitable for sales during the important holiday period.

In a study by Vijay et al. (2015), seeds dried by four different drying methods namely silica gel, saturated salt solution of lithium chloride, conc. sulphuric acid and dryer and conditioned at four different moisture content i.e. 3 ± 0.2 %, 5 ± 0.2 %, 7 ± 0.2 % and control (11 ± 0.2 %) using silica gel were evaluated for biochemical variability using SDS-PAGE profile of total soluble protein after 7 days of accelerated aging (physiological stress) at 100 % RH and 43 °C. Among the various methods used for drying seeds, silica gel was found to be the best alternative to the method using drying chamber. It not only maintained the seed quality but also provided a faster drying rate than dryer. Total soluble seed protein banding pattern of different categories samples revealed that there has been decline in band intensity, band numbers or loss of some bands in accelerated aged seed as compared to unaged seeds but highest decline was observed in acid dried and ultra-dried seeds. Optimum moisture content for storage of sorghum seeds was found to be 5–7 %. Seed survival was adversely affected below 5 % moisture content. Therefore, drying by silica gel and up to 5–7 % can be suggested for sorghum seeds.

Shere et al. (2015a, b) reported that increase in proportion of oat flour in the composite cookies greatly improved taste, flavour and textural qualities. The protein, fat, ash, fibre contents were increased with increase in oat flour in the composite cookies, however, carbohydrates decreased with increase in proportion of oat flour.

Chilli cultivars LCA-334 dried at faster rate and had maximum physiological weight loss in 5 days with less pod discolouration. The chemicals treatment with potassium carbonate and calcium carbonate accelerated drying, with quick maximum physiological weight loss. However, maximum oleoresin was observed in cultivar Indom-5. Potassium carbonate treatment recorded maximum oleoresin and capsanthin colour value (Sarada et al. 2015).

Antoniolli et al. (2012) determined the shelf life of minimally processed (MP) 'Perola' pineapples treated with ascorbic acid (AA) and citric acid (CA) based on physical, chemical, sensorial and microbiological attributes. Slices were dipped into drinking water (control) or combined solutions of AA: CA (%) (1.0:0.5 and 1.0:1.0) with sodium hypochlorite (NaClO 20 mg L^{-1}) for 30 s. The samples were conditioned in polyethylene terephtalate packages and stored at 4 ± 1 °C for 13 days.

The low peroxidase activity in the slices treated with antioxidant combinations was related to low pH values observed in these samples. The treatments 1.0:0.5 and 1.0:1.0 (AA: CA, %) favoured maintenance of the initial a* values and avoided the pulp browning. The ascorbic acid increased more than double on the 2nd day in the treated slices. By the 4th day, the CO_2 values suggested a higher respiratory activity in the slices treated with anti-browning compounds. The antioxidant treatments did not produce detectable residual flavours in the MP pineapple. Regardless of microbiological safety during the 13 days of cold storage, the control slices can be kept by 6 days, afterwards the colour and dehydration become strong enough to affect the appearance. On the other hand, MP 'Perola' pineapples treated with 1.0:0.5 (AA:CA, %) and NaClO (20 mg L^{-1}) can be stored for 8 days at 4 ± 1 °C, which represents the extension of the shelf life in 2 days. After this period, the overripe odour starts to develop.

Suhasini et al. (2015) studied the osmotic dehydration of karonda. The physico-chemical properties like moisture content, ascorbic acid, Fe content and acidity of the product were evaluated during the storage period. During storage, slight decrease in ascorbic acid, iron content, acidity and increase in moisture content of osmo-dried product of karonda was noticed. The microbiological changes of the products and a taste panel evaluated the organoleptic quality of the products during the storage period. All the products were acceptable at all the storage periods. However, osmotic pre-treatment with 5 % NaCl for 3 h found highly acceptable. Minimum microbial count (2.00) was recorded for osmosis in 5 % NaCl solution for 3 h. The product of 5%NaCl solution for 3 h when stored 4 months at room temperature secured highest sensory score (82.05).

Prathibha et al. (2015) suggested lycopene extracted from tomato processing waste can be effectively utilized to enhance the nutritive value of tomato beverages especially tomato soup. The antioxidant activity of lycopene fortified tomato soup at 30 mg 100 g^{-1} on 60th day was recorded to be 23.04 mg AAE 100 g^{-1}. Improvement in appearance and nutritional quality of the products was observed by fortification with extracted lycopene.

12.8 Post-harvest Quality

Post-harvest quality of agricultural crops is one of the important parameters for its sale or preservation. It not possible to increase the quality of crop after harvest and if a good quality of produce is received at the processing centre then only we can achieve a high quality of value added product. So it is important that a quality product be used for value addition. The quality product also fetches good price in the market. Pineapple is one of the important tropical fruit crops of Bromeliaceae family, considered as an important table fruit because of its enticing sugar acid blending and also popularly used in canning industry. Around 47 % of the total pineapple produced in India is from North Eastern Himalayan (NEH) region, which proved that it is a potential crop of this agro-eco region. Though in Mizoram, pineapple is cultivated in comparatively smaller area (1.53 thousand hectare) and the total state production is around 13.53 thousand tonnes but the crop is widely consumed as table fruit in Mizoram. Pineapples have a short post-harvest shelf life at ambient temperature and deteriorate quickly. The average minimum loss reported is 21 % and occasional instance estimated of 40–50 % and above. Thus it is of utmost importance to develop a technique for extending the shelf life of pineapple at ambient temperature by reducing the post-harvest decay and maintaining the physico-chemical qualities of fruit. The effect of nine post-harvest treatments viz., spraying of NAA at 100 mg L^{-1}, gibberellic acid at 100 mg L^{-1}, fruit dipping in salicylic acid at 5 mM, covering the fruit with perforated polythene and newspaper bag, fruit coating with wax at 60 gL^{-1}, fruit dipping in maleic hydrazide at 500 mgL^{-1}, covering of fruit with dry straw and untreated control on fruit physico-chemical qualities and shelf life of pineapple cv. Giant Kew studied has shown that GA_3 at 100 mgL^{-1} may be the best post-harvest treatment to extend the shelf life while maintaining the fruit physico-chemical qualities of pineapple cv. Giant Kew during storage at room temperature (Mandal et al. 2015).

Javaherashti et al. (2012) studied nutrition value and sensory characteristics of some different winter pumpkin genotypes. The results showed a significant difference for nutritional quality in investigated genotypes. Ziaber genotype had the highest fresh weight, dry matter, antioxidant capacity, total flavonoid and total phenolics content as compared to the other genotypes. Furthermore, a significant relationship was found between antioxidant capacity, total flavonoid and total phenolics content of fruits.

Aktas et al. (2012) reported that the harvested bunch tomatoes are sensitive to fruit abscission and desiccation of calyxes, which limit product marketability. It is concluded that pre-harvest application of ReTain containing 15 % aminoethoxyvinylglycine (AVG) (150 ppm) 7 days before commercial harvest to 'Monroe' peaches can increase financial returns to growers through increased fruit size and quality benefits as well as maintain a higher flesh firmness following post-harvest storage and export.

The impact of harvest stages and storage conditions over the post-harvest quality of mango (Langra and Samar Bahisht Chaunsa) varieties has been investigated. The fruit was harvested at 80 (early stage): 95 (mid stage) and 110 (late stage) days after the fruit setting and designated as samples I, II and III, respectively. The harvested fruit was stored under three different storage conditions till its ripening. Significant variations were observed in quality characteristics by varying harvest stage, storage conditions and their combinations. The contents of vitamin C and acidity were highest in sample I and sugar contents in sample III of the fruit. The weight loss was highest and shelf life was longest for sample I and waste percentage was lowest for sample II. The waste percentage, weight loss, pH, total soluble solids, carotenoids and total sugar increased the percentage of acidity and vitamin C was decreased with storage time or ripening process, irrespective of maturity stages. The ripening rate was increased and the shelf life was decreased with the increase in storage temperature. The skin colour, total soluble solids, sugar contents and carotenoids were well correlated (Baloch and Bibi, 2012).

The compounds of peach aroma had great difference between pre-harvest and postharvest storage. The content of lactones and esters, characteristic compounds of peach aroma, were much lower after 4 weeks of storage (Peng et al. 2012). Christen et al. (2012) determined the quality of several apricot cultivars by non-destructive VIS/NIR spectroscopy using different devices. In order to determine non-destructively the soluble solids content (SSC), titratable acidity, firmness and ground colour values of apricot, NIRs spectra and data obtained from destructive tests were subjected to PLS regressions (Partial Least Square Regression). In a first approach, a portable device was used in order to predict and to follow-up the fruit quality and maturity on tree for the cultivar 'Orange Red Reg'. Calibration models allowing the determination of soluble solids content (SSC), titratable acidity, firmness and ground colour were carried out with high precision. SSC was determined with a root mean square error of cross-validation (RMSECV) of 0.62 °Brix, titratable acidity was determined with RMSECV of 1.45 meq 100 g^{-1}, firmness with RMSECV of 6.3 DI and ground colour intensity with RMSECV of 2.4 (a*). R-values were situated between 0.49 and 0.66. External validation of the models gave similar levels of precision. In a second approach, two devices were tested for the prediction of the post-harvest quality of 40 apricot cultivars. Global models combining 40 cultivars were carried out with high precision for the non-destructive determination of SSC and with reliable precision for firmness in post-harvest. Depending on the device used, SSC was determined with RMSECV-values between 0.58 and 0.71 °Brix and R-values between 0.94 and 0.95 and firmness with RMSECV-values between 8.4 and 9.1 DI and R-values between 0.77 and 0.85. Such results show the potential of the VIS/NIR technology as a non-destructive tool for measuring apricot quality and for the follow-up of this quality from the orchard and along the whole supply chain. Ramos et al. (2012) studied three different spacing's, combined with ten diploid watermelon cultivars, eight of which were experimental hybrids, originated from the Plant Breeding Program of Embrapa Semiarido and two commercial cultivars (Sugar Baby and Smile) for agronomical performance and post-harvest fruit quality. The experimental hybrids Hyb 02, Hyb 03, Hyb 04, Hyb 05 and Hyb 07 showed the best emergency rates, with outstanding results for the hybrids 02, 03 and 06, which were among the best considering the seedling height and colon diameter besides fresh weight of shoot. The different spacing's did not influence the plant development or the physiological quality of the fruit of the genotypes assessed. Means of transportation, packaging, harvesting procedures, maturation at harvest, climatic conditions and many other variables affect the quality and post-harvest life of agricultural products. In Brazil, tomatoes are still packaged in the wooden box type 'k', first used to transport fuel in World War II. In this type of package, the fruits are placed randomly and the boxes are stacked during harvest, transport and distribution to the consumer. This handling leads to quantitative and qualitative losses. Better handling is needed to decrease mechanical damage and maintain the quality and price of tomatoes. The present work aimed to evaluate some quality characteristics of tomato fruits ('Pizzadoro') submitted to an induced compression, simulating the wooden boxes stacked up during transportation. A specific weight per unit area of the original wooden box 'k' was determined (12.5 g cm^{-2}) and used to obtain the weight needed (5 kg) to compress fifteen fruits inside small wooden boxes of 23 × 20 × 20 cm during 48 h. After this treatment, these fruits and control fruits (without treatment) were stored at 15 or 25 °C. The following variables were measured during 18 days of storage: colour, lycopene, total soluble solids and pH. It was concluded that the compression treatment significantly altered all variables studied, except for pH and total soluble solids of fruits stored at 15 °C (Oliveira et al. 2012).

Litchi (*Litchi chinensis* Sonn.) is an economically important fruit crop with great potential in the international market due to its characteristic flavour and exotic taste. Quality and safety are the two most important factors in the international marketing of horticultural commodities in the present scenario of liberalized world trade. Litchi is a crop with a limited marketing period due to the narrow harvest season and short shelf life. The application of chemicals was done when fruits were at bean-seed size and second application 2 weeks prior to harvest. Pre-harvest application of 4 % boric acid resulted in higher TSS and lower acidity content in fruits during storage (5–7 °C). Total sugars (15.92 %) and reducing sugars (11.94 %) were also enhanced with a 4 % boric acid pre-harvest application. The physical parameters of fruits (weight and diameter of fruit and pulp weight) were found to be positively influenced with the application of calcium nitrate at 1.5 % as pre-harvest spray (Alila and Achumi, 2012).

Danaee et al. (2012) reported that GA_3 50 mg l^{-1} and BA 50 mg l^{-1} were the most effective treatments based on vase life, fresh weight, solution uptake, membrane stability and total soluble solids of gerbera cut flowers. Cytokinins and gibberelins have potential to enhance post-harvest quality of cut gerbera flowers.

Sadaf et al. (2012) reported that all potassium treatments significantly increased yield characteristics as well as post-harvest quality of tomato fruit compared with untreated one (control). Potassium application significantly increased number of flowers $plant^{-1}$, fruit setting rate, number of truss $plant^{-1}$, fruits $plant^{-1}$ and yield ha^{-1}. Moreover, increased potassium levels also had positive effect on post-harvest life attributes of tomato fruit. In addition, the shelf life, quality of general appearance and taste were significantly influenced as potassium levels increased to 375 kg K_2O ha^{-1} but decreased at 450 Kg K_2O ha^{-1}. On the other hand, while increased potassium levels decreased all the undesired parameters, wilting and drying of calyx, physiological weight loss percentage and percentage of non-marketable yield. Therefore, when potassium was applied at 450 kg K_2O ha^{-1}, it increased the yield and decreased the adverse effect on marketable yield, wilting and drying of calyx and weight retention of tomato fruit. In conclusion, 375–400 Kg K_2O ha^{-1} is recommended as it produced better quality tomatoes with longer postharvest life.

Sarmento et al. (2012) recommended the utilization of the organic cultivation that produces banana of good quality free of residues of chemical products and avoids damage to the environment. Singh et al. (2012) concluded that ethrel 750 ppm was found to be the most suitable treatment in improving physico-chemical traits i.e. ripening, storage, quality and shelf-life for commercial purpose in mango. 'Williams' pear (*Pyrus communis* L.) is the most important cultivar grown in Alto Valle of Rio Negro, Argentine. 1-Methylcyclopropene (1-MCP) proved to be highly effective in inhibiting ethylene action and maintaining postharvest quality. However, due to logistical problems, the commercial application of 1-MCP is often delayed, which may reduce the efficiency of this treatment. A study was undertaken to determine the influence of 0.3 and 0.6 ml L^{-1} of 1-MCP applied after 0, 3, 7 and 10 days at 0 °C on the post-harvest quality of early and late harvested 'Williams' pears. Untreated fruit was used as control. After 60 and 90 days of storage at −0.5 °C, ethylene production and maturity indexes (flesh firmness, soluble solids, acidity, starch degradation, epidermis colour) were measured. The results showed that all 1-MCP treatments significantly reduced ethylene production and fruit ripening irrespective of the delay period. The ethylene production of fruits was undetectable during the delay period at 0 °C, meaning that ethylene was not affected by time at 0 °C. Moreover, no differences were observed between 1-MCP concentrations, with 0.3 ml L^{-1} sufficient to reduce ethylene production and ripening even in late harvested fruits. It was hypothesized that the temperature during the delay period plays a key role on the maintenance or reduction of 1-MCP efficacy. It is important to emphasize that the results obtained under experimental conditions may not be directly extrapolated to a commercial scale, since there are other factors that could reduce the effectiveness of 1-MCP treatments (Calvo and Candan, 2012).

Campos et al. (2011a)) concluded that high doses of irradiation promoted a negative effect on physical-chemical characteristics of guava 'Pedro Sato', verifying that only the lowest dose associated with modified atmosphere provided fruits with higher quality and acceptability, due to higher maturation rate and soluble solids obtained.

Chen et al. (2011) studied the changes of nutritional quality and physiological biochemistry of the spinach treated by brine in different concentrations at room temperature to explore the green and safe storage method of post-harvest spinach. The results showed that the respiratory peak was delayed and the evaporation and decay of spinach were slowed down with the treatment of appropriate concentration of brine. In addition, the kind of brine had certain inhibitory effects not only on the degradation of protein and organic acid but also on the activities of PPO, POD and CAT. The storage life of spinach was prolonged for 1–2 days, and the best treating concentration of brine for the storage of spinach was 0.9 %.

Kwon et al. (2011) elucidated the effects of mulching materials and removing time of the transparent polyethylene (PE) film on the growth of garlic at Uiseong experimental field, Korea. The experimental mulching materials comprised of transparent polyethylene film (0.025 mm) and net polyethylene (NPE). Plant height and leaf number of garlic were highest at PE treatment when the PE removing date was March 18 and the treatment also promoted the number of cloves. Length of leaf sheath and bolting rate were highest and bulb weight loss rate was lowest at PE + NPE treatment when the PE removing date was March 18. But clove number was the lowest in this treatment compared to conventional PE film treatment. Conventional mulching method accelerated secondary growth rate but bulb weight loss was vice versa. There were statistically no differences in bulb diameter among treatments but conventional treatment positively focused on bulb diameter. PE film suppressed the weeds compared to no mulching treatment. The dry weight of weeds has shown increasing trends with the delay on removal dates of PE film. Transparent PE or PE + NPE treatments can be recommended to grow best quality garlic.

Wang et al. (2011) reported that the treatment of 1-MCP was found to delay the peak time of respiration rate, inhibit decomposition of chlorophyll, decrease the superoxide anion ($O_2^{-\cdot}$) content and membrane permeability and increase protein content and antioxidative enzyme activities. However, it also inhibited the increase in titrable acidity and VC content. The effects of second time application of 1-MCP treatment were more pronounced. Two times application of 1-MCP (1 mg l^{-1}) was found to enhance the post-harvest quality of papaya fruits during its long term storage.

Singh and Solomon (2011) reported that deterioration of harvested cane could be minimized to a considerable extent by spraying of electrolyzed water and trash covering. A new deterioration indicator i.e. mannitol formation was also found to be reduced in EW treated cane. Chang et al. (2011) reported the lower incidence of Fusarium wilt in the organic system was considered to be associated with a positive change in the properties of the soil. Horticultural parameters at shooting did not show significant differences between organic and conventional banana, except for a lower number of healthy leaves in the organic banana. No differences in the post-harvest quality could be found between organic and conventional banana. A smaller bunch weight was recorded for the organic banana; however, more harvested bunches resulted in a higher overall yield in the organic farming system. Promotion of the organic farming system for adoption by more banana growers could enhance the sustainability of the Chinese Taipei banana industry. Bautista and Yang (2012) recommended moisture efficient clamshell as a better storage or packaging method for prolonging fresh blueberry's shelf life and maintaining good berry quality.

The coating cassava starch + chitosan provided the best results, with less than 6 % of loss in fruit mass, lower counts of yeast and psychrophilic microorganisms and the best appearance according to the sensory analysis (Campos et al. 2011b)

12.9 Marketing

Jan et al. (2011) analyzed the process and issues of regional agriculture product brand establishment for Taiwan farmers' associations. The research focused on farmers' associations with 'The Brand Fruit Certification System'. According to the quantitative survey, from the average score of the establishment process of regional agriculture product's brand, the ranking orders of the basis stage were 'the producing area', 'regional resources and brand establishment' and 'cooperative marketing'. The ranking orders of the formative stage were 'the major producing area', 'promotion of regional brand establishment' and 'organization marketing'. The ranking orders of the established stage were 'management of regional brand equity', 'the high major producing area' and 'integrated marketing'. Further, according to the analysis of the case data, the farmers' associations constructed process of regional agriculture's brand partly achieved the condition of the major producing area but did not reach the level of a major producing area. For the brand construction dimension, the farmers' associations interviewed also apply the mark of the regional brand. The consumers are aware of the regional brand from the knowledge of the producing area. Due to the lack of human resources, the farmers' associations could not effectively manage the marketing channels of the producing area and could not become the marketing integrator. Therefore, the farmers' associations were still in the initial stage for the regional brand. The study shows the farmers' associations have to improve the condition of the producing area diversify the marketing and strengthen the equity of the regional brand. Export of agriculture products has an important role in non-petroleum export and has specific significance among the agriculture product of pistachio, its export from Iran going to parts of the world since 2500 years ago. Pistachio figures prominently among other nuts due to its nutritious value. Pistachio trees are the second plant (after date) that can be alive for a long time without water. Until 1979, Iran was the only unique exporter of pistachio that entered into the world market through offering the pistachios product of California with competitive qualification and use of facilities. According to the latest statistics released by the World Food and Agriculture Organization (FAO) affiliated to the United Nations, the pistachio production in the world stood at 548,759 metric tons (MTs) in 2002, Iran ranking first with a production of 300,000 MTs. The pistachio gardens in Iran's rival country, the US occupies a total area of 44,000 hectares vs. 280,000 hectares in Iran. Aflatoxin is one of the serious problems in pistachio industries; some pistachios have suitable conditions for growth of fungus within the fruit, which are the basis of aflatoxin, because of different reasons, the most important of them being splitting of the external green layer of the fruit and maybe with aflatoxin. Therefore maximum levels for aflatoxin should be determined. In this study, weather conditions for pistachio planting, provinces that are planting pistachio in Iran and statistics of pistachio production in these provinces, kinds of pistachio existing in Iran, statistic study of Iran pistachio export during a few years using valid resources and strategies to increase Iran pistachio export in global markets are discussed (Nejad, 2012). Yakam et al. (2012) investigated the information of longan production and marketing and studied the market conduct of longan in Chiang Mai and Lamphun provinces. From the survey among 200 farmers, it was found that there were small size longan orchards 67.5 %. The next was medium size longan orchards 22.5 % and large size longan orchard 10.0 %. In the study of longan production, it was found that 87.5 % of farmers were producing in season longan and 6.5 % were producing off season. Both group mostly had small size orchards with some medium size orchards. The study on longan market conduct found that longan producing farmers did not set their price. Longan product pricing depends on trading means. The traders, who come to buy, will offer the price according to product grade. Longan product price offered by the traders is referred from the middleman, who buys longan in the country, and middleman in China. About competitive conduct, it was found that farmers, who produced in season, off season and both in season and off season longan throughout the year, did not compete in pricing. They had no bargaining power to negotiate about product price with the merchant. Except for farmers that sell their products wholly, they can negotiate with wholesale traders. In addition, it was found that farmers did not gather into a group to sell their products. Longan traders did not compete in pricing against one another, except for middleman that buy longan. They did not gather into a group to sell the products among traders. The result on marketing conduct due to government's longan marketing policy, it was found that most farmers and traders never participated in longan purchasing project according to the government's longan marketing project in 2004–2008. They were affected by the government's policy but not much. In 2009–2010, all three groups of longan producing farmers did not participate in the project and did not sell their products to any institute. Therefore, they did not receive the benefit from government's longan marketing policy. Sweet cherry (*Prunus avium* L.) is a model specialty crop – high potential returns per hectare, high costs of production and ephemeral in its supply. Faced with increasing competition and imminent shortages of harvest labour, the US sweet cherry industry must improve harvest efficiency without compromising consumer appeal of the fruit. Further, for research to have meaningful and sustainable impact on industry, innovation, discovery and outreach must integrate the total value chain, from genetics and breeding to processing and marketing, while partnering with stakeholders and end users. This presentation describes an innovative, multi-disciplinary research program funded through the USDA's Specialty Crop Research Initiative. The program's long-term goal is improving the sustainability of the US sweet cherry industry by developing a highly efficient, production, processing and marketing system for fresh market quality sweet cherries. Project directors address these issues this with an integrative project built around the development of mechanical, mechanical-assisted and novel hybrid harvest technologies. In this presentation, research results and a model, total systems approach integrating critical elements of a profitable, sustainable cherry industry are presented including:

*Genetics and genomics research to establish the genetic bases for abscission in cherry; utilizing marker-assisted breeding strategy to accelerate the generation of new cultivars amenable to mechanization and with high consumer appeal. * Development of high efficiency fruiting wall orchard management systems, novel compostable packaging, retail markets for mechanically harvested cherries and models of system profitability to facilitate adoption. * Engineering research to develop and deploy innovative mechanical and/or mechanical-assist cherry harvest technologies and consumer or packaging research to extend the shelf life and improve consumer appeal of fruit (Whiting et al. 2012).

Modern retail marketing of high-value agricultural commodities is of inevitable importance in the developing countries like India (Aparna and Hanumanthaiah 2012). The supply of vegetables by farmers to modern retail outlets has brought in a new form of organized marketing based on consumer demand. The study has assessed the marketing system by comparing marketing cost, marketing margins, price spread and producer's share in consumer's rupee, marketing efficiency and marketing constraints of supermarket channels and two traditional marketing channels in the Rangareddy district of Andhra Pradesh. Two vegetables, viz., brinjal and bhendi, have been selected for the study. It has been revealed that the net price received by the farmers and producer's share in consumer's rupee are higher in supermarket channel than in traditional channels. The supermarket channel has been found more efficient than the traditional channels. Rejections of low grade produce, procurement according to indent and lack of knowledge of grading have been identified as the major constraints of supermarket supply farmers. The major constraints expressed by the traditional market supply farmers include middlemen menace, higher distance to the market and high market charges. The study has observed that government intervention is required to create a policy environment that may ensure a mutually beneficial relationship between the farmers and the organized sector. Investment in infrastructure, development of extension activities and linkages with farmers are the important areas where government should give due attention to strengthen vegetable supply channels in the state.

Sarukhanyan et al. (2011) showed that in 2007–2009 the specific weight of legume in overall crop land was approximately 94 %, and about the 96 % of the gross harvest. Local production needs appropriate marketing strategy. The research of local market showed that more attention should be paid to the consumption of goods produced by the farmer households, as well as to offer them to various consumer groups. Yakam et al. (2012) found that there were small size longan orchards 67.5 %. The next was medium size longan orchards 22.5 % and large size longan orchard 10.0 %. In the study of longan production, it was found that 87.5 % of farmers were producing in season longan and 6.5 % were producing off season. Both group mostly had small size orchards with some medium size orchards. The study on longan market conduct found that longan producing farmers did not set their price. Longan product pricing depends on trading means. The traders, who come to buy, will offer the price according to product grade. Longan product price offered by the traders is referred from the middleman, who buys longan in the country, and middleman in China. About competitive conduct, it was found that farmers, who produced in season, off season and both in season and off season longan throughout the year, did not compete in pricing. They had no bargaining power to negotiate about product price with the merchant. Except for farmers that sell their products wholly, they can negotiate with wholesale traders. In addition, it was found that farmers did not gather into a group to sell their products. Longan traders did not compete in pricing against one another, except for middleman that buy longan. They did not gather into a group to sell the products among traders. The result on marketing conduct, due to government's longan marketing policy, it was found that most farmers and traders never participated in longan purchasing project according to the government's longan marketing project in 2004–2008. They were affected by the government's policy but not much. In 2009–2010, all three groups of longan producing farmers did not participate in the project and did not sell their products to any institute. Therefore, they did not receive the benefit from government's longan marketing policy.

Bhandari et al. (2012) describes the collection and marketing of *C. sinensis* [*Ophiocordyceps sinensis*], a highly priced medicinal plant in Himalayas, as a good source of income for locals. However, its over-exploitation is alarming. It was suggested that the collection of *keera jari* should be done in alternate years, and that the Government of Uttarakhand should take stringent steps to curb illegal collection and supply. Price spread refers to the difference between the price paid by the consumer and the price received by the producer for an equivalent quality of product. The study of price spread involves ascertainment of the actual price at various stages of the marketing and the costs incurred in the process of the management of cut flower from farm to the consumer and the margin of the various intermediaries. The study was based on primary data for the year 2009 collected from different flower markets in Akola district. The transaction of cut flowers and loose flowers takes place in an open condition on the road side. The business transaction takes place between 4 pm and 6 pm. Large varieties of flowers such as rose, chrysanthemum, gladiolus, gerbera, carnation, lilies and marigold area mainly brought to the market from Patur, Barshitakali, Akot tahsils of Akola districts. The Indian Government realized the value of horticulture and provided a major thrust to same by raising the budgetary allocation from a meagre amount of INR24 crores in the seventh five year plan to more than INR 2000 crores in the eleventh five year plant, besides providing a large no of concessions, subsides and incentives to the growers. During the last 10 years, taking advantage of the incentives offered by the government, a number of floriculture units were established in India for producing exportable flowers to the developed countries. Exporting centres for flowers are located near Mumbai, Bangalore, Pune and Delhi (Tayade et al. 2015).

Rao et al. (2015) studied to workout various facets of economics involved in production and marketing of jaggery, constraints faced by jaggery producers conducted during 2013–2014 in Andhra Pradesh state of India. Multistage sampling technique was adopted in selecting the sampling units. Averages, benefit cost ratio (BCR), break even output (BEO), pay back period (PBP) and response priority index (RPI) were employed as analytical tools. The prime factor in jaggery production was cost of sugarcane (74.55 %). The cost of cultivation of sugarcane per hectare was INR1,88,014 with the variable cost (Cost A) INR1,37,817 and fixed cost INR 50,197 they accounts for 73.30 % and 26.70 %, respectively. The benefit cost ratio was 1.12 in jaggery making and the breakeven output is 50.37 tonnes. The major constraints in jaggery production and marketing are lack of infrastructural facilities and insufficient market price information dissemination, respectively. These constraints should be addressed as quickly as possible to make sugarcane cultivation profitable.

12.10 Summary

Post-harvest technology is one of the important branches in handling of cash crops. As in context to India, lot of cash crops are produced, and India is the second largest producer of fruits and vegetables. But, if we look at the losses data in fruits and vegetables about 30–40 % goes waste and if that can be prevented we can feed whole Europe. Further, the processing capacity is also very low that is to the tune of 2 %. To reduce the losses, one must understand the biological and environmental factors involved in the deterioration of agricultural commodities. Since the land resources are inelastic, the potential of increasing the area under cultivation is limited but still there is scope to increase the availability of food crops by applying or using recent post-harvest technologies. Hence, to increase further the agriculture GDP of the country post-harvest technology can play a very important role.

Indian economy is agricultural based and we are the leading producers of various food crops. But then too loses are also on higher side and that due to various reasons such as mainly due to lack of infrastructural facility needed for the post-harvest management of the perishable and poor utilization of these important resources. Post-harvest technology plays an important role in the reduction of post-harvest losses, value addition, an important contributor to the economy, making availability of agriculture commodities throughout the year or in off seasons, a tool for employment generation, besides adding taste, variety, nutrition, waste utilization, etc.

There is imperative need to strengthen need based research in the areas like novel food packaging and user friendly post-harvest techniques. The country is witnessing a boom in fruit and vegetable sector but the connectivity between the producers and the consumers at the end of supply chain is yet to guarantee a fair price for the producer. Very often distress sale of agriculture commodities due to lack of proper post-harvest techniques, connectivity and marketing facilities to the urban centres, compel farmers to sell at the production point at unbelievably low rates. So there is need in future that more emphasis on problem oriented research with integrated approach can alone solve the problems faced in the post-harvest technology of agricultural crops. Development of techniques causing minimum wastage needs active attention of the post-harvest technologist or scientists. Another important component is the quality and safety of the products should continue to receive greater attention in the near future. It is also important that adequate inputs for developing infrastructure for post-harvest handling of agricultural commodities be provided. The dominance of middle man and traders in marketing of agricultural commodities should be eliminated. Further, the reduction in marketing chain will surely increase the margin of farmer and provide food to consumers at cheaper rate. It is also important that fruit collection centres and processing facilities should be provided at the farmer level also. Since the nation is losing an important revenue generator, the agricultural commodities from farm to market or even at terminal market so it's essential that proper storage facilities be created at farm and terminal markets. Further, the packaging methods for different crops be standardized and packaging material be made available. Thus, the efforts of various scientists and technologist are only meaningful if effective post-harvest technology is developed and be communicated among the farmers of the nation.

12.11 Conclusion

To achieve a significant breakthrough in post-harvest technology and management of agricultural commodities and ultimately to give farmers proper returns, we must concentrate on well-planned strategies. The agricultural commodities are the living entities so it is very important to handle them with lot of care. It is important that once the commodity is matured on the plant it should be harvested at the proper stage of maturity. After harvest, proper precooling treatments should be adopted to increase its shelf life. Not only this, these should be stored under proper storage conditions. Further, to reduce the transportation quality these should be packed in proper packaging material. The novel packaging techniques as antimicrobial films, biodegradable packaging material, temperature time indicators should also be tried and technology be transferred to the ultimate users. Not only this, the processing sector should also be strengthened so that during glut season these valuable sources be converted into valuable products. This will not only reduce the losses of the agricultural commodities but also increase the income of farmers to some extent.

References

Adarsh K, Rekha E, Kukanoor L et al (2015) Effect of pre-treatments on bio-chemical and organoleptic qualities of dehydrated onion slices. In: Compendium of abstracts of the 2nd international conference on bio-resource and stress management, ANGRAU & PJTSAU, Hyderabad, India, 7–10 January 2015

Agarwal G, Choudhary D, Singh VP et al (2012) Role of ethylene receptors during senescence and ripening in horticultural crops. Plant Signal Behav 7(7):827–46

Aktas H, Bayndr D, Dilmacunal T et al (2012) The effects of minerals, ascorbic acid, and salicylic acid on the bunch quality of tomatoes (*Solanum lycopersicum*) at high and low temperatures. HortSci 47(10):1478–1483

Alakali JS, Eze SO, Ngadi MO (2012) Influence of variety and processing methods on specific heat capacity of crude palm oil. Int J Chem Eng Appl 3(5):300–302

Alila P, Achumi I (2012) Pre harvest chemical treatments affect post-harvest quality of litchi fruit. Acta Hortic 934:755–761

Antoniolli LR, Benedetti BC, MdeSMde SF et al (2012) Shelf life of minimally processed pineapples treated with ascorbic and citric acids. Bragantia, Campinas 71(3):447–453

Aparna B, Hanumanthaiah CV (2012) Are supermarket supply channels more efficient than traditional market channels? Agric Econ Res Rev 25(2):309–316

Aruna P, Sunitha N, Sumati V (2015) Post harvest management on Tamarind by Assessment of tamarind Dehuller-Cum-Deseeder for processing. In: Compendium of abstracts of the 2nd international conference on bio-resource and stress management, ANGRAU & PJTSAU, Hyderabad, India, 7–10 January 2015

Avila JMM, Toralles RP, Cantillano RFF et al (2012) Influence of planting system and cold storage on the physical chemistries characteristics and volatile compounds development in strawberries. Ciencia Rural 42(12):2265–2271

Aviram M, Dornfeld L, Kaplan M et al (2002) Pomegranate juice flavonoids inhibit low-density lipoprotein oxidation and cardiovascular diseases: studies in atherosclerotic mice and humans. Drugs Exp Clin Res 28:49–62

Bahri MH, Niari SM, Rashidi M (2012) Effect of chemical materials application and storage periods on water content and total soluble solids of lettuce during ambient storage. Middle-East J Sci Res 12(4):479–483

Bal E, Kok D, Celik S (2011) The effects of some postharvest treatments on Kozak Siyahı grape variety. J Tekirdag Agric Fac 8(2):65–76

Baloch MK, Bibi F (2012) Effect of harvesting and storage conditions on the post harvest quality and shelf life of mango (*Mangifera indica* L.) fruit. S Afr J Bot 83:109–116

Bautista N, Yang WQ (2012) Evaluation of postharvest quality of three commercial blueberry (*Vaccinium corymbosum*) cultivars stored in two different types of clamshells. Int J Fruit Sci 12:75–82

Bethke PC, Busse JS (2010) Vine-kill treatment and harvest date have persistent effects on tuber physiology. Am J Potato Res 87:299–309

Bhandari AK, Negi JS, Bisht VK et al (2012) *Cordyceps sinensis*: fungus inhabiting the Himalayas and a source of income. Curr Sci 103(8):876

Blakey RJ, Tesfay SZ, Bertling I et al (2012) Changes in sugars, total protein, and oil in 'Hass' avocado (*Persea americana* Mill.) fruit during ripening. J Hortic Sci Biotechnol 87(4):381–387

Brackmann A, Both V, Weber A et al (2012) Changes in temperature, oxygen and CO_2 during controlled atmosphere storage of 'Royal Gala' apples. Semina: Ciências Agrárias (Londrina) 33(6):2247–2256

Byanna CN, Gowda IND (2012) Standardization of recipe for the preparation of ready-to-serve (RTS) beverage from sweet orange (*Citrus sinensis* Osbeck) var. sathgudi using sugar substitutes and its storage. Crop Res 44(3):356–362

Calvo G, Candan AP (2012) Is delay in application a factor that reduces efficacy of 1-methyl-cyclopropene response in pears? Acta Hortic 945:345–349

Campos AJ, Fujita E, Costa SM et al (2011a) Irradiation and passive modified atmosphere on the post-harvest quality of guavas 'Pedro Sato'. Agro@mbiente Online 5(3):233–239

Campos RP, Kwiatkowski A, Clemente E (2011b) Postharvest conservation of organic strawberries coated with cassava starch and chitosan. Revista Ceres 58(5): 554–560

Cantwell M, Kasmire RF (2002a) Postharvest handling systems: flower, leafy, and stem vegetables. In: Kader AA (ed) Postharvest technology of horticultural crops. University of California, Agriculture and natural resources, Publication 3311, Oakland, pp 423–433

Cantwell M, Kasmire RF (2002b) Postharvest systems: fruit vegetables. In: Kader AA (ed) Postharvest technology of horticultural crops. University of California, Agriculture and natural resources, Publication 3311, Oakland, pp 407–421

Chang CM, Chao CP, Huang SN et al (2011) Fusarium wilt incidence, growth, yield and post-harvest quality of banana as affected by organic farming in Chinese Taipei. In: Proceedings of the international ishs-promusa symposium on global perspectives on Asian challenges, Guangzhou, China, 14–18 September 2009

Chen WH, Peng MY, Zhong XJ et al (2011) Effect of brine on post-harvest nutritional quality and physiological activity of spinach. Guizhou Agric Sci 2011(09):115–129

Christen D, Camps C, Summermatter A et al (2012) Prediction of the pre- and postharvest apricot quality with different VIS/NIRs devices. Acta Hortic 966:149–153

Danaee E, Abdossi V, Mostofi Y et al (2012) Effect of GA_3 and BA on postharvest quality and vase life of gerbera cut flowers. Hortic Environ Biotechnol 52(2):140–144

Duvenhage AJ, Jooste M, Johnson SA (2012) Irradiation as a postharvest quarantine treatment for a new plot cultivar. Sci Hortic 147:64–70

Elkins RB, Meyers JM, Duraj V et al (2011) Comparison of platform versus ladders for harvest in Northern California pear orchard. Acta Hortic 909: 241–249

Fernandes A, Barreira J, Antonio AL et al (2013) Study of chemical changes and antioxidant activity variation induced by gamma-irradiation on wild mushrooms: comparative study through principal component analysis. Food Res Int 54(1):18–25

Fischer IH, Fileti MS, Palharini MCA et al (2012) Injuries in Pera oranges cultivated in organic and conventional systems marketed in Ceagesp. Citrus Res Technol 33(1):39–47

Gouble B, Reling P, Renard CMGC et al (2012) Influence of temperature on apricot fruit respiration and ethylene production. In: Proceedings of the international symposium on apricot breeding and culture, Armenia, 20–24 June 2011

Gupta MK, Sehgal VK, Arora S (2013) Optimization of drying process parameters for cauliflower drying. J Food Sci Technol 50(1):62–69

Gustafson SJ, Yousef GG, Grusak MA et al (2012) Effect of postharvest handling practices on phytochemical concentrations and bioactive potential in wild blueberry fruit. J Berry Res 2(4):215–227

Hameed U, Saeed A, Muhammad A et al (2012) Response of mango cultivars to modified atmosphere storage at an ambient temperature cv. Alphanso and Chounsa. Pak J Agric Sci 49(3):293–299

Horvitz S, Cantalejo MJ (2014) Application of ozone for the postharvest treatment of fruits and vegetables. Crit Rev Food Sci Nutr 54(3):312–339

Hu JG, Liao XF, Sun YP et al (2011) Changes in postharvest physiology and quality of Hami melon with different maturities at harvest. China Cucurbits Vegetables 2011(03)

Incedayi B, Suna S (2012) Effects of modified atmosphere packaging on the quality of minimally processed cauliflower. Acta Alimentaria 41(4):401–413. doi:10.1556/AAlim.41.2012.4.2

Jadhao SD, Waghmode KT, Hadole SS et al (2015) Effect of different gauges and ventilations of packaging material on biochemical changes of kagzi lime. In: Compendium of abstracts of the 2nd international conference on bio-resource and stress management, ANGRAU & PJTSAU, Hyderabad, India, 7–10 January 2015

Jainuddin SM, Subba Rao DV, Vijaya Kumari R et al (2015) The Prospects of Food Processing in India. In: Compendium of abstracts of the 2nd international conference on bio-resource and stress management, ANGRAU & PJTSAU, Hyderabad, India, 7–10 January 2015

Jan YH, Tsai BK, Jeans SG (2011) A study on the establishment process of regional agriculture product branding for farmer' associations. J Agri Assoc Taiwan 12(2):149–150

Javaherashti M, Ghasemnezhad M, Lahiji HS et al (2012) Comparison of nutritional value and antioxidant compounds of some winter pumpkin (*Cucurbita* sp) species fruits in Iran. Adv Environ Biol 6(10):2611–2616

Jiang N, Liu CQ, Li DJ et al (2010) Explosion puffing technology optimization for sweet potato chips. Trans Chinese Soc Agric Eng 26(11):361–367

Kandasamy P, Moitra R, Mukherjee S (2015) Diffusion channel system for enhancing shelf life of tomato. In: Compendium of abstracts of the 2nd international conference on bio-resource and stress management, ANGRAU & PJTSAU, Hyderabad, India, 7–10 January 2015

Konowalchuk J, Speirs JI (1978) Antiviral effect of apple beverages. Appl Environ Microbiol 36(6):798–801

Krishna KR, Rao DVS (2015) Chitosan coating: its effects on the physiochemical characteristics of guava (*Psidium guajava* L.) fruits during storage at room temperature. In: Compendium of abstracts of the 2nd international conference on bio-resource and stress management, ANGRAU & PJTSAU, Hyderabad, India, 7–10 January 2015

Kuhn BF, Bertelsen M, Sorensen L (2011) Optimising quality-parameters of apple cv. 'Pigeon' by adjustment of nitrogen. Sci Hortic 129:369–375

Kumar S, Beniwal BS, Sandooja JK et al (2011) Effect of packaging materials and storage conditions on flowers characters and post-harvest quality of chrysanthemum (*Chrysanthemum morifolium* Ramat.) cv. Shanti. Crop Res 41(1–3):135–138

Kweon HJ, Kim MJ, Lee CC et al (2012) Effects of aminoethoxyvinyl glycine application and heat treatment on fruit quality of 'Fuji' apples during CA storage. Korean J Hortic Sci Technol 30(5):527–533

Kwon KS, Azad MOK, Hwang JM (2011) Mulching methods and removing dates of mulch affects growth and post harvest quality of garlic (*Allium sativum* L.) cv. Uiseong. Korean J Hortic Sci Technol 29(4):293–297

Ladaniya M (2008) Citrus fruit: biology technology and evaluation. Academic, San Diego

Ledbetter C (2012) Postharvest dried apricot color degradation of three California apricot accessions. Acta Hortic 966:163–168

Lee UY, Oh KY, Moon SJ et al (2012) Effects of 1-methylcyclopropene (1-MCP) on fruit quality and occurrence of physiological disorders of Asian pear (*Pyrus pyrifolia*): 'Wonhwang' and 'Whasan', during shelf-life. Korean J Hortic Sci Technol 30(5):534–542

Li JB, Chen JY (2010) Effects of Semper fresh treatment on post-harvest physiology and storage of Newhall navel orange. Acta Agric Univ Jiangxiensis 2010:06

Liu DD, Zhu X, Wang J et al (2012) Effect of pre-harvest oligochitosan treatment on the storage qualities of apricot fruits. Modern Food Sci Technol 2012:10

Mandal D, Lalremruata, Hazarika TK et al (2015) Effect of post harvest treatments on quality and shelf life of pineapple (*Ananas comosus* [L.] Merr. 'Giant Kew') fruits at ambient storage condition. In: Compendium of abstracts of the 2nd international conference on bio-resource and stress management, ANGRAU & PJTSAU, Hyderabad, India, 7–10 January 2015

Mattiuz CFM, Mattiuz BH, Pietro J (2012) Effect of different pulse treatments on vase life of inflorescences of *Oncidium varicosum* 'Samurai'. Acta Hortic 934:465–471

Muhammad RH, Bamisheyi E, Olayemi FF (2011) The effect of stage of ripening on the shelf life of tomatoes (*Lycopersicon esculentum*) stored in the evaporative cooling system (E.C.S). J Dairying Foods Home Sci 30(4):299–301

Nag P, Panigrahi TK, Chandel N (2015) Quality enhancement by post-harvest in vegetables and fruits. In: Compendium of abstracts of the 2nd international conference on bio-resource and stress management, ANGRAU & PJTSAU, Hyderabad, India, 7–10 January 2015

Navarro S (2012) The use of modified and controlled atmospheres for the disinfestation of stored products. J Pest Sci 85(3):301–322

Nejad ASM (2012) Global pistachio production and marketing challenges in Iran. Int Symp Mycotoxins Nuts Dried Fruits 963:133–141

Nikam P, Wasnik PG, Khare A et al (2015) Studies on physico-chemical and textural Properties of burfi blended with orange pulp. In: Compendium of abstracts of the 2nd international conference on bio-resource and stress management, ANGRAU & PJTSAU, Hyderabad, India, 7–10 January 2015

Oliveira GHH, Correa PC, Finger FL et al (2012) Color and other quality aspects of tomato fruits subjected to a short period of induced compression. Acta Hortic 934:1205–1211

Pan Y, Che FB, Zhang HL et al (2010) Effect of different storage treatments on postharvest physiology and storage effectiveness of Xinjiang Lizao jujube (*Zizyphus jujuba* Mill.). Xinjiang Agric Sci 2010(05):89–104

Pandey SK, Goswami TK (2011) Modeling of respiration rate of capsicum at various temperatures. Int J Food Eng 7:5

Peng XY, Che FB, Wu ZH et al (2012) Pre- and post-harvest changes in aroma compounds of oblate peach fruit. Xinjiang Agric Sci 2012(10):1824–1833

Prathibha G, Narayana CK, Yadav TV (2015) Fortification of tomato soup with lycopene extracted from tomato peel. In: Compendium of abstracts of the 2nd international conference on bio-resource and stress management, ANGRAU & PJTSAU, Hyderabad, India, 7–10 January 2015

Puerta-Gomez AF, Moreira RG, Kim J et al (2013) Modeling the growth rates of *Escherichia coli* spp. and *Salmonella Typhimurium* LT2 in baby spinach leaves under slow cooling. Food Control 29:11–17

Ramos ARP, Dias RCS, Aragao CA et al (2012) Performance of watermelon genotypes of small fruit under several planting densities. Hortic Bras 30(2):333–338

Rao S, Rao IVYR, Rao SG (2015) Production and marketing of processed products of sugarcane in Andhra Pradesh, India. In: Compendium of abstracts of the 2nd international conference on bio-resource and stress management, ANGRAU & PJTSAU, Hyderabad, India, 7–10 January 2015

Rekha E, Kukanoor L, Adarsh K et al (2015) Effect of different storage methods on bio-chemical composition of onion during storage. In: Compendium of abstracts of the 2nd international conference on bio-resource and stress management, ANGRAU & PJTSAU, Hyderabad, India, 7–10 January 2015

Rongsennungla K, Singh VB, Maiti RK (2012) Effect of pre- and post-harvest treatments on yield, quality and post-harvest shelf life of king chilli (*Capsicum chinense* Jacquin) under foothill conditions of Nagaland (India). J Spices Aromatic Crops 21(1):42–47

Sadaf J, Khan MQ, Rahman HU et al (2012) Response of tomato yield and post harvest life to potash levels. Sarhad J Agric 28(2):227–235

Sandhu J, Bansal H, Takhar PS (2013) Experimental measurement of physical pressure in foods during frying. J Food Eng 115:272–277

Santos AEOdos, Assis JS, Berbert PA et al (2011) Influence of cassava starch and corn starch biofilms on the post harvest quality of 'Tommy Atkins' mangos. Revista Brasileira de Ciências Agrárias 6(3): 508–513

Sarada C, Rajani A, Vijayalakshmi T et al (2015) Post harvest management of chilli cultivars with chemicals and fungicides for enhancing drying and quality. In: Compendium of abstracts of the 2nd international conference on bio-resource and stress management, ANGRAU & PJTSAU, Hyderabad, India, 7–10 January 2015

Sarmento JDA, Morais PLD, Almeida MLB et al (2012) Post harvest quality of banana submitted to the organic and conventional cultivation. Revista Brasileira de Produtos Agroindustriais14(1):85–94

Sarukhanyan R, Sarukhanyan N, Torchyan V et al (2011) Study and analysis of the legume crop market in Armenia. Ratarstvo i Povrtarstvo 48(2):419–422

Sarvan V, Cde L, Verkerk DM et al (2013) Glucosinolate content of blanched cabbage (*Brassica oleracea var. capitata*) fermented by the probiotic strain *Lactobacillus paracasei* LMG-P22043. Food Res Int 54(1):706–710

Sefloo NG, Pirnia M, Shahri MK (2012) Identification of some black Aspergillus from various raisins in Khorasan-e-Razavi Province. Acta Hortic 963:189–198

Shamsudin WR, Takriff MS, Hassan O (2011) Chemical compositions and thermal properties of the Josapine variety of pineapple fruit (*Ananas comosus* L.) in different storage systems. J Food Process Eng 34(5):1558–1572

Sharma R, Sharma R (2015) Development of Aloe-Aonla based reduced calorie functional squash using stevioside and sorbitol for 'At Risk' consumers. In: Compendium of abstracts of the 2nd international conference on bio-resource and stress management, ANGRAU & PJTSAU, Hyderabad, India, 7–10 January 2015

Sharma RN, Maharshi RP, Gaur RB (2012) Biocontrol of post-harvest green mould rot (*Penicillium digitatum*) of kinnow fruits using microbial antagonists. Indian Phytopathol 65(3):276–281

Shere DM, Pawar VS, Shere PD (2015a) Process standardization and quality evaluation of oat cookies. In: Compendium of abstracts of the 2nd international conference on bio-resource and stress management, ANGRAU & PJTSAU, Hyderabad, India, 7–10 January 2015

Shere PD, Sadawarte SK, Pawar VS et al (2015b) Preparation of protein enriched low calorie oat cookies. In: Compendium of abstracts of the 2nd international conference on bio-resource and stress management, ANGRAU & PJTSAU, Hyderabad, India, 7–10 January 2015

Sidhu GK, Kumar S, Arora S (2012) Packaging and storage of minimal processed garlic (*Allium sativum* L). J Dairying Foods Home Sci 31(3):207–211

Singh P, Singh VK, Kumar M et al (2012) Effect of physico-chemical treatments on ripening behavior and post-harvest quality of Amrapali mango (*Mangifera indica* L.) during storage. J Environ Biol 33:227–232

Singh P, Solomon S (2011) Effect of spraying electrolyzed water on post-harvest quality of sugarcane at high ambient temperature. Indian J Sugarcane Technol 27(02):65–67

Sood K, Bhat A, Singh A et al (2011) Effect of harvesting methods and post harvest treatments on quality of tomato (*Lycopersicon esculentum*). Ann Food Sci Technol 12(1):58–62

Sree KB (2015) Studies on preparation and evaluation of sweet orange and pomegranate blended RTS. In: Compendium of abstracts of the 2nd international conference on bio-resource and stress management, ANGRAU & PJTSAU, Hyderabad, India, 7–10 January 2015

Sreedevi P, Rao PS (2015) Effect of high pressure processing on enzyme inactivation and microbial destruction in sugarcane juice. In: Compendium of abstracts of the 2nd international conference on bio-resource and stress management, ANGRAU & PJTSAU, Hyderabad, India, 7–10 January 2015

Suhasini L, Vanajalatha K, Padmavathamma AS et al (2015) Studies on osmotic dehydration of karonda (*Carissa carandas* L.). In: Compendium of abstracts of the 2nd international conference on bio-resource and stress management, ANGRAU & PJTSAU, Hyderabad, India, 7–10 January 2015

Supe VS, Saitwal YS (2015) Processing, evaluation and storability of tamarind jelly. In: Compendium of abstracts of the 2nd international conference on bio-resource and stress management, ANGRAU & PJTSAU, Hyderabad, India, 7–10 January 2015

Tayade PM, Deshmukh RG, Nandeshwar NS (2015) Study the price spread of flowers in Akola district. In: Compendium of abstracts of the 2nd international conference on bio-resource and stress management, ANGRAU & PJTSAU, Hyderabad, India, 7–10 January 2015

Thamaraiselvi SP, Jwaharlal M, Ganga M et al (2010) Packaging technology for long term storage of jasmine (*Jasminum sambac* Ait.) flowers. J Ornamental Hortic 13(3):171–181

Vijay MK, Pandey S, Pandey CD (2015) Total soluble protein profile of sorghum (*Sorghum bicolor* (L.) Moench) seeds conditioned by different drying methods and at different moisture contents. In: Compendium of abstracts of the 2nd international conference on bio-resource and stress management, ANGRAU & PJTSAU, Hyderabad, India, 7–10 January 2015

Wang JN, Deng DM, Li HB et al (2011) Effect of 1-methylcyclopropene (1-MCP) on post-harvest storage quality of papaya. J South Agric 42(7):719–723

Waseem S, Tahir I, Gul F (2009) Effect of dry and wet storage at cool temperatures on postharvest performance of Consolida ajacis cv. Violet Blue Spikes. J Plant Biol 36(1–2):5–9

Weber A, Brackmann A, Thewes FR et al (2012) Relative humidity and its interaction with the storage temperature of 'Gala' apples and mutants. Rural Sci 42:2159–2165

Whiting D, Oraguzie S, Pierce Z et al (2012) A total systems approach to developing a sustainable, stem-free sweet cherry production, processing and marketing system. WSHA Poster session (Abstracts): Washington State Horticultural Association Annual Meeting, Yakima, Washington, December 2010

Yadav K, Sarkar BC, Kumar P (2006) Pomegranate: recent developments in harvesting. Process Util Indian Food Ind 25(2):56–62

Yakam S, Jierwiriyapant P, Nilsamranchit S et al (2012) Market conduct of fresh longan in Chiang Mai and Lamphun provinces. J Agric Res Ext 29(3): 41–50

Yang H, Yin MA, Yu QP et al (2011) Effects of UV-C on ox horn pepper fruit shelf-life and its post-harvest physiology mechanism. J Northwest A & F Univ (Natural Science Edition) 2011(08):117–126

Yao JY, Long X, Peng HX et al (2010) Advances in post-harvest biology of litchi and its developmental status in Guangxi. Guangxi Agric Sci 2010 41(5): 475–478

Yasir N, Ahmad M, Anjum MB et al (2011) Factors affecting post harvest losses of citrus; a case study of district Sargodha. Am-Eurasian J Agric Environ Sci 10(5):934–937

Yuan MM, Gao LP, Wang Q et al (2012) Effect of chitosan coating on chilling injury in summer squash. J Henan Agric Sci 41(10):114–117

Zeuthen P, Sùrensen LB (2003) Introduction. In: Zeuthen P, Sùrensen LB (eds) Food preservation technique. CRC, New York, p 1

Zhang ZK, Gao ZY, Li M et al (2012) Hot water treatment maintains normal ripening and cell wall metabolism in mango (*Mangifera indica* L.) fruit. HortSci 47(10):1466–1471

Recent Trends in Seed Science and Technology

13

Ashok Kumar Thakur and Ratikanta Maiti

Abstract

Improvement of ovule to seed and seed to plant ratios remains the ultimate aim of all research and developmental efforts of seed science and technology. Being a connecting link between two generations of plants or crops, seed is a carrier of new technology or new improvements made by the breeders. The seed production protocols then need to be efficient enough to produce seeds that possess desired quality traits in terms of genetic, physical, physiological and health quality. Hence, the implementation of improved seed production technology of various crops will improve the ovule to seed ratio in the mother crop and seed to plant ratio in the subsequent crop raised from the seed. Additionally, the seed to plant ratio can be maintained or improved by adopting accurate and adequate seed enhancement and storage strategies. The available literature has been segregated in different heads: seed production, genetic quality, physiological quality, seed health, seed enhancement and seed storage.

13.1 Introduction

Seed is the vital component not only of agriculture but reproduction of most plants and bioresource conservation as well. Seed is crucial, critical, vital and basic input in agriculture. The quality of seed is directly and positively correlated to quality and productivity of the produce. Seed quality is, therefore, one of the cheapest and most essential inputs for improving the agricultural economy of the country. The seed quality, according to Thomson (1979), is a complex of several components and their relative importance in different sets of production systems. He categorised these components as (1) analytical quality, (2) species purity, (3) freedom from weeds, (4) cultivar purity, (5) germination capacity, (6) vigour, (7) size, (8) uniformity, (9) health and (10) moisture content. These components were categorised in more systematic manner by Nema (1988) into four major components: (1) genetic quality, (2) physical quality, (3) physiological

A.K. Thakur (✉)
Department of Seed Science and Technology, Dr Y S Parmar University of Horticulture and Forestry, Solan, Himachal Pradesh, India
e-mail: ashok.horticulture@gmail.com

R.K. Maiti
Facultad de Ciencias Forestales, Universidad Autónoma de Nuevo León, Carr. No 85, km 145, NL 67700 Linares, Mexico
e-mail: ratikanta.maiti@gmail.com

R.K. Maiti et al. (eds.), *Bioresource and Stress Management*, DOI 10.1007/978-981-10-0995-2_13

quality and (4) health quality. These components independently and their interaction at all levels sum up to constitute the seed quality. The quality of seed means that the seed is superior and pure in terms of its genetic composition, physiologically strong enough to produce a plant of desired kind, free from physical impurities and biotic stress-causing agents. The importance of quality seed is well recognised round the globe. In India, the need for quality seed was felt with the introduction of high-yielding varieties of bajra, sorghum and wheat in the 1960s. It is very much evident from the history of grain production in India and other developing countries that the availability of quality seed was one of the keys to Green Revolution. The introduction of dwarf and high-yielding varieties of wheat resulted in three to four times increase in the productivity of wheat in the country. This sudden and intense enhancement in productivity of wheat is termed as Green Revolution. In fact, it is not the development of high-yielding varieties alone that brought Green Revolution, but the production and availability of quality seeds of these novel cultivars are equally important.

In a crop production system, the quality of seed is the basic input to which the other inputs are applied to have the desired output. It ensures other agro-production and protection inputs to be productive and cost-effective. Feistritzer (1975) has mentioned that in a production system, the good quality seed is when combined with other complementary agro-inputs such as better nutrition, irrigation, plant protection measures, etc. that resulted in rapid and substantial increase in productivity. Well-planned evaluation and release of new varieties; systematic seed production and distribution, seed testing and certification; and enforcement of seed legislation, in toto, are essential to exploit the technological breakthroughs in any crop production system.

The sole and ultimate aim of seed science and technology is to bring down the ovule to seed ratio in the seed crop and seed to plant ratio in the commercial crop raised from the seed. The lower these ratios are, the lower will be the stress on the bio-resource. The recent work on research and development on seed science and technology has been compiled hereunder:

13.2 Seed Production

13.2.1 Planting Time

Planting time is crucial factor for the growth, development and yield of any crop. In cucumber, crop sown in the third week of January (20th January) gave higher seed yield per hectare (63.13 kg) with good germination (81.38 %), while crop sown on 31st July and 7th August gave higher seed yield per hectare (31st July, 41.58 kg, and 7th Aug, 39.37 kg) and germination (31st July, 93.31 %, and 7th Aug, 87.31 %) (Kishor et al. 2010).

In marigold cv. Pusa Narangi Gainda, Pramila et al. (2012) reported that number of seeds per plant (165), seed yield per plant (14.51 g) and test weight (2.63 g) were higher in *kharif* season, whereas germination (88.40 %) and seed vigour (vigour index I and vigour index II, 1102 and 88, respectively) were recorded in *rabi* season.

13.2.2 Plant Population and Geometry

Plant population and geometry is a key factor in deciding the canopy dimensions that directly affect the seed yield. In cucumber, seed yield and number of seeds per fruit increased with decrease in spacing up to 1.25 × 0.30 m (Kishor et al. 2010).

Ade et al. (2015) reported that seed cotton yield was significantly higher with closer plant spacing of 90 × 45 cm than wider spacing of 90 × 60 cm. The fertiliser application at 150 % RDF recorded significantly higher seed cotton yield.

Cluster bean (*Cyamopsis tetragonoloba* L.) planted at 45 × 15 cm spacing produced a higher graded seed yield (9.80 and 10.80 q ha^{-1}) in comparison to 45 × 30 cm (8.19 and 9.30 q ha^{-1}) and 60 × 30 cm (6.83 and 7.90 q ha^{-1}) during *kharif* and summer season, respectively (Siddaraju et al. 2010a, b).

Among the three planting methods, transplanting of 25-day-old seedling with the spacing of 180 × 75 cm recorded higher seed yields (23.5 q ha^{-1}) and net returns (INR 62,922 ha^{-1}) over dibbling and farmer method of sowing (Sujathamma et al. 2015).

The highest seed setting and seed dry weight (106.84 g) in faba bean is obtained in the highest plant density (06 plants pot^{-1}) for all varieties; however, genetic differences also existed. Genotypic differences in the growth rate were also evident (Thalji 2010).

13.2.3 Planting Material

The quality of the mother seed or planting material is the deciding factor in the final quality of the harvest. Shukla et al. (2010) concluded that among the treatments, extra large size curd of cauliflower cv. Pusa Snowball K-1 showed maximum potential for seed yield, while extra large and medium curd size were promising for seed quality parameters.

Khulbe et al. (2010) concluded that extra large curd size showed maximum potential for seed yield, while extra large and medium curd size were promising for seed quality parameters of cauliflower cv. PSBK-1.

Transfer of *Amaranthus* seeds during the early imbibitional period from 25 to 45 °C for 4, 8, 12 and 16 h resulted in the leakage of UV-absorbing substances and electrolytes up to 72 h. High-temperature treatment caused greater membrane damage by membrane lipid peroxidation and decreased ethylene formation in germinating seeds. Reduced activities of free radical scavengers like peroxidase, catalase and superoxide dismutase, with a concomitant rise in free H_2O_2 level, are also observed (Bhattacharjee and Mukherjee 1998).

The medium seed size variety JS 335 performed better than all small (NRC 37 and JS 97 52) and bold seed variety (TAMS 98 21 and TAMS 38) in respect of seed yield and yield contributing characters (Lakhote et al. 2015).

The *rabi* rice gave the highest yield when raised from 24-day-old seedlings under machine transplanting and gradual reduction in yield observed with age of seedling transplanted (Reddy et al. 2015).

The seed quality in tomato in terms of germination (92 %) and vigour was obtained in the seeds harvested from the crop raised from 30-day-old transplants and was followed by 27- and 33-day-old transplants, respectively (Singh et al. 2015b).

13.2.4 Growth Regulators

The seed yield and quality are directly influenced by the growth and development of mother plant especially during reproductive phase. In bitter gourd, three sprays of GA_3 at 50 ppm at four leaves, flower and fruit initiation stage improved the vine length (Hilli et al. 2010). NAA at 50 ppm sprayed between 85 and 100 DAS in bitter gourd (*Momordica charantia*) cv. Pusa Vishesh produced higher number of fruits per vine (14.94), fruit yield (119.68 q ha^{-1}), highest seed germination (83.25 %), seedling vigour index I (1757) and lowest electrical conductivity (0.316 dSm^{-1}), whereas boron at 4 ppm produced highest number of seeds per fruit (29.00), seed yield (6.84 q ha^{-1}), test weight (185.11 g), seedling length (21.16 cm), seedling dry weight (129.51 mg), seedling vigour index II (10749) and dehydrogenase enzyme activity (0.352 OD value) (Arvindkumar et al. 2012).

While, in pumpkin, Das and Das (1995) reported that GA_3 tends to increase the staminate/pistillate flower ratio, NAA and Ethrel decreases. A staminate/pistillate flower ratio of 6.61:1.00 was achieved by treating plants with 200 ppm Ethrel.

The foliar spray of 200 ppm ethephon significantly increased pistillate flower $plant^{-1}$ (24.7) and fruit $plant^{-1}$ (4.9) in cucumber cv. Picklingham (Sitaram et al. 1989). Foliar application of 200 ppm ethephon on the crop sown on 30th January brought a remarkable increase in fruit yield (899 g $plant^{-1}$) and seed yield (13.76 g $plant^{-1}$) of cucumber cv. Picklingham; however, the germination was generally higher in the seeds produced on December sown crop (Sitaram et al. 1989).

Kallihal et al. (2013) sprayed Planofix at 0.5 ml L^{-1} at flower initiation stage in pigeon pea which reduced flower drop (77.6 %), increased seed yield (17.4 %) and enhanced seed quality in terms of germination and vigour.

Foliar application of GA_3 at 300 ppm and at 20 and 40 DAT significantly increased the seed yield and quality in *Tagetes erecta* (Rohith et al. 2015). In another study on chrysanthemum (*Chrysanthemum coronarium* L.), application of GA_3 at 200 ppm increased number of seeds (265.33 $flower^{-1}$) and seed yield (500.00 kg ha^{-1}). The seed quality parameters like thousand seeds weight (2.14 g); germination %age (67.67%); seedling length (10.60 cm); vigour index (717) and dry weight (36.37 mg) were also higher in treatment of GA_3 @ 200 ppm (Sainath et al. 2012).

13.2.5 Harvesting

The harvesting and threshing or extraction of seeds is important operation as it determines the physical, physiological and health quality of seeds as well. Natural fermentation is more appropriate for better seed quality in cucumber than alkali and acid extractions (Chethan et al. 2013) as it gave increased germination (94.50 %): vigour index I (SVI-I and SVI-II of 2671 and 1183, respectively), field emergence (92 %) and low electrical conductivity (341 μ Sm^{-1}) (Chethan et al. 2013).

The delayed harvesting and after-ripening improves the seed yield and quality. Kortse and Oladiran (2013) reported that the best longevity of 'egusi itoo' melon (*Cucumeropsis mannii* Naudin.) seeds were obtained from the 20 days after ripened fruits were harvested when all leaves dried.

The maximum germination and normal seedlings in tomato is attained from the seeds extracted from the fruits harvested at breaker stage. The seed quality did not change if fruits are harvested there after until red ripe stage; however, the speed of germination is fastest in the seeds extracted from the fruits harvested at red ripe stage. Delay in harvest beyond red ripe stage, in general, led to deterioration in seed quality (Valdes and Gray 1998).

The marigold seeds harvested at 120 and 125 DAT recorded better seed quality in terms of germination (81.60 % and 83.2 %), 1000-seed weight (2.87 and 2.90 g) and seed vigour, i.e. vigour index I (995 and 994) and vigour index II (83 and 84), compared to earlier and later stages of harvest (Pramila et al. 2012).

13.2.6 Irrigation and Nutrition

Irrigation and nutrition of seed crop plays an important role in determining the seed yield and physiology. Irrigation increased the seed yield of mustard (*Brassica juncea* L.) by 44–57 % compared to rain-fed crop (Majid and Simpson 1999).

Application of sulphur at 45 kg ha^{-1} significantly improved the seed yield and 1000-seed weight of jute (Singh et al. 2015a).

Moklyachuk et al. (2013) recommended the implementation of zeolites, impregnated with zinc and copper ammoniates for improving the soil quality and micronutrients level in seed oat and barley. Application of zinc in the form of $ZnSO_4$ at 10 kg Zn ha^{-1} was more effective and remunerative for seed production of cereal legumes (Naaiik et al. 2015).

13.2.7 Economics

Direct seeding of rice required two irrigations lesser than the transplanting as maturity time is reduced by 7–10 days before when compared to transplanting rice (Rao et al. 2015). This technology considerably reduced the cost of cultivation by INR 10,000 ha^{-1} and improved cost-benefit ratio of *kharif* crop to 1.00: 2.94 and *rabi* crop to 1.00: 2.00 (Rao et al. 2015). It will be more suitable for dry land farmers.

Bellundagi et al. (2015) stated that the net return obtained from Bt cotton seed production was found to be higher in the case of Monsanto

Seed Company contract farmering (INR 46387), followed by Kaveri Seed Company contract farming (INR 33076). While, returns to rupee investment was found to be higher (1.48) in the case of Monsanto Seed Company contract farming, followed by JK Seed Company (1.36) contract farming.

13.2.8 Productivity

Mandal and Mohanta (2015) reported that the lateritic belt of West Bengal is highly suitable for quality seed production of *kharif* onion with an average seed yield of 10.4 q ha^{-1}, while among the cultivars, Indam Marshal (14.8 q ha^{-1}) produced the highest seed yield followed by Agrifound Dark Red (12.6 q ha^{-1}).

Patta et al. (2015) reported that higher tomato hybrid seed productivity (>80 % fruit set) of Pusa Hybrid-4 (Pusa Selection 120 × Chikoo) was achieved when the same stigma was pollinated twice for two consecutive days even up to 4 days after emasculation. Maximum fruit and seed set were observed even with 3-day stored pollen (Patta et al. 2015).

13.2.9 Organic Seed Production

The commercial agrochemical-based agriculture resulted in an ecological imbalance over the past few decades. One of the options to mitigate this imbalance is organic farming. Organic agriculture is a production system, which excludes the use of synthetically compounded fertilisers, pesticides and growth regulators. Organic agriculture is based on principles of self-sufficiency, biodiversity, crop rotation and recycling. It relies on organic manures produced from farm wastes and other biomasses. It also encompasses a conglomeration of various techniques and practices like crop rotation, intercropping, mulching, cover cropping, trap cropping, etc. Organic farming also employs various biological pest control methods, which eliminates the use of synthetic chemicals even at the storage levels. In other words, organic products are grown under a system of agriculture without the use of chemical fertilisers and pesticides with an environmentally and socially responsible approach. It works at grass-root level, preserving the reproductive and regenerative capacity of the soil, good plant nutrition and sound soil management, and producing nutritious food rich in vitality that has resistance to diseases. Organically grown food products offer the following advantages:

- Safe for the environment
- Sustains and enriches the soil
- Safe for the growers and producers
- Safe for the consumer
- More nutritious food
- Better flavour and aroma

The organic seed attains a more focused view to increase the acceptability of organic farming and organic produce as well. The organic production system is incomplete without the organic seed. Organic seed production is the need of the hour to meet out the seed needs of farming communities (Sripathy et al. 2012). As per the standards of National Programme for Organic Production, 'All seeds and plant material should be certified organic'. The standards are further extended as 'When organic seed and plant materials are available, they shall be used. The certification programme shall set time limits for the requirement of certified organic seed and other plant materials. When certified organic seed and plant materials are not available, chemically untreated conventional materials shall be used'. This relaxation is for a certain period of time because today the organic seed is not readily available. In times to come, the principle shall be strictly implemented and full compliance shall be required for the certification.

Organic seed is commercially not available for most of the crops or varieties, and if available, it is usually significantly more expensive than the same conventionally grown varieties. At present, seed companies producing certified organic seeds are limited and are often producing only regionally adapted varieties. Some organic producers have been disappointed with the seed vigour and quality of organic seed and have invested into on-farm seed propagation.

Traditionally, farmers saved a portion of their crops for their own seeds and occasionally to sell to neighbours. If the drafted regulations do pass, this practice will invariably increase and potentially become a viable source of additional income to many organic farmers. In the past, organic farmers have often relied on fields in transition to produce 'organic' seed, not only as a means towards self-sufficiency but an easy way of documenting that the transitional crop was separate from the organic crop. But, under current wording of the regulation, this practice would be prohibited, if a readily available organic seed source is present.

As of now, the work done on organic seed production is just of what is actually needed to cope with the emerging demand of organic agriculture in India. Many studies have been carried out on various aspects of organic production world over. Peacock (1990) suggested that the avoidance of disease and insect attack in organic crops could be achieved by crop rotation, adjustment of planting and harvesting time and transplanting and avoiding high pest areas on the farm.

Fernandez et al. (1998) reported that certified organic growers in the USA relied primarily on traditional organic practices such as crop rotation, animal manuring, composting, cultural or biological control of diseases or pests, water management techniques, adjustment of planting and harvesting dates and beneficial organisms.

In a comparison of conventional vegetable production system with organic system, D'Ercole and Cembalo (1999) found that a premium of at least 20 % over conventional produce was the minimum price point to make organic farming profitable.

13.3 Genetic Quality

The genetic quality is the set of parameters that the seed must possess to produce a plant with essential characteristics to confirm to be called as a particular variety or cultivar. In strict genetic terms, it must possess genetic constitution as that of the variety or cultivar of which the seed is being produced. In general, in a seed production process, the genetic quality is controlled by controlling the following factors:

- Cropping history of the field
- Seed source
- Isolation

13.3.1 Rouging and Weeding

The genetic purity can be tested by many postharvest tests ranging from the basic grow out test based on morphological markers to the applications of modern techniques like biochemical and molecular markers, DNA finger printing, ELISA, etc. The recent trends in research and development in this regard are summarised below.

13.3.2 Morphological Markers

In a study on the heterogeneity in a blend of two *Medicago sativa*, lots and a lawn grass seed mixture (*Festuca rubra* subsp. *rubra*, *F. trachyphylla*, *Poa pratensis* and *Lolium perenne*), Niffenegger (1991) described procedure for determining heterogeneity in batch-mixed commercial seed lots. The steps followed for testing heterogeneity are:

- Determining the amount of indicator seed needed
- Obtaining and marking the indicator seeds and mixing the seed lot
- Incorporating the indicator seeds as additional components of the lot
- Sampling and recording numbers of indicator seeds in samples
- Analysing the results

The test requires no specialised equipment and can be especially useful for testing the effectiveness of seed mixing systems.

Pramila et al. (2011) differentiated marigold (*Tagetes erecta* L.) cultivars based on flower and seedling characteristics like anthocyanin pigmentation, colour and incision of margin of flower.

The hybridity or genetic purity of eggplant (*Solanum melongena* L.) hybrid No. 1461734 is improved from 85.76 to 95.97 % by seed conditioning using top screen aperture, size 4.0 mm (round); middle, 3.5 mm (Round) and bottom, 0.9 mm (slotted) (Vasudevan et al. 2012).

13.3.3 Molecular Markers

Esterase isoenzyme analysis using polyacrylamide electrophoresis is an effective tool to assess the genetic purity of bent grass species or various cultivars within each species in combination with visual seed analysis or grow out tests (Freeman and Yoder 1998). The test is based on esterase banding profiles that segregate *Agrostis palustris*, *A. tenuis* and *A. canina*. The *A. tenuis* cultivars are characterised by prominent bands at Rf 0.55–0.59, whereas *A. palustris* cultivars were characterised by Rf 0.50–0.60 and *A. canina* by bands at Rf 0.57–0.66. The cultivars within species are characterised by minor banding pattern variations.

Livneh et al. (1990) observed that polymorphism is inherited as a simple codominant allelic marker, enabling the distinction between the parents themselves and between parents of the pepper (*Capsicum annuum*) hybrid cultivar Maccabi and progenies (Livneh et al. 1990). Noli et al. (1999) reported 22 polymorphic bands within cultivated *Lycopersicon esculentum* using RAPD analysis. Hybrid purity in heading Chinese cabbage (*Brassica pekinensis*) hybrid Renewed Qingza 3 is determined by DNA extraction and RAPD analysis (Meng et al. 1998).

There exist abundant polymorphic SRAPs and SSRs among closely related cultivars of *Allium fistulosum* L. However, in conformity with the morphological traits analysis, SRAP is more accurate in analysing genetic diversity, and SSR should be preferred for genetic purity test and identification of cultivars in *Allium fistulosum* L. (Li et al. 2008).

The molecular markers (two dominant RAPD primers and two codominant SSR primers) are highly efficient and reproducible for genetic purity testing of seed genetic purity of a hybrid tomato 'Hezuo906' (Liu et al. 2008).

Mishra et al. (2005) suggested that analysis of tuber protein using SDS-PAGE is a good tool for testing of genetic purity in potato (*Solanum tuberosum* L.).

13.4 Physiological Quality

13.4.1 Seed Viability

Phosphatidyl choline and phosphatidyl ethanolamine showed a consistent decline with ageing, resulting in the loss of viability in pigeon pea (Kalpana and Rao 1996). Free radical-induced lipid peroxidation did not seem to play a role in the rapid ageing of pigeon pea seeds. Superoxide dismutase activity tended to increase, while catalase and peroxidase activities decreased to various degrees depending on cultivar, indicating the probable role of hydrogen peroxide in the accelerated ageing process (Kalpana and Rao 1994).

The interaction between the PEG treatment (1.2 MPa) and the light treatment only delayed the germination process but neither inhibited the germination nor induced the secondary dormancy. High temperatures of 35 °C, 40 °C and 45 °C with duration of heating for 24 and 48 h did not induce the dormancy even though the radical emergence percentage decreased and the delayed germination value increased. However, these did not affect the viability of *Amaranthus tricolor* seed (Fatimawati and Ilyas 1996).

13.4.2 Seed Germination

The inhibitory compounds were present in the pericarp and seed coat of *Nyctanthes arbor-tristis* and cause poor germination. These phenolic compounds, leached out of imbibed seeds, interfered in germination. The removal of pericarp and seed coat increased germination from 24 % (intact seeds) to 94 % (Bhattacharyya et al. 1999). Leaching of the germination inhibitors was restricted by treating seeds prior to germination with solutions of antioxidants like polyvinylpyrrolidone (PVP) and polyvinylpolypyrrolidone (PVPP), resulting in better germination (Bhattacharyya et al. 1999).

Combined heat desiccation pulsing (scarification) of the exo- and endotestas resulted in strongly increased germination compared with controls; however, the exotesta appears to act as the primary barrier to oxygen diffusion to the embryo (Brits et al. 1999). In natural fynbos, desiccation-mediated scarification of the testal layers is the main means of regulating oxygenation and thus synchronous germination of dormant *Leucospermum* seeds after fire.

The duration of ageing required to reach a germination percentage around 75 % was identified as 5 days for black gram and 3 days for cowpea and red gram (Sujatha et al. 2012).

The bold seeds possess better seed quality parameters. The size group of 5.00 mm and 4.75 mm in French bean (*Phaseolus vulgaris* L.) cv. Arka Suvidha possessed better seed quality than the bulk seed lot (Raj et al. 2012).

13.4.3 Seed Vigour

The poor field establishment problems associated with the African yam bean [*Sphenostylis stenocarpa* (Hochst. ex A. Rich) Harms] and pigeon pea (*Cajanus cajan*) is low seed vigour and not viability (Olisa et al. 2010).

Bael [*Aegle marmelos* (L.) Corr.] fruits were categorised based on the colour into three different groups as green, greenish yellow and yellow. The recovery of yellow and greenish yellow fruits were 41 % and 33 %, respectively, from bulk. The seeds extracted from yellow fruits produced vigorous seedlings (Venudevan and Srimathi 2013).

The seed quality of dill (*Anethum graveolens* L.) depends on the position where it is borne on the umbel or inflorescence. The lower its position on the seed stalk, the lower the seed yield and germination capacity. The best seeds, in terms of their 1000 seed weight, came from the external part of the main umbel. The recorded seed embryo length from the internal part of the primary umbel was smaller than its values from the middle and external parts of that umbel (Holubowicz and Morozowska 2011).

Sreenivas et al. (2015) reported that response of seeds of different cultivars of paddy varied to the EC of water; however, cultivar cv. MTU-1010 performed well on an average of all the EC levels of water ranging from 0 to 20 dS m^{-1}.

13.5 Seed Health

13.5.1 Seed Treatment

One of the most important research focuses in seed science and technology nowadays is looking for new methods to improve seed health quality. Seed treatment with plant-based ethereal oils is one option. Chen and Houbowicz (2010) reported that 5 % alcoholic solution of camphor tree (*Cinnamomum camphora* L.) ethereal oil and nine different treatments of 5 % alcoholic solution of patchouli plant (*Pogostemon cablin* Benth.) ethereal oil lowered the amount of identified *Alternaria alternata* and *Cladosporium sphaerospermum* fungi on the seeds and percentage of dead seeds with no effect on the germination of lettuce seeds; however, the length of seedling is lowered.

Khusro et al. (2013) reported that the aqueous extracts of *Allium sativum* is a potential spice for controlling *Bacillus licheniformis* strain 018 compared to *Bacillus tequilensis* strain ARMATI. *Clonostachys* spp. and grapefruit extract significantly reduced seed infestation with *A. radicina* and improved seed germination capacity in carrot (Szopinska et al. 2010).

Seed treatment with methyl jasmonate (MeJA) or benzothiadiazole (BTH) can induce plant defence against insect pests and bacterial diseases in flowering Chinese cabbage (*Brassica campestris* L. ssp. *chinensis*), owing to elevated expression of the PDF1.2, PR1 and WRKY33 genes (for insect pests, *Phyllotreta striolata* resistance) and upregulation of PDF1.2 (for postharvest stem decay) (Zhang et al. 2013).

Beneficial bacteria and fungi provide promising alternatives or supplements to chemicals as seed treatments against soil-borne pathogens. Early results also suggest variable levels of BCA compatibility with chemical seed treatments and other microbial applications. Progress in storability, ease of application and economy of biological control agent's production will be needed to move biologicals beyond their niche position in the seed treatment market (Bennett 1998).

13.5.2 Seed Health Testing

Introduction of ELISA in place of chloroplast agglutination test in plants reduced the incidence of mosaics, resulting in corresponding increase in yield (Sharma and Venkatasalam 2011).

Agar plate method was found to be suitable, as even under lesser incubation, there was higher observed incidence of seed mycoflora. The other finding depicted that the most common fungi in all crop seed as well as in treated and untreated seeds were observed were *A. niger*, *A. flavus*, *P. rubrum* and *P. citrinum*. In all stored seed, *R. stolonifer* was also observed as a common fungus (Singh et al. 2015c).

13.6 Seed Enhancement

13.6.1 Irradiation Seed Treatments

The germination and vigour of common bean (*Phaseolus vulgaris*) seeds were improved by exposure to microwaves in a microwave oven (650 W, 2450 MHz) for 15, 30, 45, 60, 90 and 120 s. Microwaves diminished the presence of *Penicillium* spp., both on the seed surface and in the inner seed tissues, but were not effective in controlling *A. alternata* and *Fusarium* spp. (Tylkowska et al. 2010).

Nargis et al. (1998) reported that tomato cv. PKM-1 seeds exposed to 10 and 20 Kr of gamma radiation showed an increased speed of germination and vigour but did not show significant increase at 30 Kr. However, seeds treated with vitamin C or vitamin E resulted in increased seedling length even at 30 Kr. No damage was observed on the seed coat due to gamma radiations.

The finger millet cultivars had different response to gamma ray irradiation. The LD50 values determined based on the germination percentage was 1.20 KGy for Vr-708, Vr-900, saptagiri and kalyani; 1.40 KGy for Vr-762, Vr-847 and padmavati; and 1.60 KGy for Vr-(w)-936 (Gali and Maheshwari 2015).

13.6.2 Seed Scarification

Solanum viarum seeds scarified for 10 min using commercial sulphuric acid at 25 ml kg^{-1} of seed recorded the highest germination percentage (80 %), followed by those scarified for 15 min (60 %). The root and shoot length (4.3 and 4.5 cm, respectively), dry matter production (0.034 mg) and vigour index values (704) were also the highest with 10 min scarification. Sulphuric acid scarification beyond 15 min was injurious and caused reduction in germination and seedling growth (Ramamoorthy et al. 2010).

Treatment with H_2SO_4 for 60 and 120 s and scarification with sandpaper gave germination percentages of 86.46, 86.0 and 80.78, respectively, compared with 28.76 % on controls (Tomer and Kumari 1991). In a further experiment, the percentage of hard seeds declined from 50 % immediately after harvest to 7 % after 12 weeks of storage under ambient conditions (Tomer and Kumari 1991).

Takaki and Gama (1998) reported that the seeds of lettuce cv. Grand Rapids scarified with sodium hypochlorite germinated in both dark and continuous light. The scarified seeds pre-incubated at 36 °C behaved similar to non-scarified seeds. The chemical scarification changed the control of seed germination from low fluence response to the very low fluence response, and pre-incubation at 36 °C returned the control to the low fluence response of phytochrome action.

13.6.3 Seed Priming

Osmopriming of kohlrabi seeds improved the seed germination and vigour despite the presence of *A. brassicola*; however, it is conducive to penetration of *A. brassicola* into kohlrabi seeds. The osmopriming is of limited usefulness for seeds heavily infected with *A. brassicola* as it leads to increased infected seedlings (Dorna et al. 2010).

Three successive soaking-draining cycles (soaking for 24 h then draining) of carrot cv. Zino seeds increased germination from 30 to 63 % and also enhanced the speed of germination and vigour (Sundaralingam et al. 1998).

Priming of cantaloupe (*Cucumis melo*) seed sample with 20 % of infestation of fungi (*Alternaria, Cladosporium, Epicoccum* and *Stemphylium* spp.) for 6 days in darkness at 25 °C in a $KNO_3 + KH_2PO_4$ (1.5 + 1.5 %) aerated solution increased the microorganism incidence to 60 % on Captan 50 WP (3 g kg^{-1} seeds) treated seeds and 94 % on non-treated seeds; however, in non-primed seeds, fungicide treatment eliminated microorganism incidence (Nascimento and West 1998).

Black gram (cv. GOBG 282/1) and cowpea (cv. CO 6) seeds can be soaked up to 5 h and 3 h in 33 or 50 % volume of water (V/V of seed), respectively, without any soaking injury (Vijaya and Ponnuswamy 1997b).

Germination of PEG-primed seeds of carrot, leek, and onion was not affected by drying and storage; however, the number of abnormal seedlings increased with storage duration. The slow-germinating seed lots benefited more from priming than faster ones (Drew et al. 1997).

Priming is a tool to improve the performance of chilli seed, known to have poor germination in the field at temperatures near 15 °C. Priming in solutions of 0.3 M NaCl and 0.3 M NaCl + 4 μg ProGibb T g^{-1} seed for 5 days at 23 °C improved the germination rate of seeds of chilli cvs. Tam Veracruz and Early Jalapeno. Under these conditions, ProGibb T cannot be used alone as 70 % of the seeds germinated during priming process (Carter 1997).

At 0.7 MPa, the increase in germination percentage, compared with the control, determined by osmopriming ranged between 13 % (priming in PEG6000 for 8 days) and 40 % (priming in $KNO_3 + K_3PO_4$ for 8 days). The most effective treatment ($KNO_3 + K_3PO_4$ for 8 days) reduced MTG by about 4 days compared with the control where it was 9.5 days (Mauromicale and Cavallaro 1995).

Seed priming, particularly hydro-priming, is a low-cost technology in which controlled hydration of seeds followed by re-drying is done to break dormancy, improve seed germination and stand establishment. Sowmya et al. (2013a) reported that hydro-priming at 25 ± 1 °C for 48 h was found to be optimum for priming in cucumber (*Cucumis sativus* L.).

Priming is one of the most common methods of improving seed quality, which many a time significantly affecting their storability. Dorna et al. (2013) found that for maximal seed viability and germination rate after 6 and 12 months storage, both hydro- and osmoprimed seeds of onion should to be stored at 4 °C rather than 20 °C. Osmopriming significantly increased seed infestation with *Penicillium* spp., which remained stable up to 12 months of storage. The number of seeds infested with *Botrytis* spp. significantly decreased after priming and storage, especially at 20 °C.

Solid matrix priming offers an effective means of raising seed performance in sweet corn (*Zea mays*) carrying the shrunken-2 (sh-2) gene. Nevertheless, the storability of primed sh-2 seeds is still unknown. The priming effects on sh-2 sweet corn seeds can be maintained after storage for 6months at less than 10 °C. However, the primed sh-2 seeds deteriorated more rapidly than non-primed sh-2 seeds when stored at 25 °C for 6 months (Chang and Sung 1998).

Seed moisturization by increasing moisture up to 14–16 % of *Phaseolus vulgaris* seed improved emergence, while seedling dry weight continued to increase until 18 % (Demir et al. 1998).

Grzesik and Nowak (1998) stated that matriconditioning of seeds for 6 days with Micro-Cel E in the presence of kinetin (20 mg L^{-1}), Busz (15,000 mg L^{-1} a mixture of N, P and K), ethephon (1500 mg L^{-1}) and GA_3 (300 mg L^{-1}) improved seed performance and seedling emergence. Such treatment also increased seedling frost resistance and decreased the harmful effect of water stress on seed performance of *Helichrysum bracteatum*.

Seed quality enhancement of cucumber (*Cucumis sativus* L.) shall be done at 25 ± 1 °C temperature by priming with 100 ppm ethrel or 1 % KNO_3, followed by 100 ppm GA_3 (Sowmya et al. 2013b).

P. fluorescens is an effective antagonist for *G. lucidum* with 85.91 % inhibition and 10 mm of inhibition zone in oil palm (*Elaeis guineensis* Jacq.). In the bio-priming pot culture experiment, combination of *P. fluorescens* + *T. viride* proved effective with 23.56 PDI as compared to control with 100 PDI. The same treatment took 8 weeks of incubation period for onset of symptoms (Deepthi and Rao 2015).

The primary seed priming of red bean (*Phaseolus calcaratus*) with water followed by drying and complementary priming with GA_3 significantly increased dry matter (53.06 g) and grain yield (Rastin et al. 2013).

The germination index, mean germination rate and total fresh weight significantly increased in primed tomato seeds compared with unprimed seeds under salinity stress (100 mM NaCl conditions). Seed priming improves tomato seed vigour under stress conditions associated with a decrease in seed lipid peroxidation (Zhang et al. 2012).

Onion seeds hydro-primed with 80 % sand for 24 h improved the seed quality characteristics throughout the period of storage up to 4 months (Selvarani et al. 2011).

In order to harness maximum potential of seed priming, the most suitable method and the methodology should be adopted, specific to each crop species. For onion, sand matric priming (24 h in 80 % WHC of sand) recorded the highest improvement in term of percent and speed of germination. For carrot, hydro-priming (24 h in water at double the volume of seed) recorded the highest improvement (Selvarani and Umarani 2011).

Priming is a treatment that partially hydrates seeds so that germination processes begin, but radical emergence does not occur. Experimentally, priming treatments are limited only to conditions that do not result in premature radical extension and may include equilibration under conditions of high humidity, soaking in water or osmotic solution, equilibration with a matric potential control surface, intermixture with a porous matrix material and moisture addition to seed water content less than required for germination. Welbaum et al. (1998) emphasised that priming treatments must be determined empirically. The optimal water potential for priming varies with the species in question and so is the effect of priming.

Priming of chick pea seeds with custard apple leaf extract at 3 % improved the seed quality with respect to germination, vigour, dehydrogenase and α-amylase activity irrespective of containers, at the end of tenth months of storage period (Shakuntala et al. 2015).

Soaking of okra seeds in solution of GA3 at 100 ppm and foliar sprays Cycocel at 750 and 1000 ppm at 30 and 45 DAS, respectively, were found to be beneficial in terms of early germination (2.75 days); the highest germination percentage (99.5); reduced height of plant (86.65 cm) and length of internodes (5.10 cm); and increased number of leaves (43), number of internodes (15.90), number of branches (3.15) and leaf area (1249.5 cm2). It also resulted in early flowering (34 days); increased number of flowers (23.40), fruit set (87.54 %), number of fruits (20.46) and yield per plant (201.30 g $plant^{-1}$) of okra (Bhagure and Tambe 2015).

Seed priming is a pre-sowing strategy that influences the seedling development by modulating pregermination metabolic activity prior to emergence of the radical and generally germinates more rapidly leading to better crop performance.

Among the organics used for seed treatment, custard apple leaf extract at 3 % concentration recorded significant improvement in seed germination and vigour of sorghum (Vasudevan et al. 2015). The genotypic response to various organic priming treatments varied and the sorghum cultivar M 35 1 responded very well followed by the cultivars Muguti and BJV 44 (Vasudevan et al. 2015).

Pre-sowing treatments of mango (*Mangifera indica* L.) stones of GA_3 at 100 ppm increased the germination and enhanced the vigour (Kolekar et al. 2015).

Vaddinakatti et al. (2015) reported that sunflower seeds priming with custard apple leaf extract at 3 % and stored in polyethylene bag (700 gauge) increased the germination and vigour. However, various genotypes responded variably to the priming and storage treatments. Seeds of cultivar KBSH-44 primed with custard apple leaf extract at 3 % and stored in polyethylene bag (700 gauge) recorded significantly highest seed quality parameters than KBSH-53 and KBSH-41 genotypes at the end of tenth month of storage period.

Organic seed treatment with *beejamruta* at 50 % (Shankrayya et al. 2015) and panchagavya at 3 % (Nagarajaiah et al. 2015) improved quality in terms of higher germination, vigour, dehydrogenase enzyme activity and lower electrical conductivity of seed leachate in paddy seeds.

Cucumber seeds primed with PEG 6000 1.5 MPa for 24 h exhibited improvement in quality in terms of speed of germination, total germination and vigour, up to 6 months of storage; however, after 6 months of storage, there was a sharp decline in germination and vigour of primed seeds (Shukla et al. 2015).

13.6.4 Magnetic Seed Treatment

The seeds treated with south pole of magnet showed earlier flowering, a higher inflorescences plant^{-1} and a higher reproductive efficiency index (RE) over north pole treated cauliflower seeds (Samy 1998); however, the seeds treated with both poles of magnet failed to give significant increases in siliqua setting percentage and days to 50 % flowering from bolting. In general, south pole treatments were better than north pole treatments. Samy (1998) further observed that exposure for 8 h was most beneficial, followed by spraying with south pole treated water and soaking in south pole treated water for 16 h.

Gurusamy (1998) reported significant increase in curd diameter, length and yield from the plants raised from cauliflower cv. Kibo Giant seeds exposed to a south magnetic pole for 8 h.

13.6.5 Seed Upgradation

Carrot cv. Zino seeds air blown for 2 min and retained on a 12 × 12 BSS sieve after removing the bristles increased the germination (71 %), 1000-seed weight and vigour index; however, the seed recovery was 52.2 % (Sundaralingam and Karivaratharaju 1998).

Taylor et al. (1991) developed a system to effectively upgrade *Brassica* seed (cabbage cvs. Danish Ballhead, King Cole, cauliflower cv. Snowball, broccoli cv. Citation F_1, oilseed rape cv. Westar and the ornamental species *Erysimum hieraciifolium* cv. Orange Bedder) quality by exploiting sinapine leakage. Seeds were first hydrated and then coated with an absorbent (10 % Pelgel) to trap the leaking sinapine. The coated seeds were dried and sorted into fluorescent and nonfluorescent categories using UV light.

In paddy, grain discolouration is considered as one of the important problem affecting quality of the paddy seed used for planting. Field experiment on discoloured seed at different levels with quintal (iprodione + carbendazim) and carbendazim has resulted in increased seedling growth rate and speed of germination (Durga et al. 2015).

13.6.6 Seed Hardening

'Hardening' of seeds has been defined as 'a treatment preliminary to sowing during which seeds are moistened and dried back (once or a number of times) to activate certain physiological mechanisms which will enable the resulting plants to withstand adverse environmental conditions'. It can be done by using organic and inorganic products. This technique is mainly followed for pulses, viz. chick pea, green gram, groundnut, etc. The pre-sowing seed hardening with chemicals, viz. hypertonic solutions of salts (e.g. KNO_3, $MnSO_4$) or of osmotically active but chemically inert substances (e.g. polyethylene glycols of high molecular weight), is one of the simple technique being employed to modify the morpho-physiological and biochemical nature of seed. It has been reported to induce drought resistance capacity in plants and such seeds have the capacity to withstand moisture stress. Seed hardening induce the absorption of more water due to the development of stronger and efficient root system and increased dry matter production. It is similar process of 'advancing' that has the restricted aim of enabling seed to pass through the first stages of germination, short of radicle emergence, before the seed is sown. It is possible that these procedures might improve seed quality, because fully imbibed seeds have the capacity to repair damaged tissues (Janaki et al. 2015).

13.6.7 Seed Fortification

The germination and field emergence of black gram seeds was significantly increased with treatment of $ZnSO_4$ at 100 mg kg^{-1} seed; however, seed germination was not affected by the trace elements, but field emergence was highest with the combined application of $ZnSO_4$, $MnSO_4$ or Na_2Mo_4 at 100 mg kg^{-1} seed (Vijaya and Ponnuswamy 1997a).

Seed fortification is one of the important seed invigoration treatments. Jayanthi et al. (2013) recommended seed fortification of rice (cv. Co.43) with 2 % horse gram sprout extracts at seed moisture of 12 ± 1 % by soaking for 12 h for enhancement in seedling quality characteristics.

13.6.8 Seed Pelleting

Water activity measurements were a versatile method of assessing water status of coated and non-coated seeds on a routine basis. It provided accurate determinations of water status of pelleted and coated seeds of lettuce (*Lactuca sativa*) lot and one onion (*Allium cepa*) (from six pellet sources and three film coating formulations) for two crops. The seed moisture content could be calculated from the water activity values with equations from the moisture isotherm for each crop (Taylor et al. 1997).

Rhizobium phaseoli pre-inoculated pelleted seed of *P. vulgaris* cv. Contender contained sufficient bacteria per seed for effective nodulation after 56 day storage at 25 °C in study conducted by Jansen et al. (1994). The pre-inoculation was done by incorporating the inoculums to the filler media of pellet. Liquid inoculum of *Rhizobium* failed to produce the same effects (Jansen et al. 1994).

Srimathi et al. (2013) reported that Jatropha seeds pelleted with pungam leaf powder recorded 169 % higher germination and *P. pinnata* seeds pelleted with *A. indica* leaf powder registered 92 % higher germination than control after 9 months of storage. The botanical leaf powder seed pelleting not only improve the longevity of seeds through protecting the seeds from fungal and insect attack but also improve the seed and soil relationships through enriching the rhizosphere region of seed to produce better growth and development.

Pelleting of tomato cv. PKM-1 seeds with $ZnSO_4$ (300 mg kg^{-1}) showed improvement in seed emergence and subsequent plant growth and yield (Shashibhaskar et al. 2011).

13.6.9 Electric Seed Treatments

Kumar et al. (1990) reported that bitter gourd cv. Faizabadi seeds treated with electric current 250 mA for 3 min produced plants with a significantly increased number of female flowers (from 10.5 to 39.0), female/male ratio (from 0.038 to 0.165), number of fruit $plant^{-1}$ (from 6.75 to 20.5) and fruit yield $plant^{-1}$ (from 204.43 to 383.16 g).

13.6.10 Seed Coating

The seed quality parameters, viz. 100 seed weight, germination percentage, seedling length, seedling dry weight, vigour index and electrical conductivity, were found to be higher in pigeon pea (*Cajanus cajan* (L.) Millsp) seeds obtained and treated with deltamethrin 2.8 EC at 0.3 ml kg^{-1} seeds + vitavax powder at 3 g kg^{-1} seed + polymer seed coating at 5 ml kg^{-1} seeds compared to other treatments (Vinodkumar et al. 2013).

Tian et al. (2013) suggested that labelling with rhodamine B dye by coating at 1 Kg per

20–30 Kg pea seeds can be used as an anti-counterfeiting technique in pea seeds. Moreover, the vascular bundles of stem, roots and aerial parts of seedlings treated with rhodamine B dye emitted brilliant fluorescence for a long time, which could be used as a marker in seedlings.

Polymer 'Quick Roots' seed treatment increased the tolerance of cotton seed to both high as well as low moisture content and improved germination and seedling establishment (Deepika et al. 2015).

13.6.11 Seed Nanotechnology

Nanoparticles are small-size (between 1 and 100 nm) atoms having more surface area that facilitate the plants to speed up the growth and development process. Agronomic efficiency of N, P and K was recorded highest with 80 % recommended dose of fertiliser (RDF) plus TiO_2 nanoparticles (Kumar et al. 2015). Apparent recovery efficiency of N and P was found highest with RDF plus ZnO, whereas in the case of K, it was highest with RDF plus TiO_2 nanoparticles. From this experiment, it can be concluded that seed treatment with ZnO and TiO_2 nanoparticles are found best in terms of wheat yield and NUE (Kumar et al. 2015).

13.7 Synthetic Seed

Synthetic seeds are defined as somatic embryos engineered to be of use in commercial plant production (Gray and Purohit 1991). Depending on the crop, synthetic seeds:

1. May or may not have a synthetic seed coat
2. May be hydrated or dehydrated
3. May be quiescent or not

The first basal section of cotyledons results in 58 % embryogenic callus formation and five regenerated plantlets in cucumbers cv. Poinsettia76 (Guedes and Jennings 1999).

The embryo encapsulated with an artificial endosperm containing gibberellic acid (GA_3) improved the recovery of plantlets from somatic embryos in *Citrus reticulata* Blanco. Encapsulation with GA_3 was also useful for storage of somatic embryos at 4 °C for 1 month (Antonietta et al. 1998).

Supplementation of embryo production medium with 10 6 M abscisic acid (ABA) or 1.0 mM proline improved desiccation tolerance and survival in celery (Kim and Janick 1990).

13.8 Seed Storage

13.8.1 Prestorage Treatments

The large-sized wheat seed are more suitable as a planting material and prestorage aspirin, and *Trigonella* seed powder treatments are very effective for improvement of storability of wheat (De et al. 2012).

The prestorage dry treatment with bleaching powder at 2 g kg^{-1} of seed of large-sized okra seed improved storability and field performance (Guha and Mandal 2011). Phostoxin had a remarkable effect on germination (78.4 %) and vigour (80.67 %) of okra seeds (Akintobi 2009).

Prestorage dry seed treatments with bleaching powder at 2 g kg^{-1} of seed and red chilli powder at 1 g kg^{-1} of seed for high vigour seed and wet treatment (soaking for 2 h followed by drying) for medium vigour seed improve germinability during storage of onion seeds (Biswas et al. 2011).

13.8.2 Mid-storage Treatments

The tomato and eggplant crop raised from seeds treated with disodium phosphate showed the best growth and yield (812 and 2163 g $plant^{-1}$, for tomato and eggplant, respectively, compared with control values of 529 and 1192 g $plant^{-1}$, respectively) (Geetharani and Karivaratharaju 1996).

13.8.3 Post-storage Treatments

The germination and field performance of 16-month-old naturally aged seeds of okra cv. Pusa Sawani and Parbhani Kranti were increased by soaking in aqueous solutions of 50 or 100 ppm GA_3, IBA or thiourea for 24 h at 25 °C. GA_3 which increased the seed yield by 18 % (Kumar et al. 1997).

13.9 Summary

The research and development is progressing continuously in all sheers of seed science and technology. The emphasis among the seed production technologies is on plant population and geometry, use of plant growth regulators and efficient methods of seed threshing or extraction to improve seed yield and quality. The quality management, the other important aspect of seed technology, records new methods and techniques of testing seed quality in terms of genetic purity, germination and vigour and seed health. The new morphological markers using seed size variation in parental and hybrid seed has been described. The use of biotechnological tools in combination with traditional visual assessment has been devised in many crops. The molecular techniques have been developed to test the hybridity or hybrid purity in major crops. The biochemical basis of problems related to low germination and vigour has been discovered and the strategies for improvement have been suggested in various crops. The insight of recent literature reveals considerable advances in terms of diagnosis of seed heath problems and their preventive and corrective measures.

The recent trends show that the quantum of research in seed enhancement technologies is more compared to all other fields. Through the major emphasis on various techniques of seed priming, many new techniques have been described including magnetic and electrical seed treatments; irradiation with gamma rays, microwave, etc.; and seed scarification, hardening, upgradation, fortification, pelleting and coating. The application of synthetic seed and nanotechnology has also been employed to improve the performance of seed. The research in the field of seed storage is also prominent to lower down the seed deterioration.

13.10 Conclusion

Seed is the vital component of plant life cycle and so is for the agriculture. The performance of plant or crop entirely revolves around the inherent quality of seed from which it is raised. Seed even plays a pivotal role in the productivity of other agricultural inputs. Moreover, improved seed is the easiest way of transfer of technology in an effective manner. The applications of modern technologies like space breeding, nanotechnology, and physical treatments using magnet, plasma, microwaves, etc. provide new avenues about seed to scientists and technologists. Though a tremendous progress has been made in the field of seed science and technology to improve the performance of seed, however, the zenith will be achieved when every seed harvested will produce a healthy plant.

References

Ade BD, Chorey AB, Deotalu AS et al (2015) Effect of plant geometry and fertilizer response to hirsutum hybrids on productivity dynamic of seed cotton under rainfed condition. In: Compendium of abstracts of the 2nd international conference on bio-resource and stress management, ANGRAU & PJTSAU, Hyderabad, India, 7–10 January 2015

Akintobi DCA (2009) Effect of seed treatment on seed longevity, seedling growth, and yield of okra. Moor J Agric Res 10:43–52

Antonietta GM, Emanuele P, Alvaro S (1998) Effects of encapsulation on *Citrus reticulata* Blanco somatic embryo conversion. Plant Cell Tiss Org Cult 55(3):235–237

Arvindkumar PR, Vasudevan SN, Patil MG et al (2012) Influence of NAA, triacontanol and boron spray on seed yield and quality of bitter gourd (*Momordica charantia*) cv. Pusa Visesh. Asian J Hortic 7(1):36–39

Bennett MA (1998) The use of biologicals to enhance vegetable seed quality. Seed Technol (USA) 20(2):198–208

Bellundagi V, Patil SS, Vijaya BW et al (2015) Constraints Faced by Farmers in Bt Cotton seed Production in Karnataka. In: compendium of abstracts of the 2nd international conference on bio-resource and stress management, ANGRAU & PJTSAU, Hyderabad, India, 7–10 January 2015

Bhagure YL, Tambe TB (2015) Effect of seed soaking and foliar sprays of plant growth regulators on germination, growth and yield of okra [*Abelmoschus esculentus* (L.) Moench] Var. Parbhani Kranti. In: Compendium of abstracts of the 2nd international conference on bio-resource and stress management, ANGRAU & PJTSAU, Hyderabad, India, 7–10 January 2015

Bhattacharjee S, Mukherjee AK (1998) The deleterious effects of high temperature during early germination on membrane integrity and subsequent germination of *Amaranthus lividus*. Seed Sci Technol 26(1):1–8

Bhattacharyya S, Das B, Ghose TK et al (1999) Investigation on seed germination of *Nyctanthes arbor tristis* (Oleaceae) in relation to the total phenol content. Seed Sci Technol 27(1):321–327

Biswas J, Guha P, Mondal AK (2011) Responsiveness of high and medium vigour seed to seed treatment for improved storability of stored onion (*Allium cepa* L.). Indian Agric 55(1/2):51–56

Brits GJ, Calitz FJ, Brown NAC (1999) Heat desiccation as a seed scarifying agent in *Leucospermum* spp. (Proteaceae) and its effects on the testa, viability and germination. Seed Sci Technol 27(1):163–176

Carter AK (1997) Effect of NaCl and Pro Gibb T priming treatments on germination of 'Tam Veracruz' and 'Early Jalapeno' Chile (*Capsicum annuum*) seed. Seed Technol 19(1):16–23

Chang SM, Sung JM (1998) Deteriorative changes in primed sweet corn seeds during storage. Seed Sci Technol 26(3):613–626

Chen WJ, Houbowicz R (2010) Effect of treating lettuce (*Lactuca sativa* L.) seeds with solutions of ethereal oils from camphor tree (*Cinnamomum camphora* L.) and patchouli plant (*Pogostemon cablin* Benth.) on their germination. Acta Sci Polonorum Hortorum Cultus 9(3):69–83

Chethan HN, Devaraju PJ, Dinesh HB et al (2013) Influence of different methods of seed extraction on seed quality in cucumber [*Cucumis sativus* (L.)] cv. Hassan Local. Int J Agric Sci 9(1):323–325

D'Ercole E, Cembalo L (1999) Organic vegetable production a compromise between profit and environment. Italus Hortus 6(1–2):22–27

Das BC, Das TK (1995) Efficacy of GA_3, NAA and Ethrel on sex expression in pumpkin (*Cucurbita moschata* Poir) cv. Guamala Local. Orissa J Hortic 23(1&2):87–91

De BK, Guha P, Mandal AK (2012) Wet and dry dressing treatments on different seed sizes of wheat (*Triticum aestivum* L.) for extended storability and improved field performance. Indian Agric 56(3/4):97–105

Deepika S, Manonmani V, Ambika S (2015) Germination and vigour of polymer coated cotton seeds under different water holding capacities. In: Compendium of abstracts of the 2nd international conference on bio-resource and stress management, ANGRAU & PJTSAU, Hyderabad, India, 7–10 January 2015

Deepthi KP, Rao MA (2015) Bio-priming of oil palm sprouts: a preventive biological control strategy for management of basal stem rot caused by *Ganoderma lucidum* Karst. In: Compendium of abstracts of the 2nd international conference on bio-resource and stress management, ANGRAU & PJTSAU, Hyderabad, India, 7–10 January 2015

Demir I, Gunay A, Ceylan Y (1998) Seed moisturization as an enhancement treatment for emergence and seedling growth in bean (*Phaseolus vulgaris* L.). Seed Sci Technol 26(2):281–288

Dorna H, Szopinska D, Tylkowska K (2010) Germination and vigour of kohlrabi seeds subjected to priming in the presence of *Alternaria brassicicola* (Schw.) Wiltshire. Acta Sci Polonorum Hortorum Cultus 9(1):45–54

Dorna H, Jarosz M, Szopinska D et al (2013) Germination, vigour and health of primed *Allium cepa* L. seeds after storage. Acta Sci Polonorum Hortorum Cultus 12(4):43–58

Drew RLK, Hands LJ, Gray D (1997) Relating the effects of priming to germination of unprimed seeds. Seed Sci Technol 25(3):537–548

Durga KK, Bharathi V, Sudha Rani M et al (2015) Influence of seed discolouration on seed quality in paddy. In: Compendium of abstracts of the 2nd international conference on bio-resource and stress management, ANGRAU & PJTSAU, Hyderabad, India, 7–10 January 2015

Fatimawati FI, Ilyas S (1996) Study of the possibility to induce secondary dormancy of *Amaranthus tricolor* L. seed. Keluarga Benih (Indonesia) 7(1):99–105

Feistritzer WP (1975) The role of seed technology for agricultural development. Seed Sci Technol 3:415–429

Fernandez CJ, Greene C, Penn R et al (1998) Organic vegetable production in the U.S.: certified growers and their practices. Am J Altern Agric 13(2):69–78

Freeman GW, Yoder FA Jr (1998) Electrophoresis of esterases to identify bentgrass cultivars. J Seed Technol 12(2):120–126

Gali R, Maheshwari KU (2015) Effect of gamma radiation on germination and seedling parameters of finger millet (*Eleusine coracana* (L.) Gaertn). In: Compendium of abstracts of the 2nd international conference on bio-resource and stress management, ANGRAU & PJTSAU, Hyderabad, India, 7–10 January 2015

Geetharani PAS, Karivaratharaju (1996) Influence of mid storage seed treatment on the production potential of stored seeds of tomato and brinjal. Madras Agric J 83(11):734–735

Gray DJ, Purohit A (1991) Somatic embryogenesis and development of synthetic seed technology. Crit Rev Plant Sci 10(1):33–61

Grzesik M, Nowak J (1998) Effects of matriconditioning and hydropriming on *Helichrysum bracteatum* L. seed germination, seedling emergence and stress tolerance. Seed Sci Technol 26(2):363–376

Guedes NMP, Jennings PH (1999) Somatic embryogenesis using *Cucumis sativus* L. cotyledons. Seed Technol 21(1):72–76

Guha P, Mandal AK (2011) Pre storage dry physiological treatments on different seed sizes of okra (*Abelmoschus esculentus* L.) for improved storability and field performance and productivity. Indian Biol 43(1/2):39–50

Gurusamy C (1998) A note on the effect of magnetic seed treatment on curd production of cauliflower. Orissa J Hortic 26(1):112–113

Hilli JS, Vyakarnahal BS, Biradar DP et al (2010) Effect of growth regulators and stages of spray on growth, fruit set and seed yield of ridgegourd (*Luffa acutangula* L. Roxb). Karnataka J Agric Sci 23(2):239–242

Holubowicz R, Morozowska M (2011) Effect of umbel position on dill (*Anethum graveolens* L.) plants growing in field stands on selected seed stalk features. Folia Hortic 23(2):157–163

Janaki N, Naidu LN, Ramana CV (2015) Seed hardening technique-A tool for drought resistance. In: Compendium of abstracts of the 2nd international conference on bio-resource and stress management, ANGRAU & PJTSAU, Hyderabad, India, 7–10 January 2015

Jansen S, Mathot P, Brakel J (1994) Use of bacteria protections in pelleting for preinoculation of bean seeds (*Phaseolus vulgaris* L.). Seed Sci Technol 22(2):329–336

Jayanthi M, Umarani R, Vijayalakshmi V (2013) The effects of organic fortification with pulse sprout extracts from horse gram and cowpea on the seedling quality characteristics of rice variety Co. 43. Afr J Agric Res 8(31):4179–4183

Kallihal MS, Saxena KB, Kumar V (2013) Effect of growth regulators spray at different growth stages on flower drop, pod setting, seed yield and quality of female parent of hybrid pigeon pea cv. ICPH 2740. Int J Agric Stat Sci 9(1):197–204

Kalpana R, Rao KVM (1994) Absence of the role of lipid peroxidation during accelerated ageing of seeds of pigeonpea (*Cajanus cajan* (L.) Millsp.) cultivars. Seed Sci Technol 22(2):253–260

Kalpana R, Rao KVM (1996) Lipid changes during accelerated ageing of seeds of pigeonpea (*Cajanus cajan* (L.) Millsp.) cultivars. Seed Sci Technol 24(3):475–483

Khulbe H, Prabha SS, Khulbe D et al (2010) Effect of curd size on seed yield and seed quality parameters of cauliflower (*Brassica oleracea* var. *botrytis* L.). Trends Biosci 3(2):130–132

Khusro A, Aarti C, Preetamraj JP et al (2013) *In-vitro* studies on antibacterial activity of aqueous extracts of spices and vegetables against *Bacillus licheniformis* strain 018 and *Bacillus tequilensis* strain ARMATI. Int J Curr Microbiol Appl Sci 2(9):79–88

Kim YH, Janick J (1990) Synthetic seed technology: improving desiccation tolerance of somatic embryos of celery. Acta Hortic 280:23–28

Kishor S, Tomar BS, Singh B et al (2010) Effect of season, spacing and planting time on seed yield and quality in cucumber. Indian J Hortic 67(1):66–69

Kolekar SN, Kadam AS, Ulemale PH et al (2015) Effect of different pre soaking treatments on germination and success of grafting in Mango (*Mangifera indica* L.). In: Compendium of abstracts of the 2nd international conference on bio-resource and stress management, ANGRAU & PJTSAU, Hyderabad, India, 7–10 January 2015

Kortse PA, Oladiran AJ (2013) The effects of leaf colour at fruit harvest and fruit after ripening duration on (*Cucumeropsis mannii* Naudin.) seed quality. J Biol Agric Healthcare 3(2):190–200

Kumar R, Vyas RP, Goyal RD (1990) Effect of electric current on bitter gourd (*Momordica charantia*) seeds. Seed Res 18(2):166–168

Kumar S, Singh P, Katiyar RP et al (1997) Beneficial effect of some plant growth regulators on aged seeds of okra (*Abelmoschus esculentus* (L.) Moench) under field conditions. Seed Res 24(1):11–14

Kumar R, Pandey DS, Singh VP et al (2015) Wheat growth and nutrient use efficiency in relation to nanoparticles treated seed. In: Compendium of abstracts of the 2nd international conference on bio-resource and stress management, ANGRAU & PJTSAU, Hyderabad, India, 7–10 January 2015

Lakhote JS, Patil EH, Ajay K (2015) Effect of seed size on yield and seed quality traits in soybean (*Glycine max* L. Merrill). In: Compendium of abstracts of the 2nd international conference on bio-resource and stress management, ANGRAU & PJTSAU, Hyderabad, India, 7–10 January 2015

Li HZ, Yin YP, Zhang CQ et al (2008) Comparison of characteristics of SRAP and SSR markers in genetic diversity analysis of cultivars in *Allium fistulosum* L. Seed Sci Technol 36(2):423–434

Liu L, Wang Y, Gong Y et al (2008) Genetic purity test of F_1 hybrid tomato using molecular marker analysis. Acta Hortic 771:231–238

Livneh N, Tal HG, Beckmann JS et al (1990) RFLP analysis of a hybrid cultivar of pepper (*Capsicum annuum*) and its use in distinguishing between parental lines and in hybrid identification. Seed Sci Technol 18(2):209–214

Majid A, Simpson GM (1999) Yield and quality of mustard (*Brassica juncea* L.) as affected by irrigation pattern. Pak J Soil Sci 16(1&2):5–10

Mandal J, Mohanta S (2015) *Kharif* onion seed production as influenced by cultivars in lateritic belt of eastern India. In: Compendium of abstracts of the 2nd international conference on bio-resource and stress management, ANGRAU & PJTSAU, Hyderabad, India, 7–10 January 2015

Mauromicale G, Cavallaro V (1995) Effects of seed osmopriming on germination of tomato at different water potential. Seed Sci Technol 23(2):393–403

Meng XD, Hong M, Zhang WH et al (1998) A fast procedure for genetic purity determination of head Chinese cabbage hybrid seed based on RAPD markers. Seed Sci Technol 26(3):829–833

Mishra AC, Singh NP, Mishra JP (2005) Tuber protein electrophoregram as a tool of genetic purity test in potato (*Solanum tuberosum* L.). Seed Res 33(2):129–133

Moklyachuk F, Strelko M, Shynkarenko L et al (2013) The use of natural zeolites impregnated with zinc and copper ammoniates as carriers of micronutrients in growing vegetables. Emirates J Food Agric 25(12): 980–985

Naaiik RVTB, Shanti, M, Chandrika V et al (2015) Enhancement of seed and forage production under cereal legume based cropping system. In: Compendium of abstracts of the 2nd international conference on bio-resource and stress management, ANGRAU & PJTSAU, Hyderabad, India, 7–10 January 2015

Nagarajaiah LS, Shakuntala NM, Vasudevan SN et al (2015) Influence of seed biopriming on seed health of paddy during storage In: Compendium of abstracts of the 2nd international conference on bio-resource and stress management, ANGRAU & PJTSAU, Hyderabad, India, 7–10 January 2015

Nargis S, Gunasekaran M, Lakshmi S et al (1998) Effect of gamma irradiation of seed germination and vigour of tomato (*Lycopersicon esculentum* Mill.). Orissa J Hortic 26(2):47–49

Nascimento WM, West SH (1998) Microorganism growth during muskmelon seed priming. Seed Sci Technol 26(2):531–534

Nema NP (1988) Principles of seed certification and testing. Allied Publishers, New Delhi

Niffenegger D (1991) Using indicator seeds to test for seed lot heterogeneity. J Seed Technol 15(1):1–11

Noli E, Conti S, Maccaferri M (1999) Molecular characterization of tomato cultivars. Seed Sci Technol 27(1):1–10

Olisa BS, Ajayi SA, Akande SR (2010) Physiological quality of seeds of promising African yam bean (*Sphenostylis stenocarpa* (Hochst. ex A. Rich) Harms) and pigeon pea (*Cajanus cajan* L. Mill sp.) landraces. Res J Seed Sci 3(2):93–101

Patta S, Vari AK, Tomar BS et al (2015) Low cost standard technology for tomato hybrid seed crop productivity enhancement. In: Compendium of abstracts of the 2nd international conference on bio-resource and stress management, ANGRAU & PJTSAU, Hyderabad, India, 7–10 January 2015

Peacock L (1990) Practical constraints and opportunities for improving crop protection in organic vegetable production. Monogr Br Crop Protect Coun 45:157–162

Pramila CK, Prasanna KPR, Jayanthi R (2011) Assessment of marigold (*Tagetes crecta* L.) genotypes for morphological characters. Mysore J Agric Sci 45(3):544–550

Pramila CK, Prasanna KPR, Jayanthi R et al (2012) Influence of seasons and maturity stages on seed yield and quality in marigold cv. Pusa Narangi. Mysore J Agric Sci 46(3):510–516

Raj MSP, Devraju PJ, Gowda KRR et al (2012) Effect of seed size and threshing methods on seed quality parameters of French bean (*Phaseolus vulgaris* L.) cv. Arka Suvidha. Crop Res Hisar 44(1/2):51–54

Ramamoorthy K, Geetha R, Sujatha K (2010) Germination improvement through alleviation of seed hardness in *Solanum viarum* Dunal. Int J Plant Sci Muzaffarnagar 5(1):393–394

Rao NM, Cheralu CK, Rao NK et al (2015) Direct seeding with drum seeder a climate resilient and low cost technology in rice. In: Compendium of abstracts of the 2nd international conference on bio-resource and stress management, ANGRAU & PJTSAU, Hyderabad, India, 7–10 January 2015

Rastin S, Madani H, Shoaei S (2013) Effect of seed priming on red bean (*Phaseolus calcaratus*) growth and yield. Ann Biol Res 4(2):292–296

Reddy PRR, Varma NRG, Jagadeeshwar R et al (2015) Standardization of seedling age for *rabi* machine transplanted rice. In: Compendium of abstracts of the 2nd international conference on bio-resource and stress management, ANGRAU & PJTSAU, Hyderabad, India, 7–10 January 2015

Rohith CH, Chandrasekhararao C, Ushakumari K et al (2015) Effect of micronutrients and GA_3 foliar sprays on growth, flowering, flower yield, seed yield and quality parameters in tagetes erecta under coastal conditions.). In: Compendium of abstracts of the 2nd international conference on bio-resource and stress management, ANGRAU & PJTSAU, Hyderabad, India, 7–10 January 2015

Sainath, Uppar DS, Meena MK (2012) Effect of different plant growth regulators on growth, quality, yield and yield components in chrysanthemum (*Chrysanthemum coronarium* L.). Int J Plant Sci Muzaffarnagar 7(1):10–17

Samy CG (1998) Magnetic seed treatment. I. Influence on flowering, siliqua and seed characters of cauliflower. Orissa J Hortic 26(2):68–69

Selvarani K, Umarani R (2011) Evaluation of seed priming methods to improve seed vigour of onion (*Allium cepa* cv. *aggregatum*) and carrot (*Daucus carota*). Int J Agric Technol 7(3):857–867

Selvarani K, Umarani R, Sivasubramaniam K (2011) Effect of priming on stored onion seeds. Madras Agric J 98(4/6):124–128

Shakuntala NM, Vasudevan, SN, Hareesh M et al (2015) Seed quality enhancement through organic priming technique in chick pea genotypes. In: Compendium of abstracts of the 2nd international conference on bio-resource and stress management, ANGRAU & PJTSAU, Hyderabad, India, 7–10 January 2015

Shankrayya, Shakuntala NM, Vasudevan SN (2015) Studies on the effect of organic seed treatment on old seeds of paddy cv. BPT 5204 (*Sona masuri*). In: Compendium of abstracts of the 2nd international conference on bio-resource and stress management, ANGRAU & PJTSAU, Hyderabad, India, 7–10 January 2015

Sharma AK, Venkatasalam EP (2011) Impact of replacing the chloroplast agglutination test with enzyme linked immunosorbent assay (ELISA) on the health status and productivity of seed potatoes. Arch Phytopathol Plant Protect 44(20):1999–2003

Shashibhaskar MS, Vasudevan SN, Nagabhushan et al (2011) Effect of seed pelleting treatment on growth, seed yield and quality of tomato (*Lycopersicum esculentum* Mill.) cv. PKM. Plant Arch 11(1):443–445

Shukla PS, Kulbe H, Prasad S (2010) Correlation of yield contributing characters with seed vigour parameters in cauliflower under mid Himalayan region of Uttarakhand. Trends Biosci 3(1):58–59

Shukla YS, Raturi HC, Mehta DK et al (2015) Seed priming and its impact on vigour and storage of cucumber seed under temperature stress. In: Compendium of abstracts of the 2nd international conference on bio-resource and stress management, ANGRAU & PJTSAU, Hyderabad, India, 7–10 January 2015

Siddaraju R, Narayanaswamy S, Ramegowda et al (2010a) Studies on growth, seed yield and yield attributes as influenced by varieties and row spacings in cluster bean (*Cyamopsis tetragonoloba* L.). Mysore J Agric Sci 44(1):16–21

Siddaraju R, Narayanaswamy S, Ramegowda et al (2010b) Influence of varieties and row spacings on seed yield and quality of cluster bean (*Cyamopsis tetragonoloba* L.). Mysore J Agric Sci 44(3):679–683

Singh A, Kumar M, Mazumdar SP et al (2015a) Yield and quality of jute seed as affected by sulphur and potassium application in Burdwan district of west Bengal. In: Compendium of abstracts of the 2nd international conference on bio-resource and stress management, ANGRAU & PJTSAU, Hyderabad, India, 7–10 January 2015

Singh DP, Singh S, Kumar (2015b) Effect of age of transplants on seed yield and quality of tomato. In: Compendium of abstracts of the 2nd international conference on bio-resource and stress management, ANGRAU & PJTSAU, Hyderabad, India, 7–10 January 2015

Singh S, Sinha A, Kumar S (2015c) Dynamics myco flora of cereal and legumes seed under treated and untreated condition by Agar and Blotter methods. In: Compendium of abstracts of the 2nd international conference on bio-resource and stress management, ANGRAU & PJTSAU, Hyderabad, India, 7–10 January 2015

Sitaram, Habib AF, Kulkarni GN (1989) Effect of growth regulators on seed production and quality in hybrid cucumber (*Cucumis sativus* L.). Seed Res 17(1):6–10

Sowmya KJ, Gowda R, Bhanuprakash K et al (2013a) Influence of hydropriming on seed quality attributes in cucumber (*Cucumis sativus* L.). Mysore J Agric Sci 47(2):285–291

Sowmya KJ, Gowda R, Bhanuprakash K et al (2013b) Enhancement of seed quality through chemopriming in cucumber (*Cucumis sativus* L.). Mysore J Agric Sci 47(1):22–30

Sreenivas C, Rani TU, Sireesha A (2015) Effect of different levels of E.C water on germination of Paddy seeds. In: Compendium of abstracts of the 2nd international conference on bio-resource and stress management, ANGRAU & PJTSAU, Hyderabad, India, 7–10 January 2015

Srimathi P, Mariappan N, Sundaramoorthy L et al (2013) Effect of organic seed pelleting on seed storability and quality seedling production in biofuel tree species. J Hortic For 5(5):68–73

Sripathy KV, Hosamani J, Bellundagi A et al (2012) Organic seed production. Environ Ecol 30(1):102–105

Sujatha K, Ramamoorthy K, Sivasubramaniam K (2012) Assessment of accelerated ageing in pulses. Int J Agric Sci 8(2):505–508

Sujathamma P, Vineetha U, Reddy CR et al (2015) Response of Pigeon pea [*Cajanus cajan* (L.) Millspaugh] hybrid ICPH 2740 to planting methods. In: Compendium of abstracts of the 2nd international conference on bio-resource and stress management, ANGRAU & PJTSAU, Hyderabad, India, 7–10 January 2015

Sundaralingam K, Karivaratharaju TV (1998) Upgradation of seed quality in carrot cv. Zino. S Indian Hortic 46(1/2):85–87

Sundaralingam K, Raju TVK, Srimathi P (1998) Maximization of seed germination in carrot. Madras Agric J 85(5/6):319–320

Szopinska D, Jensen B, Knudsen IMB et al (2010) Non-chemical methods for controlling seedborne fungi in carrot with special reference to *Alternaria radicina*. J Plant Protect Res 50(2):184–192

Takaki M, Gama LHP (1998) The role of the seed coat in phytochrome controlled seed germination in *Lactuca sativa* L. cv. Grand Rapids. Seed Sci Technol 26(2):355–362

Taylor AG, Min TG, Mallaber CA (1991) Seed coating system to upgrade Brassicaceae seed quality by exploiting sinapine leakage. Seed Sci Technol 19(2):423–433

Taylor AG, Grabe DF, Paine DH (1997) Moisture content and water activity determination of pelleted and film coated seeds. Seed Technol 19(1):24–32

Thalji T (2010) Effect of plant density on seed yield and agronomic characters of faba bean (*Vicia faba* L.) under green house conditions. Biosci Res 7(1):22–25

Thomson JR (1979) An introduction to seed technology. Wiley, New York, p 252

Tian XY, Wang QT, Hu J et al (2013) Application of fluorescent dyes for falsification preventing of pea seeds (*Pisum sativum* L.). Aust J Crop Sci 7(1):147–151

Tomer RPS, Kumari P (1991) Hard seed studies in black gram (*Vigna mungo* L.). Seed Sci Technol 19(1):51–56

Tylkowska K, Turek M, Prieto RB (2010) Health, germination and vigour of common bean seeds in relation to microwave irradiation. Phytopathol Pol 55:5–12

Vaddinakatti S, Shakuntala NM, Vasudevan SN et al (2015) Effect of different seed priming treatments on seed quality of sunflower (*Helianthus annuus* L.) during storage. In: compendium of abstracts of the 2nd international conference on bio-resource and stress management, ANGRAU & PJTSAU, Hyderabad, India, 7–10 January 2015

Valdes VM, Gray D (1998) The influence of stage of fruit maturation on seed quality in tomato (*Lycopersicon lycopersicum* (L.) Karsten). Seed Sci Technol 26(2):309–318

Vasudevan M, Patil, Lokeshappa BL et al (2012) Influence of seed conditioning to improve genetic purity in eggplant. Int Proc Chem Biol Environ Eng IPCBEE 37:100–103

Vasudevan SN, Shakuntala NM, Nagarajaiah LS et al (2015) Seed quality enhancement through orgopriming technique in sorghum genotypes. In: Compendium of abstracts of the 2nd international conference on bio-resource and stress management, ANGRAU & PJTSAU, Hyderabad, India, 7–10 January 2015

Venudevan B, Srimathi P (2013) Influence of fruit colour variation on physical, physiological, biochemical and nursery characters of endangered medicinal tree bael (*Aegle marmelos* (L.) Corr.,). J Med Plants Res 7(26):1951–1956

Vijaya A, Ponnuswamy AS (1997a) Effect of seed fortification with micronutrients on seed quality in black gram and cowpea. Madras Agric J 84(4): 223–225

Vijaya J, Ponnuswamy AS (1997b) Effect of soaking duration and volume of water on seed quality in black gram and cowpea. Madras Agric J 84(4):226–228

Vinodkumar SB, Vyakaranahal BS, Dhananjaya P et al (2013) Effect of seed polymer coating on field performance and quality of pigeon pea [*Cajanus cajan* (L.) Millsp]. Environ Ecol 31(1):43–46

Welbaum GE, Shen Z, Oluoch MO et al (1998) The evolution and effects of priming vegetable seeds. Seed Technol (USA) 20(2):209–235

Zhang W, Yuan Y, Cheng W et al (2012) Osmopriming improves tomato seed vigor under aging and salinity stress. Afr J Biotechnol 11(23):6305–6311

Zhang W, Zhang FM, Zhu (2013) Soaking seeds in methyl jasmonate or benzothiadiazole induces resistance to an insect pest and stem decay in *Brassica campestris* L. ssp. *chinensis*. J Hortic Sci Biotechnol 88(6):715–720

MIX
Papier aus verantwortungsvollen Quellen
Paper from responsible sources
FSC® C105338

If you have any concerns about our products,
you can contact us on
ProductSafety@springernature.com

In case Publisher is established outside the EU,
the EU authorized representative is:
Springer Nature Customer Service Center GmbH
Europaplatz 3, 69115 Heidelberg, Germany

Printed by Libri Plureos GmbH
in Hamburg, Germany